AF770091

SCIENCE:
IMAGe IN AcTION

THE SCIENCE AND CULTURE SERIES – ASTROPHYSICS

Series Editor: A. Zichichi

Proceedings of the 7th International Workshop on Data Analysis in Astronomy "Livio Scarsi and Vito DiGesù"

SCIENCE: IMAGe IN ACTION

Erice, Sicily, Italy 15 – 21 April 2011

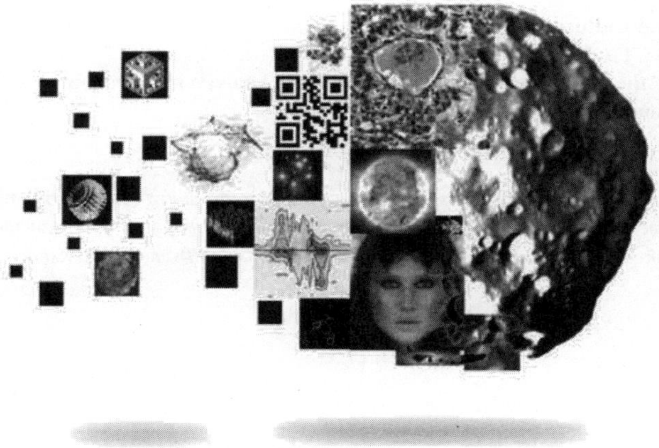

Edited by

Bertrand Zavidovique
Université Paris-Sud XI, France

Giosue' Lo Bosco
Università degli Studi di Palermo, Italy

World Scientific

NEW JERSEY • LONDON • SINGAPORE • BEIJING • SHANGHAI • HONG KONG • TAIPEI • CHENNAI

Published by

World Scientific Publishing Co. Pte. Ltd.
5 Toh Tuck Link, Singapore 596224
USA office: 27 Warren Street, Suite 401-402, Hackensack, NJ 07601
UK office: 57 Shelton Street, Covent Garden, London WC2H 9HE

British Library Cataloguing-in-Publication Data
A catalogue record for this book is available from the British Library.

The Science and Culture Series – Astrophysics
IMAGE IN ACTION
Proceedings of the 7th International Workshop Data Analysis in Astronomy
"Livio Scarsi and Vito DiGesù"

Copyright © 2012 by World Scientific Publishing Co. Pte. Ltd.

All rights reserved. This book, or parts thereof, may not be reproduced in any form or by any means, electronic or mechanical, including photocopying, recording or any information storage and retrieval system now known or to be invented, without written permission from the Publisher.

For photocopying of material in this volume, please pay a copying fee through the Copyright Clearance Center, Inc., 222 Rosewood Drive, Danvers, MA 01923, USA. In this case permission to photocopy is not required from the publisher.

ISBN-13 978-981-4383-28-8
ISBN-10 981-4383-28-7

Printed in Singapore by World Scientific Printers.

"Science: IMAGe IN AcTION"
Seventh International Workshop of the
"Data Analysis in Astronomy ≪Livio Scarsi and Vito Di Gesù≫" Series
EMFCSC, Erice, Italy
15-21 April 2011

ORGANIZING COMMITTEES

DIRECTOR OF THE WORKSHOP

Bertrand Zavidovique	– I.E.F. Université Paris Sud, Orsay, France

INTERNATIONAL STEERING COMMITTEE

D. Chakrabarty	– University of Warwick, Coventry, UK
M. Concetta Maccarone	– Ist. Astrofisica Spaziale e Fisica Cosmica/INAF, Palermo, Italy
G. Bernroider	– University of Salzburg, Austria
L. Boi	– Ec. des Ht. Et. en Sciences Sociales, Paris, France
U. Bottazzini	– Universitá degli Studi di Milano, Italy
M. Kumar Kundu	– Indian Statistical Institute, Kolkata, India
S. Roy	– Indian Statistical Institute, Kolkata, India
H. Yeshurun	– Tel Aviv University, Israel.

LOCAL SECRETARIAT

G. Lo Bosco	– DMI, Universitá di Palermo, Palermo, Italy
EMFCSC Staff	– Ettore Majorana Foundation and Center for Scientific Culture, Erice, Italy

PROCEEDINGS EDITORS

B. Zavidovique	– I.E.F. Université Paris Sud, Orsay, France
G. Lo Bosco	– DMI, Universitá di Palermo, Palermo, Italy

PREFACE

The book gathers articles that were exposed during the seventh edition of the Workshop "Data Analysis in Astronomy". It is interesting to frame the present workshop and its contents within the evolution along the former editions. Let us understand how the topics and concerns changed with the joint progresses of 1) information technology and 2) problem complexity in physical observation. The enlargement of the theme of the workshop to cover the data representation and the data analysis in other fields like biology then appears as just a logical next step towards more abstraction. The trend to search for common expressions or models capable of transcending usual disciplines might be associated with some lack in the Mathematics required to tackle current difficulties in modelling complex systems, whether gigantic or minuscule. In that sense, data analysis would be at the epicentre and a key facilitator of some current integrative phase of Science, as it was underlined collectively during the final panel session.
Speaking of the Data devil, it should be no wonder any more that few clicks on active expressions of Wikipedia lead ineluctably to the word "Philosophy", a field once known for being the mother of all others!

The Data Analysis in Astronomy Workshop series

The "Data Analysis in Astronomy" Workshop series started in 1984 aiming to provide an updated overview of advanced methods and related applications to astronomy and astrophysics. It is enough to list the topics covered over the years to show how the workshop progressively enlarged its scope toward simulation and modelling, opening to three major branches of Science (Astrophysics, Biology and Earth Environment):

- 1984: emerging systems for Data Analysis were presented; a stress was put on methodologies in image and signal analysis, while Parallel Processing and Machine Vision were beginning to be studied.
- 1986: data handling systems intended for major satellites were reviewed; the importance of physical interpretation beyond data analysis was underlined; Artificial Intelligence applications to astronomy were proposed.
- 1988: novel data analysis (chaotic processes, galaxy chains via clustering, burst with adaptive growing) were discussed to tackle frontiers in astrophysics.
- 1991: proof of the maturation of the field, a review of large working experiments was the key contribution of this edition, including comparison of data analysis

systems in light of those conjectured seven years earlier; results on experimental data were founding new goals and problems.
- 1996: data analysis problems were considered in all fields, the multi-wavelength approach was paralleled with advances in data fusion, information retrieval, high-speed computing ..., all techniques motivated by robotics and information systems.
- 2007: simulation and modelling, up to virtuality, were the main concerns of this instance, a proof this time of the huge progress in computer visualisation; evolutionary computing and learning were illustrations of cross fertilization between disciplines, motivating deeper analysis of interactions in astronomy and biology; data mining was discussed in front of the burst of experimental materials.

The series strongly contributed to stimulate and enforce the scientific interaction between astrophysicists - and then other scientists - and experts in data analysis, who debated and compared results and methods, theory and experiments. The analysis of the interrelation together with the impact of new technologies and innovative design on fields such as Cosmology, Biology or Climatology cannot be else than a beneficial fallout of the initiative. The International Consortium for Interdisciplinary Science (http://icis.unipa.it/organization.html) which was contributing significantly to the present edition was created by Vito Di Gesù after the successful 2007's edition.

The present edition of the workshop: IMAGe IN AcTION

The summary of the evolution of the workshop through its key topics shows that it followed the major trends in Data Analysis, i.e. more generally in Pattern Recognition and Optimization. Indeed, if one separates the actual applications to astronomy and astrophysics that include system descriptions and evaluations, the history of both the field and the workshop is all in the process of inventing models or borrowing them from other fields. The phenomenon is all the more sensible as more complex and cumbersome data asks for more abstract and constricting models. It is enough to check that along the editions were successively reviewed:

- conventional techniques and methodologies in tool box systems.
- attempts through new theories as chaos in mathematics or artificial intelligence in computer science.
- data fusion with the emergence of non monotonic logics or information retrieval as an abstract inverse projection.
- evolutionary computing or learning with their flavour of bio or psycho behavior.

So it seemed that it was time to put forward the notion of model in confronting modern Physics and Biology to the service of Astronomy by means of Data analysis. The process is not new. Optimization, and thus Pattern Recognition in general and Data Analysis in particular, started by a Mechanics inspiration, whether directly with gradient descents and calculus of variations or in its statistical form through the universal concept of inertia (e.g. Principal Component Analysis). Key variants were coming from the data representation and adapted distances (e.g. Dynamic Programming and the Levenstein's distance). But very soon, partly because of new kinds of complexity coming along with discrete spaces - genuine or sampled - and partly because some decision spaces ask for more than just adapting a distance (e.g. multi-partite graphs), analogies were borrowing to larger Physics like Simulated Annealing to master random scans of the data or Graph Cuts for organized spreading. Decision processes were starting to mimic naturally-converging phenomena from the leaving realm as Genetic evolution or Ant colonies, and from the societal behaviour too as Stable Marriages or College admission.

Beyond such examples of analogies in data analysis, not much was studied in our fields about the way representations are elaborated, and even less on their manner to evolve in the degree of abstraction or to merge into larger ones, from one support or one scale to another. It is admitted they are necessary for humans to understanding phenomena and they may help elaborating theories, thus condensing sparse images into a single explanatory one. They are likely key features of the leaving, where programs (acting models) of systems are transmitted, having thus been elaborated as representations, together with the systems themselves which are then "ante-replication" representations of their avatars. The importance of images for system behaviors and their understanding, for instance in control theory, puts forward the role of representations as information supports: same as conductors convey energy or aging systems convey the time, representations transmit abstracted versions of systems, for them to be better perceived and acted. Conversely, a same image gets different meanings (triggers different actions) to different perceiving entities even equally equipped. That in turns introduces the importance of noise, already latent in biological growth, and sends back to the compressing role of representations through the notion of robustness. Eventually, representation as an imaging and abstracting process appears central to any intelligent activity. Among them, Science is a swing between conjectures (i.e. abstract representations of a problem), and understanding (i.e. effective representation of solutions), through experimentations (i.e. concrete representation of both in terms of actions). And the technological burst of "informatics" images spreads them every where and at every stage of any current scientific steps: Process flowcharts, blue prints and system sketches before experiment, Signals, images, matrices or graphics and dependency graphs after. So that a problem to solve in parallel with most other scientific studies turns out to be "how to master the huge amount of data generated by the least observation". Indeed, one is far here from the unique image mentioned above and intermediate representations are rather dense than sparse!

So, in many senses one can say that everything starts and ends with images. Of course, some kill-joys will object with chicken-egg considerations, notwithstanding that the circular "start-end" makes both chicken (reality, observer included) and egg (image of both) in between the other. In data analysis, following the swing again between experiments and theories, the latter circularity sends back to the common saying "a clear problem is half way solved". And then Hadamard comes to tell us that if it is not well posed a problem may only be solved in adding constraints to restrict the solution space. Note that, in the matter, data analysis is not all about clustering or extracting information, it can be model driven (even directly so and not only through the decision process). As a consequence, models are involved in several ways: for instance they can act like supervisors, but they can also modify the topology or structure of the decision space. The change gets several forms: introducing preferences or other bias, operating in a quotient space (e.g. indiscernible elements inside granules of rough sets), or introducing a measure like uncertainty, complexity, stability etc. in a different way from through equivalences. Models then ask the question of their representation, i.e. the image of the system at stake: a close form seemed to be the Grail until computers made it doable to elaborate simplified forms, possibly from equations or with operators (the so called "agents") free of any physical likeliness! But the main point here is that both favor simulation by computers, all the more as it is likely the only way to get around uncertainty principles ..., and simulation sends back to optimization (e.g. identification in the control theory terminology). Yet another form of circularity, simulation can be directly that of the system but also can "materialize" an analogy with some other system in some other field. In other words, a representation of a system may be very different and is actually independent from the representation of-and-by its productions. In that case, do models inspire Pattern Recognition first, and then only after, through it, applications in other field (Astro-X for the present workshop)? Or are they involved as prior knowledge in a different way?

As a logical fall-out of trends in data analysis towards system understanding, the present edition of this workshop was all devoted to the question of "representation in Science". It was trying to kill a few big scientific birds with this single small iconic stone, whence its name, IMAGe IN AcTION, and its main thrusts

CIII: Communication, Image, Information and Intuition,
From perception to virtuality: mastering realism,
Structure and representation: mastering complexity,
System and data organization: mastering uncertainty,

all considered from three possible points of view: fundamentals, modeling or use. Let us underline that this classification makes concepts as "complexity" or "dynamics" appear like transverse notions: a measure among others as already said or a dimensional feature among others. It could be of course that the classification of the talks was not perfect, yet another instance of cobbler's children going barefoot! But this may well come from the intrinsic complexity of clustering a set, even a small one, of excellent researches in this dynamic field!

The content of the book

In light of the above introduction and of the last questions, the organization of the book is slightly different from that of the workshop, but still willingly ignoring the usual split into scientific fields. Papers are clustered into three parts:

Part A: Information: data organization and communication,
Part B: System: structure and behavior,
Part C: Data - System representation.

Part A broadly discusses a dialogue between experiments and information, should information be extracted-from or brought-to experiments. The concept is fundamental in statistics and it is also the crux of the matter to tailor the emergence of collective behaviors. Such examples illustrate different facets of this couple, depending for instance on whether the elaboration of one thanks to the other is considered, or the transmission from one to the other. Communication then asks for uncertainty considerations, noise to begin with, and its wider impact on the couple perception-action. Whichever its definition, noise is not the only source of uncertainty in a communication. Indeterminacy or approximation need to be accounted for: fuzzy and rough sets are among the efficient candidates to render that. Eventually, clustering can be claimed all about uncertainty handling: data set representation is so proved not to be the only solution. Introducing hierarchies with adapted metrics, a priori pre-improving the data resolution with or without fusing several sources are other methods in need of evaluation. The latter is nothing but an abstract metaphor of improving the signal-to-noise ratio by accumulation or cross-reference, except that the technology together with the higher semantic level makes it possible to involve synthetic data as simulation results for the multiplication of sources.

Part B plays with another couple important for complex systems: state vs. transition (it showed up already in Part A, for data, with the difference between representing sets or runs in clustering). It is conjectured that state-first descriptions would rather characterize physics, while transition-first would fit biology better. That could stem from life producing dynamical systems in essence i.e. from the importance of experience. Uncertainty joins causality here: for example, would local changes in interactions disrupt or not a global behavior. The system geometry can bring answers. That is interestingly paralleled with a different use of geometry in the study of wave-particle duality (electromagnetism): stable patterns in the state space involve constraints on separate degrees of freedom in regards of some dynamics consistency. Stable patterns are also characterizing biological systems expressed as neuronal networks: these are patterns of activity. It is proven that they are formed under the conjunction of multiple instabilities of auto waves. Whether transition or state is put forward, dynamics sends back to the time scale. In the living world, the complexity - i.e. a global measure on both states and transitions - does not stop increasing towards consciousness: this is conjectured for a principle of evolution. Measures get readily integral forms: the Lagrangian formalism with Eulerian accounting of locality turns adequate to deal with causality in general, reintroducing the observer. Beside geometry or measures, operators and topology have supporters too as tools for reporting on dynamical systems. An example of operator is the (self) assembly of building blocks to reconstruct a structure. The complexity is recovered afterward under the form of the simplest set of blocks to do so. Eventually and targeting universality, the category theory of topological thermodynamics is proposed as a foundation of dynamical system understanding.

Part C details examples of actual relations between data and systems within the frame of explicit applications and subsequent experiments. It starts in showing how pure computer display and animation techniques are able to link models and representations to "reality" in some "concrete", although virtual, manner. Indeed these techniques are inspired from artificial life and so with no connection to physical, biological or physiological phenomena, The Virtual Observatory is the second example given to support the evidence that simulation helps Science not only in giving access to more flexible parameter variability, but also due to the associated data and method storing-capabilities. It fosters interoperability, statistics on bulky corpuses, efficient data mining possibly through the web etc. in short a reuse of resources in general, including novel ideas and competencies. Visualization is not the only premise of virtuality, especially in astrophysics where many phenomena are perceived from very indirect observation. Galactic complexity may be one of the kind: in order to question it, the chaos is estimated in models of a galaxy as a possible hint of phase non-linearities. Since usual models assume smooth parametric phase space densities, a Bayes non-parametric approach accounting for measure uncertainties is applied to external observations, showing a non linear phase space density. The process ends into inverse modeling. Other reconstructions are more classical, explicating some type of evolving shape to be reconstructed from complex data acquisition as fluorescence tomography. Again, reconstruction from models and other projections is one among the many recognition processes. Features like symmetry or isotropy are more of the gestalt fashion; moreover they depend on the scale. Illustrations of that include the perception of texture symmetry by humans together with a physics metaphor of the same, or a correlative scheme designed for isolating anisotropic data against isotropic background. The fractal behavior depends on the scale too, so when it is involved in the model, a multi scale process can be preferred: an application to the identification of air showers is presented. Eventually, Bayesian curve fitting is frequently revisited. A mixture analysis is evaluated to establish a relationship between apparent magnitude and red shift for quasars towards clustering.

Bertrand Zavidovique
Giosuè Lo Bosco

Acknowledgements

The meeting could not have been such successful without the joint efforts of a dozen "good people": from members of the scientific committee who proposed and contacted speakers following the general scientific line decided for the workshop [Maria Concetta Maccarone (Italy) who was also the guarantor for Astrophysics and Astronomy, Dalia Chakrabarty (U.K.), Gustav Bernroider (Austria), Luciano Boi (France), Sisir Roy (India), Hezy Yeshurun (Israel)] to the Local Secretary [Giosuè Lo Bosco (Italy)] and all participants who delivered their speech or took part in the discussions. Special thanks to Marco Tabacchi who was helping organization and communication at all stages. We are also glad to thank the Italian National Institute for Astrophysics INAF, the Université Paris Sud XI, the Università degli Studi di Palermo through the CITC (Centro Interdipartimentale di Teconlogie della Conoscenza) and the GIRPR (Gruppo Italiano Ricercatori in Pattern Recognition) for their supports. Eventually, we thank the entire staff of the Ettore Majorana Foundation and Centre for Scientific Culture for support and invaluable help in organizing a successful workshop.

Special Mention

The best student's paper award was attributed to Gaetano Scandariato for his work on "The Extinction Map of the Orion Molecular Cloud" sponsored by GIRPR (Gruppo Italiano Ricercatori in Pattern Recognition).

On the behalf of Prof. A. Zichichi (President of the EMFCSC) the "DAA - Data Analysis in Astronomy" Workshops are from now dedicated to "Prof. Livio Scarsi and Vito DiGesù" who were the enthusiastic inspirer of the series.

UNIVERSITÉ
PARIS-SUD 11

UNIVERSITÀ
DEGLI STUDI
DI PALERMO

INAF

ISTITUTO NAZIONALE
DI ASTROFISICA
NATIONAL INSTITUTE
FOR ASTROPHYSICS

Centro
Interdipartimentale
per le Tecnologie
della Conoscenza

Gruppo Italiano
Ricercatori
in Pattern
Recognition

Vito Di Gesù from his first to his last disciple
(March 25, 1945 Torino - March 15, 2009 Palermo)

This workshop is dedicated to Vito.
When remembering him, it is not easy to figure out so polyhedral a personality, and so human at the same time. Only a series of images can relate the many aspects of the character. Let me spend some words on my long period, about 20 years, of research activity with Vito.

I knew Vito in 1975; he was assistant of the course on "Analytical Mechanics" at the Department of Physics, University of Palermo, a few years before he found his future at the Department of Mathematics. Although at the time I was a beginning student, we soon shared the common interest for the analysis methods applied to experimental data; then, in 1978, Vito was rapporteur of my thesis dissertation for the degree in Physics. The topic of my experimental thesis was on gamma-spectrometry in environmental radioecology and on new data analysis techniques.

These topics were at the basis of our common work that very soon was addressed to problems coming from astronomy and astrophysics. In fact, in 1981 was born the Institute which I belong, the IFCAI, Istituto di Fisica Cosmica e Applicazioni dell'Informatica (Institute of Cosmic Physics and Applied Computer Science).

The Institute, established and headed by Prof. Livio Scarsi, comprised mainly researchers in astrophysics and Vito as Collaborator. Vito, assistant professor at the Department of Mathematics, was member of the Scientific Council of IFCAI for several years and he was appointed "Head of the Computer Science and Data Analysis"; in such a position, Vito defined at IFCAI the group of "Informatics applied to astrophysics" in which Vito and I worked together for a long time.

In the field of signal and image analysis, the strength was the study and development of new methods and methodologies: Vito never stopped on what was already known; he tried alternative solutions and points of view different from the "classical" ones; it was so that our activities in this field were directed to the Bayesian deconvolution, cluster analysis and related uniformity test, mathematical morphology and new operators for gray-level images, possibility theory, fuzzy sets, and more, in terms of both theoretical and applied study.

I must stress that at the time, as always, new methods found no easy way into the astrophysics community; as an example, statistics and probability theory remained more prominent than the almost-newborn possibility and fuzzy sets theories; it was difficult to convince the community of the applicability and effectiveness of these methods, but Vito did not give up and finally won (our last work together was on fuzzy-sets, too).

Together we explored also the field of the data analysis systems and developed the basis for a decision support system in astronomical environment. During the Eighties, driven by Vito and Livio, my director, I spent several months at ESO in Garching with the MIDAS team, then in Stanford at the Department of Statistics with Jerry Friedman, and also at the Philips Labs in Bruxelles with Pierre Devijver, acquiring experience in various fields related to my scientific activity, and not only.

All this was part of a wider program that included the establishment of international working groups, scientific networks and workshops, as well as the "training" of fellows, some of them spent their first period, one year or more, at IFCAI before to move permanently to the Department of Mathematics in Palermo.

The first concrete evidence of such a program was the Data Analysis in Astronomy workshop series started on 1984, defined and directed by Livio Scarsi and Vito. And I was obviously and strongly involved in it since its beginning.

The second success was the constitution, in 1987, of a new Technical Committee of the International Association of Pattern Recognition, IAPR, the so-called TC13-PRAA on Pattern Recognition in Astronomy and Astrophysics, born from an idea of Vito who was its president for many years. Fionn Murtagh, Rafael Molina and myself we were, and are today, part of the team, together with other scientists with interests in the field of data analysis in astronomy.

As a natural corollary, few years later, in 1995, was activated the European scientific network CCMA "Converging Computing Methodology in Astronomy" of ESF (European Science Foundation), headed by me and Fionn Murtagh. Vito was a key-component of the CCMA Scientific Council; he always attended with new creative ideas and with his usual humour (I remember the meetings between Vito and Albert Bijaoui that always ended in an ironic historical-political discussion on the proper "owner" of Nice: Italy or France?).

And it was from the CCMA that born the idea of iAstro, activated at the end of 2001. Vito wanted that I belonged too, asked me often trying to convince, as he had repeatedly tried to convince me to move to the University; but I have always declared loyalty to my Institute and the commitments and responsibilities that I got there, prevented me to join to the new enterprise. In the same period my Institute changed its denomination in IASF, Institute of Astrophysics and Cosmic Physics, partially losing its Computer Science component.

So, in the late Nineties our scientific interests had to follow more and more different paths, and we had no many possibilities to take back our study on new data analysis methods.

But our friendship has never been affected by the events and Vito and I, we always stayed in touch till his demise.

At the beginning of the Nineties, Vito had prepared at the Department of Mathematics what today is his further legacy in Palermo. With his scientific skill, Vito was able to instil in young people the passion for research, accompanied by humour and humanity; so, in such a way, Vito was preparing what would become his group of "boys" (named "picciotti" in Sicilian dialect). Among his "picciotti", Giosuè is the one who, perhaps more than any other, and as for me in the past, has followed Vito mainly in the field of cluster analysis, fuzzy sets and shape classification techniques.

Therefore, to continue the story, I leave my notes as "first spiritual daughter" of Vito and pass the word to Giosuè, his "last spiritual son", as Bertrand Zavidovique defined us; knowing Vito, I am honoured for such a definition.

M. Cettina Maccarone

Astrophysics gave the possibility to Vito to approach one of the most engaging problems of computer science: pattern recognition. And he became so involved in such field, that in 1993 Vito won the Mahalanobis prize for scientific activity in pattern recognition, and in 1994 the IAPR fellow for his research in Pattern recognition, clustering and architecture for Vision.

I met Vito for the first time during those years. I was a student in Math and I was particularly attracted by computer science. Vito was associate professor at the Department of Mathematics, and he taught a subject that he called M.I.T, that was an acronym to indicate "Methods for Treatment of information". But since he was also a great joker, he said: "I have brought the M.I.T in Palermo". And most of his colleagues replied "hey Vito, who do you believe you are, ... the M.I.T.".

Anyway, in that subject Vito told us, students, about digital image analysis and processing, fuzzy logic and parallel computer architecture, giving us the possibility to observe concrete applications of computer science, differently from theoretical computer science that, at that time, was widely celebrated in the math course series where Vito worked. In particular, I found his way of teaching really unique: differently from the others, he taught many topics regarding his research activity, and this was the strong point of his teaching activity.

Many students felt fascinated by the topics he taught, adopting one of them as subject of their graduation thesis. Some of them were so struck by the subject that they started to feel how fascinating it would be to do research in those fields. I was one of them.

Vito, gave me immediately the possibility to work on research, when I was a student, not yet graduated. I like to imagine that he felt I would and could.

I met many of his collaborators and colleagues, and thanks to them too, I understood many of the contributions he had already gave at that time, e.g. the fuzzy operators he defined to perform mathematical morphology on gray level images, and the operators he introduced to detect and measure local and global symmetry in complex scenes. Moreover, in the contest of parallel computer architectures, I had heard he was one of the designers of the PAPIA-machine (Pyramid Array of Processors for Image Analysis Machine) and the leader scientist of the project HERMIA-machine (Heterogeneous and Reconfigurable Machine for Image Analysis).

Thus, my first research experience occurred when he decided to get me involved in a project named DAISY (Distributed Architecture for Intelligent System) whose main goal was autonomous robot navigation. Because of that project, we started to think to the application of evolutionary approaches on path finding problems.

At the end, my graduation thesis subject was the application of genetic algorithms to the path finding problem. Such application involved me very much, and together we developed a simulator application able to simulate a prey-predator system in an unknown environment with obstacles, and a real application of autonomous robot navigation by a LEGO robot.

After these very exciting experiences, I decided to start a PhD in computer science under his supervision.

The topic of my PhD concerned the use of evolutionary algorithms in Data Analysis. So, we started to do research on data analysis methodologies, with application to multidimensional data. We had started working on that from 2000. With Vito, I have developed a clustering methodology based on a genetic algorithm that we have applied to digital images, doing segmentation, and also to microarray data; Then, a methodology to combine classifiers, based again on an evolutionary paradigm; And also a new one-class classifier that we have been applying to computer security.

Back to 2000, that year Vito realized a dream: he established the first course of Computer science at the faculty of Science. Finally, Vito had the possibility to grow up a new generation of pure computer scientists, differently from the previous generation of mathematicians using a computer.

Topics grew up, and Vito taught on several classes dealing with data analysis. And what had occurred in the past for someone, like me, occurred for many other young pure computer science students. That is: many young students decided to be involved on Vito's research topics.

Because of that, our research group got bigger, thanks to our leader Vito, to assistant professors like me and others, and young PhD students. I and Vito continued to work on data analysis methodologies, with application to bioinformatics.

The last years of his life have been devoted to the study of a challenging problem in Biology: the nucleosome positioning. We have worked very hard on that problem, developing a methodology able to discover patterns corresponding to nucleosomes in biological data. This has been our last common adventure in research.

What I have observed most of the time during the common works with Vito, was his ability to always find alternative and new ways to approach the research problems. I think that this make the difference between a normal, and a clever researcher. This is something regarding creativity. And from my point of view, Vito was a creative.

Giosué Lo Bosco

CONTENTS

PART A

Information: Data Organization and Communication

STATISTICAL INFORMATION: A BAYESIAN PERSPECTIVE

R.B. STERN and C.A. de B. PEREIRA*

Department of Statistics, IME-USP,
São Paulo, SP, Brazil
**E-mail: cpereira@ime.usp.br*

We explore the concept of information in statistics: information about unknown quantities of interest, the parameters. We discuss intuitive ideas of what should be information in statistics. Our approach on information is divided in two scenarios: observed data and planning of an experiment. On the first scenario, we discuss the Sufficiency Principle, the Conditionality Principle, the Likelihood Principle and their relationship with trivial experiments. We also provide applications of some measures of information to an intuitive example. On the second scenario, the definition and new applications of Blackwell Sufficiency are presented. We discuss a new relationship between Blackwell Equivalence and the Likelihood Principle. Finally, the expected values of some measures of information are calculated.

Keywords: Blackwell Equivalence, Blackwell Sufficiency, Conditionality Principle, Likelihood Principle, Sufficiency Principle.

1. Introduction

The concept of information is perhaps one of the most controversial in Statistics. One can find innumerous different measures of the amount of information an experiment or a given data set brings about unknown quantities of interest — parameters. It is important to explore these measures because, after all, extracting information seems to be the ultimate goal of Statistics. More precisely, the goal is to extract — from an observed data or from an experiment to be performed — information about unknown quantities of interest. The intuitive definition of information which will guide this paper is given by Ref. 1 and is as follows:

"Information is what it does for you, it changes your opinion".

This conceptual definition leads one to the following four questions:

- Information about what?
- Where is the information?
- How is information extracted?
- How much information is used?

We are interested in defining information about a parameter, θ, which assumes possible outcomes in Θ, the parameter space. Hence, the answer to the first question is straightforward.

A parameter represents a state of nature which we are uncertain of. For example, one can be interested in the number of days in which it will rain next year. For instance, θ can be this number and Θ all natural numbers smaller or equal to 366.

Next, we try to answer the second question. It is important to note that when defining Θ we are already using some previous knowledge about θ. In the example, we have informed that any year has at most 366 days and, therefore, θ must be smaller than this number. Besides stating Θ, one might also think that some values are more probable than others. This kind of knowledge is used to elicitate the prior distribution for θ. Here the prior distribution represents our description of our present state of uncertainty about θ. Mainly, the statisticians' goal is to decrease her/his uncertainty about this unknown quantity of interest. In order to reach such objective, (s)he observes data that, in his/her opinion, is related to the parameter. Consequently, one expects that there is information about θ in the data to be observed. That is, answering the second question, statistical (expected) information is contained in the collected data set (experiment to be performed).

It seems natural at this point to ask: How to extract information contained in the observed data? In order to answer this question properly, the "scientist" considers a global probability space involving a prior distribution on θ and the experimental distribution for every possible θ. Next, the "scientist" uses the Bayes operation to obtain the posterior distribution. The posterior distribution describes the uncertainty about the parameter after calibrating the prior by the observed data. Thus, we could say that the new information also depends on the statistical framework. The act of observing the data corresponds to a mechanism of transforming unknown quantities in known ones. We also call such a mechanism an *experiment.*

The last question which is important in design of experiments is: "How much information is extracted after the experiment is performed (after the data is observed)?". After obtaining an adequate answer, it is possible to modify the question to a *pre-posterior analysis*: "How much information do we expect to obtain in a specific experiment to be performed?". The heart of this paper is to explore possible answers for both questions.

Section 3 analyzes how informative a particular data set is. Section 3.1 introduces common principles in Statistics and their relationship with the Likelihood Principle. Section 3.2 presents a simple example and three information functions compatible with the principles of Section 3.1.

Section 4 is related to experimental design and tries to answer the following question: Among the possible alternative experiments, which is our best choice? Blackwell Sufficiency is a possible criterion to compare experiments. The definition of Blackwell Sufficiency, with examples, is presented in Section 4.1. The Likelihood Principle and its relationship with Blackwell's Equivalence are discussed in Section 4.2. We argue that Blackwell Sufficiency is the best criterion whenever the experiments are comparable in that sense. Finally, since not all experiments are comparable in Blackwell's sense, Section 4.3 explores the metrics exposed in Section 3.2 using the framework of decision theory to compare experiments.

2. Definitions

In the context of experimental information we will always be concerned with a probability space, that is, a triple $(\Omega, \Im, P)$ in which Ω is a set, $\Im$ is a σ-algebra on Ω and $P : \Im \mapsto [0,1]$ is a probability function.

A random quantity R corresponds to a function from Ω to some set $\Re$. We define the probability space induced by R, $(\Re, \Im_R, P_R)$, where $\Im_R = \{M \subset \Re : R^{-1}[M] \in \Im\}$ and $P_R(M) = P\left(R^{-1}[M]\right)$. Finally, the σ-algebra induced on Ω by a random quantity R is called $\Im_{|R}$ and corresponds to $\{R^{-1}[M] : M \in \Im_R\}$.

We call an experiment any random quantity which can be observed, that is, which can be known. The realization of an experiment corresponds to the observation of this random quantity. It is important to observe that classifying a random variable as an experiment has nothing to do with the probability space, but with the limitations which exist in the world.

On the other hand, a parameter is a random quantity of interest. If the parameter were an experiment, its value could be known and the work of the statistician would be easy. Nevertheless, in many cases the parameter is not an experiment. Therefore, it is necessary to learn about it in an indirect manner, that is, observing those random quantities which are experiments and applying the Bayes Theorem. Again, classifying a random variable as a parameter has nothing to do with the probability space, but with an aspect of the world, our interest.

Therefore, in our representation, experiments and parameters are treated essentially in the same way: both are random quantities. The given names only reflect some aspects of the world not contemplated in the probability space, that is, the observability of the random quantity and our interest in it.

Let X be an experiment in $\mathcal{X}$. A function $T : \mathcal{X} \mapsto \tau$ is considered a statistic of X. Therefore, $T(X)$ is also an experiment. Whenever there is no confusion, we will use the letter T both to indicate the statistic T and the experiment $T(X)$.

From now on, we restrict ourselves to random quantities whose probability distributions are absolutely continuous or discrete. Following this restriction, $p_X(x|\theta)$ is the conditional probability (density) function of the random quantity X given θ. After the experiment is performed we write $L(\theta|X = x)$ for the likelihood function of X at point x. Whenever clear in the context, we write $p(x|\theta)$ and $L(\theta|x)$ for the former functions.

Finally, we say that an experiment $X : \Omega \mapsto \mathcal{X}$ is trivial for $\theta : \Omega \mapsto \Theta$ if $\Im_{|X}$ is independent of $\Im_{|\theta}$. This condition is equivalent to the assertion that, $\forall \theta' \in \Theta, \forall x \in \mathcal{X}$, $p(x|\theta') = p(x)$. We use the word trivial to emphasize that X and θ are not associated. Consequently, X by itself does not carry "information" about θ.

3. Data Information

3.1. *Statistical principles and information*

At this point, we hope to have convinced the reader that it is an important task to define how informative is the observed data of an experiment. Therefore it seems reasonable to assume the existence of an information function which has as arguments the experiment, its observed data, and the parameter. Although this function is still not defined, to make reference to it we will use the notation $Inf(X, x, \theta)$, where X is an experiment, x the observed data and θ the parameter.

Before establishing an exact form for such a function, we look for properties, as in Ref. 1 and Ref. 2, which follow from common statistical principles.

A statistic $T : \mathcal{X} \mapsto \tau$ is called sufficient if X and θ are conditionally independent given T, that is, X is a trivial experiment for θ given T. Previously, it was discussed that there are reasonable reasons to believe that a trivial experiment for θ does not bring information about the parameter. Therefore, since X is a trivial experiment for θ given T, it seems that all the information in X is gained by observing only T. This is the heart of the Sufficiency Principle. The Sufficiency Principle states that for any sufficient statistic T, for any x and y in $\mathcal{X}$, if $T(x) = T(x')$ then $Inf(X, x, \theta) = Inf(X, x', \theta)$. This principle is usually followed by all statisticians, although not always explicitly mentioned: for inference about θ the statistician only needs to consider a sufficient statistic.

The Conditionality Principle is another important statistical principle: it might be seen as the reciprocal of the Sufficiency one. In the latter we say that the trivial experiment realized after T does not bring information about θ. Considering X_1 and X_2 as possible experiments, the Conditionality Principle states that a trivial experiment for θ, X_1 and X_2 does not bring information about θ. Let Y be some trivial experiment for (θ, X_1, X_2) assuming values in $\{1, 2\}$ and X_Y a mixture of X_1 and X_2. X_Y is observed in the following way: If the result of Y is 1 then we observe the result of X_1, if it is 2 then we observe the result of X_2. The Conditionality Principle states that $Inf((Y, X_Y), (i, x), \theta) = Inf(X_i, x, \theta)$, $\forall i \in \{1, 2\}$. In the sequel we show that this principle is more controversial than that of Sufficiency.

The Likelihood Principle is the object of the last part of this section. The Likelihood Principle states that any two possible outcomes having proportional likelihood functions must provide the same information about the parameter. Therefore, for any experiments X_1 and X_2 and any $x_1 \in \mathcal{X}_1$ and $x_2 \in \mathcal{X}_2$, if $L(\theta|x_1) \propto L(\theta|x_2)$, then $Inf(X_1, x_1, \theta) = Inf(X_2, x_2, \theta)$. It is possible to prove that this principle is stronger than the Sufficiency Principle and than the Conditionality Principle. Below, we prove that both the Sufficiency and the Conditionality Principles imply the Likelihood Principle. Ref. 2 and Ref. 3 also presented proofs of this important result.

Theorem 3.1. *The Sufficiency and the Conditionality Principles imply the Likelihood Principle.*

Proof. Consider $x_1 \in \mathcal{X}_1$ and $x_2 \in \mathcal{X}_2$ outcomes of, respectively, experiments X_1 and X_2, such that $L(\theta|x_1) \propto L(\theta|x_2)$. The statistic $T : \{1,2\} \times (\mathcal{X}_1 \cup \mathcal{X}_2) \mapsto (\{1,2\} \times (\mathcal{X}_1 \cup \mathcal{X}_2)) \cup \{0\}$ is such that: 1) $T(i,z) = (i,z)$, if $(i,z) \neq (1,x)$ and $(i,z) \neq (2,y)$; 2) $T(i,z) = 0$, if $(i,z) = (1,x)$ or $(i,z) = (2,y)$. Define $Y \in \{1,2\}$ in such a way that $P(Y=1) = P(Y=2)$ and Y is trivial for (θ, X_1, X_2). Y is used to define a mixture X_Y. Note that T is a sufficient statistic for (Y, X_Y). Therefore, using the Sufficiency Principle, $Inf((Y,X_Y),(1,x),\theta) = Inf((Y,X_Y),(2,y),\theta)$. Next, by Conditionality, $Inf(X_1,x,\theta) = Inf((Y,X_Y),(1,x),\theta)$ and $Inf(X_2,y,\theta) = Inf((Y,X_Y),(2,y),\theta)$. The Likelihood Principle follows straightforward.

□

A "scientist" that follows the Likelihood Principle can perform inference about the parameter solely based on the likelihood function. However, this principle is not followed by many statisticians. Ref. 4 and Ref. 5 provide interesting examples in which some classical statistical methods violate the Likelihood Principle. In one example, someone observes 3 failures out of 10 trials on two experiments. On the first one, the person would realize 10 trials and therefore, the distribution of the total of failures is binomial. On the second one the person would realize trials until he observed 3 failures and, therefore, the distribution of the total of failures is the negative binomial. Even though the unbiased estimation produces different results, posterior distributions are the same. This way, since classical statisticians follow the Sufficiency Principle and do not follow the Likelihood Principle, it is possible to conclude that they do not follow the Conditionality Principle. On the other hand, as was also shown in Ref. 6, bayesian statisticians usually follow all three principles.

In our opinion, it is reasonable to follow both the Sufficiency and Conditionality Principles whenever performing statistical analysis. Therefore, when looking for a more rigorous definition of the information function in the remaining part of this paper, the Likelihood Principle will be assumed. That is, the information given by the observation of some data will not give different values for data with proportional likelihood functions.

3.2. *Information on observed data*

Our interest in this section is on a function of the information about θ contained in a particular observed data. In the last section it was seen that such a function must not give different values for points with proportional likelihoods. Nevertheless, this property only gives a vague idea about how such a function should be. Therefore, we will go back to the definition given in the introduction in search for intuition: "Information is what it does for you, it changes your opinion".

Recall that the opinion one has about the parameter before (s)he realizes the experiment is given by his/her prior distribution for it. On the other hand, the opinion one has after an experiment is realized becomes, by the use of Bayes Theorem, his/her posterior distribution. Hence, since the information should represent the changing in opinion, it seems reasonable for it to compare prior and posterior distributions. If prior and posterior distributions are equal, it seems reasonable to assume there is no gain of information. We will also define that the information given when the prior distribution is equal to that of the posterior is 0. Since for any posterior different to the prior distribution there is some change in opinion, it seems reasonable to assume that information is always greater or equal to 0.

As discussed before, a trivial experiment should not bring any information about the parameter. Therefore, it should be less or equally informative than any other experiment. Note that prior and posterior are equal for trivial experiments, and the properties expected of these experiments are consistent with statement above.

Consider the following simple example: Someone chooses 3 balls among 4 balls, 2 black ones and 2 white. The 3 balls are put in an box. Another person is offered the possibility of observing one of three experiments; Experiment 1 consists of taking only one ball out of the urn; Experiment 2, two balls with replacement and; Experiment 3, two balls without replacement. The objective is to guess the number of white balls in the urn, 1 or 2. The person, a priori, does not believe any combination of balls is more likely than others, a uniform prior. Finally, the person assumes that all balls in the urn have equal probability of being selected.

Let θ be the number of white balls in the urn and X_i the number of white balls observed in the i-th experiment, denote by $P(\theta|X_i = x_i) = (a, b)$, for $P(\theta = 1|X_i = x_i) = a$ and $P(\theta = 2|X_i = x_i) = b$. The posterior probabilities are as follows:

(1) $P(\theta|X_1 = 0) = \left(\frac{2}{3}; \frac{1}{3}\right)$, $P(\theta|X_1 = 1) = \left(\frac{1}{3}; \frac{2}{3}\right)$;
(2) $P(\theta|X_2 = 0) = \left(\frac{4}{5}; \frac{1}{5}\right)$, $P(\theta|X_2 = 1) = \left(\frac{1}{2}; \frac{1}{2}\right)$, $P(\theta|X_2 = 2) = \left(\frac{1}{5}; \frac{4}{5}\right)$;
(3) $P(\theta|X_3 = 0) = (1; 0)$, $P(\theta|X_3 = 1) = \left(\frac{1}{2}; \frac{1}{2}\right)$, $P(\theta|X_3 = 2) = (0; 1)$.

Some common information functions applied to these experiments are:

(1) The Euclidean distance:
$$Inf_E(X_i, x_i, \theta) = \sqrt{\sum_j [P(\theta = j) - P(\theta = j|X_i = x_i)]^2}.$$
(2) $Inf_V(X_i, x_i, \theta) = [E(\theta|X_i = x_i) - E(\theta)]^2$.
(3) Kullback-Leibler divergence:
$$Inf_{KL}(X_i, x_i, \theta) = \sum_j P(\theta = j|X_i = x_i) \log\left(\frac{P(\theta=j|X_i=x_i)}{P(\theta=j)}\right).$$

For experiment 1, the information given by these functions for each vector of possible outcomes is, respectively, $\left(18^{-1/2}; 18^{-1/2}\right)$, $\left(36^{-1}; 36^{-1}\right)$ and $(0.024; 0.024)$. For experiment 2, $(0.424; 0; 0.424)$, $(0.09; 0; 0.09)$ and $(0.084; 0; 0.084)$. Finally, for experiment 3, $\left(2^{-1/2}; 0; 2^{-1/2}\right)$, $(0.25; 0; 0.25)$ and $(0.301; 0; 0.301)$.

These measures have interesting properties. Nevertheless, they do not offer a straightforward way of comparing experiments. Next section, this problem will be discussed in detail.

4. Information in an Experiment

4.1. *Blackwell sufficiency*

A celebrated definition in Statistical Inference is that of Sufficiency. According to the Sufficiency Principle exposed in Section 3.1, any inference based on data or on a sufficient statistic should be the same. Nevertheless, this principle is only useful to compare statistics inside the same space. Consider two experiments, X and Y, that depend on the parameter θ. One usually wants to choose between X and Y for inferences about θ based solely on their marginal distributions — the conditional distributions of X given θ and Y given θ. In this case, clearly the Sufficiency Principle is useless. In this section we expose the concept of Blackwell Sufficiency [7] and show that it is a natural generalization of the Sufficiency Principle for comparison of experiments.

From Section 3.1, a statistic T is sufficient for an experiment X, if X and θ are conditionally independent given T. Consequently, T is sufficient if and only if $p(x|\theta) = p(t|\theta)p(x|t)$. After performing T, there exists a simple "randomization exercise" which produces an experiment like X.

Let $X \in \mathcal{X}(X)$ and $Y \in \mathcal{X}(Y)$ be two statistical experiments. X is Blackwell Sufficient for Y if there exists a map $H : \mathcal{X}(X) \times \mathcal{X}(Y) \mapsto [0,1]$, a transition function, satisfying the following properties:

- For any $y \in \mathcal{X}(Y)$, $H(\cdot, y)$ is measurable on the σ-algebra induced by X, $\Im_{|X}$.
- For any $x \in \mathcal{X}(X)$, $H(x, \cdot)$ is a probability (density) function defined on $(\mathcal{X}(Y), \Im_{|Y})$.
- For any $y \in \mathcal{X}(Y)$, $p(y|\theta) = E(H(X,y)|\theta)$, the conditional expectation of $H(X,y)$ given θ.

Let $\mathcal{X}(X)$ and $\mathcal{X}(Y)$ be countable sets and define for all $x \in \mathcal{X}(X)$, $Z_x \in \mathcal{X}(Y)$ as a trivial experiment such that $P(Z_x = y) = H(x,y)$. From the definition of Blackwell Sufficiency, one can see that the random quantities (Z_X, θ) and (Y, θ) are equally distributed: X is Blackwell Sufficient for Y if and only if one can obtain an experiment with the same distribution as Y by observing X and, after that, performing a simple randomization exercise, Z_x.

The structure presented above leads us to conclude that Blackwell Sufficiency is the version of Sufficient Statistics for the comparison of experiments. Hence, it seems reasonable to state that is better to realize X than Y. It is not difficult to see that every Sufficient Statistic is also Blackwell Sufficient.

Having presented the overall aspects of Blackwell Sufficiency, a few examples are needed. These are related to those found in Ref. 8 and seem to give support to well known beliefs.

Example 4.1. Let X and Y be two experiments, π a parameter in $[0,1]$ and q and p known constants in $[0,1]$. Representing the Bernoulli distribution with parameter p by $\text{Ber}(p)$, consider also that the conditional distributions of X and Y given π

are, respectively:

$$X \sim \text{Ber}(\pi) \text{ and } Y \sim \text{Ber}(q\pi + (1-q)p).$$

If $A \sim \text{Ber}(q)$ and $B \sim \text{Ber}(p)$, either one independent of all variables considered in this problem, then for $Y' = AX + (1-A)B$ we have that (Y', π) and (Y, π) are equally distributed. Therefore, X is Blackwell Sufficient for Y.

Example 4.2. Let X and Y be two experiments, π a parameter in $[0,1]$, q and p known constants in $[0,1]$ and $n \in N$. Consider also the following. Representing the Binomial distribution with parameters n and p by $\text{Bin}(n,p)$, consider also that the conditional distributions of X and Y given π are, respectively:

$$X \sim \text{Bin}(n, \pi) \text{ and } Y \sim \text{Bin}(n, q\pi + (1-q)p).$$

We know that X has the same distribution as a sufficient statistic for a sequence $X_1, \ldots, X_n$ conditionally independent identically distributed given π with $X_1 \sim \text{Ber}(\pi)$. Thus X is Blackwell Sufficient for $X_1, \ldots, X_n$.

From example 4.1, we know that the sequence previously considered is Blackwell Sufficient for a sequence $Y_1, \ldots, Y_n$ conditionally independent given π and such that, conditionally to π, $Y_1 \sim \text{Ber}(q\pi + (1-q)p)$.

Finally, $Y_1, \ldots, Y_n$ is Blackwell Sufficient for Y, since it has the same distribution as a sufficient statistic for $Y_1, \ldots, Y_n$. Thus, since Blackwell Sufficiency is a transitive relation, X is Blackwell Sufficient for Y.

Example 4.3. Next, we generalize the example of Section 3.2. Consider an urn with N balls. θ of these balls are black and $N - \theta$ are white. n $(\leq N)$ balls are selected from the urn.

By stating that $(X_1, \ldots, X_n)$ is a sample with replacement from the urn, we mean:

(1) Conditionally to θ, $X_1 \sim \text{Ber}\left(\frac{\theta}{N}\right)$;
(2) Conditionally to θ, $X_1, \ldots, X_n$ are identically distributed;
(3) X_{i+1} is conditionally independent of $(X_i, \ldots, X_1)$ given θ, $\forall i \in \{1, \ldots, n-1\}$.

Analogously, $(Y_1, \ldots, Y_n)$ corresponds to a sample without replacement, that is:

(1) Conditionally to θ, $Y_1 \sim \text{Ber}\left(\frac{\theta}{N}\right)$;
(2) $Y_{i+1}|(y_i, \ldots, y_1, \theta) \sim \text{Ber}\left(\frac{\theta - \sum_{j=1}^{i} y_j}{N-i}\right)$,
$\forall i \in \{1, \ldots, n-1\}$, $\forall (y_i, \ldots, y_1) \in \{0,1\}^i$.

Next, we prove that $(Y_1, \ldots, Y_n)$ is Blackwell Sufficient for $(X_1, \ldots, X_n)$.

Define $X_1^* = Y_1$, $t_i = \sum_{j=1}^{i} y_j$ and $\forall i \in \{1, \ldots, n-1\}$ two random quantities A_{i+1} and B_{i+1}. These two quantities are such that:

(1) $A_{i+1} \sim \text{Ber}\left(\frac{N-i}{N}\right)$, and is independent of all other variables;
(2) $B_{i+1}|t_i \sim \text{Ber}\left(\frac{t_i}{i}\right)$;

(3) $\forall i \in \{1, \ldots, n\}$, B_i, conditionally to t_i, is independent of $((A_1, \ldots, A_n); (B_1, \ldots, B_{i-1}); (Y_{i+1}, \ldots, Y_n); \theta)$.

Define:

$$X_{i+1}^* = A_{i+1}Y_{i+1} + (1 - A_{i+1})B_{i+1}.$$

We obtain that, conditionally to θ, $X_{i+1}^*|t_i \sim \text{Ber}(\theta/N)$, $\forall t_i \in \{0, \ldots, i\}$. Therefore, $X_{i+1}^* \sim \text{Ber}(\theta/N)$ and is conditionally independent of $(Y_i, \ldots, Y_1)$ given θ. Finally, since $(X_i^*, \ldots, X_1^*)$ is a function of $(Y_i, \ldots, Y_1)$, $(A_i, \ldots, A_2)$ and $(B_i, \ldots, B_2)$, we conclude that X_{i+1}^* is independent of $(X_i^*, \ldots, X_1^*)$ given θ.

By the previous conclusions, we know that $(X_1^*, \ldots, X_n^*, \theta)$ is identically distributed to $(X_1, \ldots, X_n, \theta)$. By construction, we also know that $(X_1^*, \ldots, X_n^*)|(Y_1 = y_1, \ldots, Y_n = y_n)$ is trivial, $\forall (y_1, \ldots, y_n) \in \{0, 1\}^n$. Therefore, it is proved that sampling without replacement is Blackwell Sufficient for sampling with replacement.

Example 4.4. A simple corollary of the above result is now presented. First we recall that T_n is a sufficient statistic for $(Y_1, \ldots, Y_n)$, the sampling without replacement presented in Example 3, and that, conditionally to θ, T_n is Hypergeometric with parameter (N, n, θ); Consequently, T_n is Blackwell Sufficient for $(Y_1, \ldots, Y_n)$. We know that $(Y_1, \ldots, Y_n)$ is Blackwell Sufficient for $(X_1, \ldots, X_n)$, , the sampling with replacement presented in Example 3. Also $(X_1, \ldots, X_n)$ is Blackwell Sufficient for $\sum_{i=1}^n X_i = S_n$ and, conditionally to θ, $S_n \sim \text{Bin}(n, \theta/N)$. Using the transitive property of Blackwell Sufficiency we conclude that T_n is Blackwell Sufficient for S_n.

Example 4.5. In many inferential problems we are concerned with an experiment Y and a parameter μ such that $Y|\mu \sim N(\mu, s)$, a normal random quantity. We commonly think of this model as $Y = \mu + \epsilon$, $\epsilon \sim N(0, s)$. We interpret ϵ as some random noise on a measure of μ. The larger the value of s, the more intense the noise is. Therefore, it seems reasonable that if $X \sim N(\mu, s')$ and $s < s'$, then Y is more informative for μ than X. It is possible to prove that this intuitive idea is preserved by Blackwell Sufficiency. Let $Z \sim N(0, s' - s)$ independent of Y, then $(Y + Z, \theta)$ is equally distributed to (X, θ).

Example 4.6. Other two important distributions are the Poisson and Exponential. We show that if, conditionally to θ, $X_1 \sim \text{Poisson}(\theta)$ and $X_2 \sim \text{Poisson}(p\theta + k)$, $k \in R_+$, $p \in [0, 1]$, then X_1 is Blackwell Sufficient for X_2. Using the same technique we will also prove that, conditionally to θ, if $Y_1 \sim \text{Exp}(\theta)$ and $Y_2 \sim \text{Exp}(p\theta + k)$, $k \in R_+$, $p \in [0, 1]$, then Y_1 is Blackwell Sufficient for Y_2.

Conditionally to θ, let Z' and A be independent Poisson Processes in $[0, 1]$ with rates respectively θ and k. Define a new process, Z, in which we choose, independently and with probability $(1 - p)$, points of Z' to be discarded. Conditionally to θ, Z is still a Poisson Process in $[0, 1]$ with rate $p\theta$ and $Z \cup A$ is a Poisson Process in $[0, 1]$ with rate $p\theta + k$. If $n(Z'), n(Z), n(A)$ and $n(Z \cup A)$ are the number of occurrences in Z', Z, A and $Z \cup A$, then $n(Z) + n(A) = n(Z \cup A)$. Since $(n(Z'), \theta)$ has the same distribution of (X, θ) and $(n(Z \cup A), \theta)$ has the same distribution of (X_2, θ), then Blackwell Sufficiency is proven.

Here, conditionally to θ, let Z' and A be independent Poisson Processes in R_+ with rates respectively θ and k. Again, Z is the process which is obtained by disregarding points with probability $(1-p)$: $Z \cup A$ is a Poisson Process in R with rate $p\theta + k$. If T'_Z, T_Z, T_A and $T_{Z \cup A}$ are the waiting times until the first occurrence, respectively, in Z', Z, A and $Z \cup A$, then $T'_Z \sim \text{Exp}(\theta)$, $T_Z \sim \text{Exp}(p\theta)$, $T_A \sim \text{Exp}(k)$ and $T_{Z \cup A} \sim \text{Exp}(p\theta + k)$. Since, $T_{Z \cup A} = \min(T_Z, T_A)$, it is proven that $\text{Exp}(\theta)$ is Blackwell Sufficient for $\text{Exp}(p\theta + k)$.

4.2. *Equivalence relation in experiment information*

In this section, Ω is a countable set and $F = \wp(\Omega)$, that is, the set of all subsets of Ω. Therefore, for any random quantity $X : \Omega \mapsto \mathcal{X}$, $F_X = \wp(\mathcal{X})$. It will also be assumed that $\forall \theta \in \Theta, \exists x \in \mathcal{X}, P(x|\theta) > 0$.

Blackwell Sufficiency was introduced in the last section. Using it, it is possible to define an equivalence relation between experiments: X and Y are Blackwell Equivalent if any one is Blackwell Sufficient for the other, $X \approx Y$. We show that this equivalence relates to the Likelihood Principle, Section 3.1. In the sequel, we prove the following result:

Theorem 4.1 (Blackwell likelihood (BL)). *Let $X \in \mathcal{X}_1$ and $Y \in \mathcal{X}_2$ be two experiments. $X \approx Y$ if and only if $P(\{x \in \mathcal{X}_1 : L_X(\cdot|x) \propto L(\cdot)\}|\theta') = P(\{y \in \mathcal{X}_2 : L_Y(\cdot|y) \propto L(\cdot)\}|\theta')$, $\forall \theta' \in \Theta$ and all likelihood function $L(\theta)$ derived from either X or Y.*

For simplicity, we use a special notation that reduces the algebra involved. Since all sets are countable, we consider an ordering inside them. Let, $\forall \theta \in \Theta$, $P(X = x|\theta)$ be a probability function, then we define that $p(.|\theta)$ is a vector such that in its i-th position the value assumed is $P(x_i|\theta)$; x_i is the i-th element of the ordering assumed in the set of values of X. If F is a map from $\xi_1 \times \xi_2$ into $[0, 1]$, then we also use the symbol F as the matrix which has in its j-th row and i-th column position the value of $F(x_i, y_j)$; x_i is the i-th element of the ordering in ξ_1 and y_j is the j-th element of that in ξ_2. Finally, we recall that a transition (transposed) matrix is a matrix in which all elements are greater or equal to 0 and for any column the sum of its elements is equal to 1.

Proof. $(\Leftarrow)$ Let $S : \mathcal{X}_1 \mapsto [0,1]^\Theta$ and $T : \mathcal{X}_2 \mapsto [0,1]^\Theta$, such that $S(x)$ and $T(y)$ are likelihood nuclei of x and y - a likelihood nucleus is a chosen likelihood between all of those which are proportional. Recall Ref. 3 that S and T are, respectively, minimal sufficient statistics for X and Y. Therefore, $S \approx X$ and $T \approx Y$. By the hypothesis, (S, θ) and (T, θ) are identically distributed, therefore they are Blackwell Equivalent. By the transitive property of Blackwell Equivalence $S \approx T$, since $S \approx X \approx Y \approx T$.

$(\Rightarrow)$ Consider the above statistics, S and T. For simplicity, we call $S(X(\Omega)) = \xi_X$ and $T(Y(\Omega)) = \xi_Y$. We also call $P(S(X) = l_x|\theta) = p_X(l_x|\theta)$ and $P(T(Y) = l_y|\theta) =$

$p_Y(l_y|\theta)$. Clearly, by construction, for every two points in ξ_X or in ξ_Y, if their likelihood functions are proportional, then they are the same point. Since S and T are minimal sufficient statistics, $S \approx X$, $T \approx Y$ and, therefore, $S \approx T$.

Since S is Blackwell Sufficient for T, there exists a map $A : \xi_X \times \xi_Y \mapsto [0,1]$ such that A is a transition matrix and:

$$Ap_X(.|\theta) = p_Y(.|\theta), \forall\theta \in \Theta.$$

On the other hand, T is also Blackwell Sufficient for S and, similarly, there exists a map $B : \xi_Y \times \xi_X \mapsto [0,1]$ such that B is a transition matrix and:

$$Bp_Y(.|\theta) = p_X(.|\theta), \forall\theta \in \Theta.$$

From these two equations, there exist two other transition matrices, $M = BA$ and $N = AB$, such that:

$$Mp_X(.|\theta) = p_X(.|\theta), \forall\theta \in \Theta,$$
$$Np_Y(.|\theta) = p_Y(.|\theta), \forall\theta \in \Theta.$$

Since M and N are transition matrices, respectively, from ξ_X to ξ_X and from ξ_Y to ξ_Y, we consider the Markov Chains associated to them. All probability functions in the family $\{p_X(.|\theta) : \theta \in \Theta\}$ are invariant measures for M. Note that there are no transient states in M. If there were, let x be a transient state in M, consequently $P(x|\theta) = 0$, $\forall\theta \in \Theta$. However, such a state cannot occur by the definition in the beginning of this section; there is no transient state in M.

To proceed with the proof we use the following result stated in Ref. 9:

Lemma 4.1. *Consider a Markov Chain on a countable space $\mathcal{X}$ with a transition matrix A and no transient states. Let A have irreducible components $C(1)$, ..., $C(n)$, ... Then, there exists an unique set of probability functions $\{p_j(\cdot) : j \in N\}$, with $p_j(x)$ defined in $\{1, \ldots, |C(j)|\}$, such that all invariant measures (μ) of A can be written as the following:*

If $c_{k,i}$ is the i-th element of $C(k)$, then $\mu(c_{k,i}) = p_k(i).q(k)$ and q is a probability function in N.

To interpret the above result, we consider the sub-matrix A_k associated to $C(k)$. Since the A_k is irreducible, it only has one invariant measure, p_k. Now suppose that at the initial position (X_0) of the Chain each component $C(k)$ has probability q_k of being chosen. As n increases the law of X_n converges to the one provided by the lemma.

Using the lemma, since $C(1), \ldots, C(n), \ldots$ are irreducible components of M and $c(k,i)$ is the element of number i of $C(k)$, then $p_1(c(k,i)|\theta) = p_k(i)q_{k,\theta}$. Consequently,

$$p_1(c(k,i)|\theta) = p_1(c(k,j)|\theta)\left(\frac{p_k(i)}{p_k(j)}\right)$$

If two states are in the same irreducible component then their likelihood functions are proportional. The same proof holds to matrix N.

The i-th element of ξ_X is said to connect to the j-th element of ξ_Y if $A(i,j) > 0$. Similarly, the i-th element of ξ_Y is said to connect to the j-th element of ξ_X if $B(i,j) > 0$. Note that every state in ξ_X connects to at least one state in ξ_Y and vice-versa. This is true because A and B are transition matrices.

For all $x_1 \in \xi_X$, if x_1 connects to $y \in \xi_Y$ then y only connects to x_1. If there were a state $x_2 \in \xi_X$ such that y connected to x_2, then x_1 and x_2 would be on the same irreducible component of M. Therefore x_1 and x_2 would yield proportional likelihood functions and, by the definition of S, $x_1 = x_2$. Similarly, if a state $y \in \xi_Y$ connects to a state $x \in \xi_X$ then x connects solely to y.

Finally, we conclude that every state in ξ_X only connects to one state in ξ_Y and vice-versa. Also, if $x \in \xi_X$ connects to $y \in \xi_Y$, then y connects to x and vice-versa. This implies that if x connects to y, then $P(X = x|\theta) = P(Y = y|\theta)$, $\forall \theta \in \Theta$. Since S and T are sufficient the Theorem is proved.

□

Applying the above Theorem and the Likelihood Principle one obtains the following result: if X is Blackwell Equivalent to Y,

$$A_e = \{x : Inf(X, x, \theta) = e\} \subset \mathcal{X}_1; B_e = \{y : Inf(Y, y, \theta) = e\} \subset \mathcal{X}_2,$$

then $P(A_e|\theta) = P(B_e|\theta), \forall \theta \in \Theta$, for all possible e — the value of information.

> For any information function, Inf, satisfying the Likelihood Principle — if x and y yield proportional likelihood functions, then $Inf(X, x, \theta) = Inf(Y, y, \theta)$ —, X is Blackwell Equivalent to Y, if and only if, the distribution of (Inf, θ) for X and Y are the same.

4.3. *Experiment information function*

In the last section, we did not define any information function but a representation of it, Inf. Now, we are interested possible functions capable of describing the information of a statistical experiment. A possible approach to this problem is considering that the information gained is an utility function [10] which the researcher wants to maximize. This way, by the results in decision theory $Inf(X, \theta) = E(Inf(X, x, \theta))$. Since we consider the data information function as non-negative, the utility function is concave, see Ref. 11 for instance.

Proceeding with this approach, we compare the different information functions presented in Section 3.2. Recall the example in Section 3.2 and remember that, for any of the information functions introduced, the maximum possible value is obtained when the posterior distribution is such that $P(\theta = 0|x) = 0$ or $P(\theta = 0|x) = 1$. Therefore, to compare those information functions, we divide all of them by these maxima.

For the three possible experiments described in Section 3.2, X_1, X_2, X_3, the information vectors using euclidean distance are: $\left(18^{-1/2}; 18^{-1/2}\right)$, $(0.424; 0; 0.424)$ and $\left(2^{-1/2}; 0; 2^{-1/2}\right)$. Since the maximum possible information is $\frac{1}{\sqrt{2}}$, in the first experiment, with probability 1 the gain of information is 33%. That is, a small gain with a small risk. On the second experiment, with probability 56% the gain is 60% of the maximum and with probability 44% it is 0% of the maximum, moderate gain with moderate risk. In the third experiment one can get 100% of the maximum possible information with probability 33% and can get 0% of the maximum possible information with probability 67%, maximum gain with great risk. In conclusion, if one uses the Euclidian's "utility", then he/she would have no preference among the three experimnents, since, for all of them, the expected information gain is of 33%. This is surprising as the third experiment is Blackwell Sufficient for both the others.

Consider the second information function, $Inf(X, x, \theta) = [E(\theta) - E(\theta|X = x)]^2$. In the three possible experiments, the information vectors using this function are, $(\frac{1}{36}; \frac{1}{36})$, $(0.09; 0; 0.09)$ and $(0.25; 0; 0.25)$. Dividing by the maximum, we obtain the following vectors: $(\frac{1}{9}; \frac{1}{9})$, $(0.36; 0; 0.36)$ and $(1; 0; 1)$. Therefore, the expected information gain is, respectively, 11%, 20% and 33%. Thus, the third experiment is more informative than the second which is more informative than the first. It is interesting to note that the information of an experiment using this metric is: $Inf(X, \theta) = V(E(\theta|X))$.

Finally, considering the Kullback-Leibler divergence, the informations vectors are: $(0.024; 0.024)$, $(0.084; 0; 0.084)$ and $(0.301; 0; 0.301)$. Dividing by the maximum, .301, we obtain: $(0.081; 0.081)$, $(0.278; 0; 0.278)$ and $(1; 0; 1)$. The expected gain of information is respectively, 2.4%, 4.6% and 33%. Again, the informativeness order, X_1, X_2, X_3, is in complete agreement with the ordering induced by Blackwell Sufficiency.

5. Final Considerations

Our main objective in writing this note was to reflect on the concept of information. It seems that there is no universal concept of information. The reader might have noticed that we follow an approach different from the one provided by Information Theory. As a matter of fact, we were not concerned with the old and important concept of information of a (prior) distribution. Another possible approach is related to Fischer Information. In this case, the value of the information depends on the value of the parameter, which is unknown and the object of the inference. It must be recognized that Fischer Information is one of the most important tools for the construction of Modern Statistical Inference. What could be questioned is the title given for this measure: Information. In fact, we do not know how to answer the questions presented in Section 1 when using this measure.

We used Basu's concept of information to develop our reflection. To operationalize Basu's concept, we had to strongly use the Statistical Principles. We alert the reader that we presented these principles with Bayesian "eyes" while, frequently, those principles are presented under a frequentist perspective. For instance, the definition of the Conditionality Principle that we presented is slightly different from that of Ref. 2 and Ref. 3. We are convinced that trivial experiments (or ancillary statistics) should not bring information about the parameter and, therefore, one should follow the Sufficiency and Conditionality Principles. Consequently, one must follow the Likelihood Principle and abandon approaches based on sample spaces: the frequentist way. One of the most celebrated hypothesis test is the asymptotic likelihood ratio test that does not depend on stopping rules. Consequently, it also does not depend on the sample spaces. The three Statistical Principles, although created by frequentist statisticians, are intrinsically considered whenever using the Bayesian operation.

On trying to understand the concept of information, we were leaded to the problem of comparison of experiments — a pre-posterior analysis. When comparing statistics in different sample spaces, a known alternative to the classical sufficiency definition is that of Blackwell Sufficiency. A principle can be naturally induced by Blackwell Sufficiency. Let X and Y be two experiments such that X is Blackwell Sufficient for Y. If you are restricted to chose only one, it should be X! This principle was used in some important cases, for example, to show that sampling without replacement is preferable than that with replacement. Blackwell Sufficiency is also useful for characterization of distributions, for instance Ref. 12.

The main achievements of this paper are the examples of the use of Blackwell Sufficiency and the theorem of Section 4.2. Looking at the examples, some of the preferences were expected — the ones in the sampling and the normal distribution. Other examples, such as 1, 2 and 6, surprise the authors. An important question related to all of these examples is: Are the experiments Blackwell Equivalent? The answer is no: for Example 1, see Ref. 8. The general proof for the negative answer uses the BL Theorem, discussed below.

BL Theorem states that two experiments are Blackwell Equivalence, if and only if, the two likelihood-function statistics are equally distributed conditionally to θ. Two applications of this Theorem are as follows. i. If one believes in the Likelihood Principle and that two experiments are equally informative if the distribution of the information functions are equal, then the person believes in the information equivalence between experiments induced by Blackwell Equivalence. ii. To prove that an experiment is not Blackwell Sufficient for another is, in general, difficult: one must show that there is no transition function from one to the other. However, if X if Blackwell Sufficient for Y, using BL Theorem, if the likelihood-function statistics, conditionally to θ, are not equally distributed, then Y is not Blackwell Sufficient for X. We leave to the reader to check that, in all the examples of Section 4.1, Blackwell Equivalence does not hold. Recall that we did not prove BL Theorem for the absolutely continuous cases.

In the first example, discussed in Section 3.2, we try three measures of information. The first one, based on the Euclidean distance, did not differentiate the experiments. This seems incoherent since there is a strict order of preference induced by Blackwell Sufficiency. The second and third measures of information both satisfy the strict order of preference induced by Blackwell Sufficiency. An important distinction between these two measures is that, while the Kullback-Leibler divergence does not take into account the possible values of the parameter space, the variance of the posterior mean does strongly depend on these values. Another difference is that the latter may not be defined, although the former always can be computed.

We end this paper by evoking the memory of D. Basu who, among other teachings, inspires the authors with the illuminating concept of information: "Information is what it does for you, it changes your opinion".

Acknowledgments

The authors are grateful to Professor Basu for his legacy on Foundations of Statistics. We are grateful to Adriano Polpo, Estéfano Alves de Souza, Fernando V. Bonassi, Luis G. Esteves, Julio M. Stern, Paulo C. Marques and Sergio Wechsler for the insightful discussions and suggestions. The authors of this paper have benefited from the support of CNPq and FAPESP.

References

[1] D. Basu, A note on likelihood, in *Lecture Notes in Statistics* **45**, (Springer-Verlag, Berlin, 1988)
[2] A. Birnbaum, *Journal of the American Statistical Association* **57**, 269 (1962).
[3] D. Basu, Statistical information & likelihood, in *Lecture Notes in Statistics* **45**, (Springer-Verlag, Berlin, 1988)
[4] D. V. Lindley and L. D. Philips, *The American Statistician* **30**, 112 (1976).
[5] C. A. B. Pereira and D. V. Lindley, *The Stastician* **36**, 15 (1987).
[6] S. Wechsler, C. A. B. Pereira and P. C. Marques, *AIP Conference Proceedings* **1073**, 96 (2008).
[7] D. Blackwell, *Proceedings of the 2nd Berkeley Symposium* **57**, 93 (1971).
[8] D. Basu and C. A. B. Pereira, *Brazilian Journal of Probability and Statistics* **4**, 137 (1990).
[9] P. A. Ferrari and A. Galves, *Coupling and Regeneration for Stochastic Processes* (Sociedad Venezoelana de Matematicas, Caracas, 2000).
[10] M. H. DeGroot, *Optimal Statistical Decisions* (Wiley, New York, 1970).
[11] M. H. DeGroot, *Annals of Mathematical Statistics* **33**, 404 (1971).
[12] D. Basu and C. A. B. Pereira, *Sankhya: The Indian Journal of Statistics, Series A* **45**, 99 (1983).

MULTI-A(GE)NT GRAPH PATROLLING AND PARTITIONING

Y. ELOR and A.M. BRUCKSTEIN
Department of Computer Science,
Technion, Haifa 32000, Israel

We introduce a novel multi agent patrolling algorithm inspired by the behavior of gas filled balloons. Very low capability ant-like agents are considered with the task of patrolling an unknown area modeled as a graph. While executing the proposed algorithm, the agents dynamically partition the graph between them using simple local interactions, every agent assuming the responsibility for patrolling his subgraph. Balanced graph partition is an emergent behavior due to the local interactions between the agents in the swarm. Extensive simulations on various graphs (environments) showed that the average time to reach a balanced partition is linear with the graph size. The simulations yielded a convincing argument for conjecturing that if the graph being patrolled contains a balanced partition, the agents will find it. However, we could not prove this. Nevertheless, we have proved that if a balanced partition is reached, the maximum time lag between two successive visits to any vertex using the proposed strategy is at most twice the optimal so the patrol quality is at least half the optimal. In case of weighted graphs the patrol quality is at least $\frac{1}{2}l_{min}/l_{max}$ of the optimal where l_{max} (l_{min}) is the longest (shortest) edge in the graph.

Keywords: Patrol, Graph Partition, Emergent Behavior, Nature-Inspired Paradigms, Swarm Intelligence.

1. Introduction

To patrol is to continuously travel through an area. The purpose of patrolling is to visit every point in the area as often as possible. "Informally, a good (patrol) strategy is one that minimizes the time lag between two passages to the same place and for all places" [1]. It is convenient to model the area being patrolled as an undirected graph. In this model, time is discrete and will be measured in cycles. Using the undirected graph model we can loosely define the patrolling task as continuously visiting all the vertices of a graph. The *idleness* of a vertex is defined as the time since the last visit to this vertex by an agent. In this work we will address the *worst idleness* evaluation criterion defined by Machado and colleagues [1, 2].

We present the Balloon DFS algorithm (BDFS) as a patrolling strategy. The algorithm addresses a simplified version of the problem in which all edge lengths are assumed equal (a variant addressing weighted graphs is described in Section 7). When k agents are patrolling a graph G using BDFS, they dynamically partition G into k connected subgraphs. Experiments show that in time the algorithm achieves a stable partition. We prove that the largest subgraph in any stable partition is of

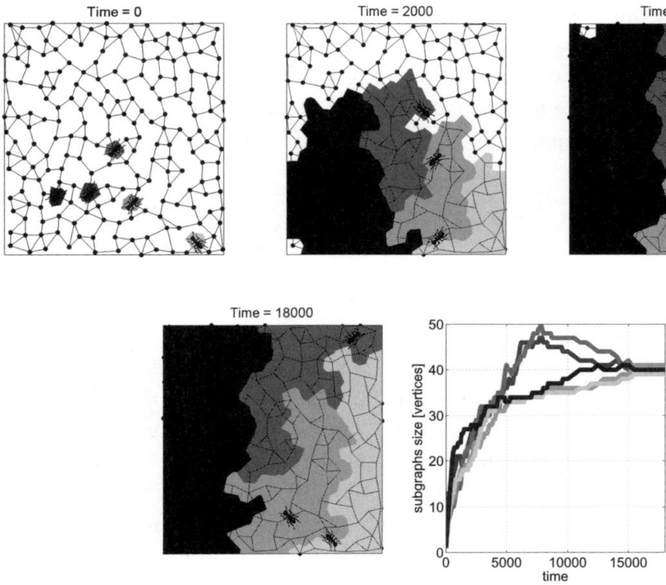

Fig. 1: A typical run of BDFS on a random spatial graph with 5 agents. The different colors represent areas controlled by different agents. The graph was created by randomizing locations for 201 vertices on the plane and then connecting every vertex to its three closets neighbors. The number of vertices controlled by every agent as a function of time is presented in the rightmost figure. The resulting configuration after about 15,000 time cycles is balanced.

size at most $\frac{|G|}{k} + \frac{k}{2}$. Every agent patrols his subgraph using DFS hence the *worst idleness* is $2\frac{|G|}{k} + k$. In every time cycle every agent can visit at most one vertex so every multi-agent strategy (including the optimal) yields a *worst idleness* of at least $|G|/k$. Assuming $|G| \gg k$, the *worst idleness* of BDFS is about $2|G|/k$ which is at most twice the optimal.

Some previously suggested partition based strategies work in two stages, see e.g. [3, 4]. In the first stage, the graph is partitioned between the agents and in the second the agents are sent to patrol the graph. The first stage complexity is high and requires some kind of "coordinator agent" who knows the graph in advance. In contrast, in our approach, the graph partitioning is executed "on the run", there is no need to a-priory know the graph and there is no need for any preprocessing before starting the patrol. Furthermore, if the graph changes during the run, or some agents break down, the agents automatically repartition the graph without any external intervention. In case of an agent break down the malfunctioning agent domain will be conquered and patrolled by other agents. On the downside, BDFS cover time is large and reaching a good partition may take a long time.

The algorithm can be classified as a local search algorithm. The agents continuously change the graph partition toward (hopefully) better configurations. Under the heavy restrictions of the lack of global information and the low communication and memory abilities of the agents a local search algorithm is a natural choice.

A short version of this paper appeared in ANTS2010 conference [5].

2. Previous Work

When facing any multi agent task one must carefully address the assumed agents' capabilities. Obviously the strategy selected will be based on the assumed agent capabilities.

Given a graph and single agent to patrol it we can find a minimal length closed path which covers all vertices of the graph. Finding that kind of path is the well known traveling sales person problem (TSP). Chevaleyre *et al.* shown that an optimal single agent strategy would be to follow the best TSP circuit [3]. Considering a multi agent scenario, we can make all agents follow the same cycle while maintaining a constant gap between them or alternatively partition the graph into subgraphs, each subgraph being patrolled by an agent performing an optimal circuit in it.

The scenario where the agents can sense the *idleness* of all the vertices of the graph lead to several *heuristic* approaches [1, 2]. Two different assumptions were made: either each agent can sense only the vertices' *idleness* relative to it own visits or each agent sense the vertices' *idleness* with respect to all agents. Note that in the second case the *idleness* of the vertices serves as a communication method. Among other strategies considered, a straightforward one was to choose the shortest route to the vertex with the highest *idleness*. A distributed sensing variation proposed by Wagner *et al.* is to choose the vertex with the highest *idleness* only from the adjacent vertices as inspired by ants [6]. We shown that a similar process can be used to construct patrolling cycles [7]. Yanovski *et al.* suggested another distributed sensing variation in which the agent marks the graph edges with pheromone marks (instead of the vertices) which leads to patrolling the graph using an Euler cycle [8].

The graph partitioning problem was widely studied, see a survey of the work up to 1997 by Elsner [9]. The most common variant of the problem is the k-cut problem. k-cut is an NP-hard problem which consists of finding a partition of a graph into k size balanced parts so that the sum of weights of all edges cut is minimized. Note that in the k-cut problem the partition subgraphs are not required to be connected. On the other hand, in our problem, we are not required to minimize the edges cut weight.

Recently, Ahmadi and Stone [10] proposed a patrolling strategy based on dynamic graph partition algorithm in which the agents exchange their boundary vertices using negotiating process. The algorithm they propose requires direct communication and computationally stronger agents. Furthermore, they do not have guarantees on the quality of the emergent partition.

In the heuristic algorithm ANTS proposed by Comellas and Sapena [11] some agents are working together to color a graph in k colors such that the weight of the cut between the colors is minimal. The subgraphs created using this approach are not connected. Another approach is constructing a "vertex-market" [2, 12]. Each agent owns some vertices and it is his responsibility to patrol his vertices. Agents can trade vertices in the market, for example an agent can trade a far vertex with a closer one and by that decrease its route length. In this scheme, a vertex can belong to several agent routes.

In the bubble scheme [13] k vertices are chosen as seeds and the seeds expand BFS-style (or "diffuse") to create a k-partition of the graph. Afterward, k new seed vertices are chosen based on the resulting partition and the process is repeated until the seeds do not change their location. Meyerhenke *et al.* proved convergence for a variant of the algorithm [14], however with no analytic guarantees on the partition quality.

Some recent works have suggested a new evaluation criteria for the patrolling problem [4, 15, 16]. Assuming some intruder model the new evaluation criterion is, roughly speaking, the ability to stop the intruder. The main theme of these patrolling strategies is to use probabilistic algorithms in order to avoid static patrolling patterns which can be exploited by intruders. Paruchuri *et al.* modeled the patrolling problem as a game between the patrolling agent and an Intruder [15]. A heuristic strategy was developed using this model. Sak *et al.* defined three types of intruders and examined the performance of different patrolling strategies against these intruders [4]. Agmon *et al.* addressed a variant of the patrolling problem in which the agents are required to patrol around a perimeter against smart intruders [16]. However, stopping intruders is only one aspect of patrolling. For applications such as cleaning, maintenance, surveillance or guarding against weak intruders, the *worst idleness* criterion remains very important. Hence in this work, our goal is to minimize *worst idleness.*

3. High-Level description of the Algorithm and the Physical Intuition

Our algorithm is inspired from gas filled rubber balloons. Consider a space volume V with N balloons inside it. The balloons are all filled with the same quantity of gas and they are infinitely elastic. Given enough time, the system will converge to a balanced state in which the balloons fill the whole space V and the pressure inside the balloons is equal. Since the amount of gas in every balloon is the same, the balloons will have a volume of V/N each. The algorithm we propose mimics the balloons' behavior. Every agent controls a connected part of the graph. There are some border edges i.e. edges who connect vertices belonging to different agents. The agents sense those border edges and apply some "artificial pressure" over them - similarly to the pressure the gas molecules apply on the balloon surface. When the pressure applied by an agent is higher than the pressure applied from the other side, the agent can conquer the vertex on the other side. Since agents who control relatively small area hit the border edges more often than agents who control larger areas, they apply more pressure and the process described will hopefully converge to a balanced partition of the graph.

In physical balloon systems the space is continuous hence there is always a perfectly balanced partition of the space. In our system $|G|/k$ is generally not an integer so such a partition does not always exist. The most balanced partition is a partition in which some agents control subgraphs of size $\lfloor |G|/k \rfloor$ and some of size $\lceil |G|/k \rceil$. It is desired that such a partition will be stable i.e. no more conquests will happen. In order to achieve this we set the rule that an agent can conquer a vertex from another agent only if the size difference of the agents' subgraphs is (strictly) larger than 1. As a result, every partition in which the size difference between every two adjacent subgraphs is at most 1 is stable. We show that in a stable partition every subgraph is of size between $\frac{|G|}{k} - \frac{k}{2}$ and $\frac{|G|}{k} + \frac{k}{2}$ (see Lemma 5.3).

4. The Algorithm

4.1. *Model*

Ant-like agents having limited capabilities are considered here. The agents are assumed to have little memory (a finite number of registers with $O(log_2 |G|)$ bits), they all execute the same algorithm and no direct communication is allowed between them. However, the agents can mark their neighborhood with pheromone stamps which can later be sensed. These markings are used as a primitive form of distributed memory and communication. The number of markings on every vertex or edge is fixed where the size of each marking is $O(log_2 |G|)$ bits.

For simplicity, we use a *strong synchronous* model. We assume that agents are operating in rounds. In every round the agents have slightly different timing phase so they never work simultaneously. In this model every passage between adjacent vertices takes one time cycle. Other operations like marking or sensing marks are assumed to be instantaneous. These *strong synchronous* assumptions can be relaxed later.

4.2. *Multi level DFS*

Depth First Search (DFS) can be used in order to travel a graph. A variant of DFS called Multi Level DFS (MLDFS) was introduced by Wagner *et al.* [17]. MLDFS is carried out by using marks on the graph edges and vertices. An important property of the MLDFS is its noise robustness. If at some point the agent loses its backtracking path he just starts a new search. In MLDFS, a time cycle is lost every time a new search starts, hence our MLDFS cover time is $2|G|-1$ as opposed to $2(|G|-1)$ in a regular DFS.

4.3. *The balloon mechanism*

In BDFS every agent controls a connected subgraph and is responsible for patrolling it. We introduce the balloon mechanism which eventually balances the graph partition. The balloon mechanism requires an additional mark on each edge: the pressure mark. The *pressure* on an edge is a *vector mark* i.e. a mark which gives opposite readings when being sensed from different sides of the edge. For example, consider the edge uv. Assuming $P(uv) = p$ i.e. the pressure of edge uv when sensed from vertex u is p; then $P(vu) = -p$ i.e. the pressure when sensed from vertex v is $-p$.

A *border edge* is an edge connecting two vertices belonging to different agents. Every time an agent enters a vertex for the first time in the current search, it senses the pressure of all border edges emanating from the vertex. Assuming the agent is on vertex u, let uv be a border edge and denote the sizes of the subgraphs containing u and v by $|G_u|$ and $|G_v|$ respectively. Upon sensing the pressure of edge uv the agent acts as follows:

(1) $P(uv) \leq 0$ and $|G_u| \neq -P(uv) - 1$. In this case the agent announces his area size by setting $P(uv) \leftarrow |G_u|$.
(2) $P(uv) \leq 0$ and $|G_u| = -P(uv) - 1$ implies with high probability that $|G_u| = |G_v| - 1$ so the agent will set $P(uv) \leftarrow 0$.
(3) $P(uv) > 0$ implies with high probability that $|G_u| \leq |G_v| - 2$ so the agent will conquer vertex v with probability r.

We can show that most of the time the agent knows what his subgraph size is (see Lemma 5.1). A high-level pseudo code of BDFS can be found in Figure 2.

5. Analysis

The proposed algorithm behavior is analyzed by defining system configurations and deriving the transition probabilities between them. A full description of the system includes the areas controlled by agents, the agents locations, the agents routes and the pressure over border edges. Instead of using the full description, we will use a partial one including only the areas controlled by agents. Formally, a system configuration c is a mapping of vertices to markings, $c : v \in V \to \{id_0, ..., id_k\}$.

5.1. *Observations*

We first introduce some notations. The system comprised k agents patrolling undirected graph G with $|G|$ vertices. $G_u(t)$ is the subgraph controlled by agent u at time t and G_u^c is the subgraph controlled by agent u in configuration c. This notation implies that vertex $u \in G_u$. G_u is a connected subgraph of G. All the vertices of G_u are marked by the pheromone mark of agent u denoted by id_u. However it may happen during the execution of the algorithm that some vertices are marked by id_u but do not belong to G_u, after a *balloon explosion* which is discussed in section 6. Time periods are denoted by $[T_0, T_1]$ where $t \in [T_0, T_1]$ implies that $T_0 \leq t < T_1$.

```
// the agent is on vertex u
if (u is not marked with AgentId) {
  go to a neighbor vertex marked by AgentId (return)
}
if (u is not marked with AgentId) { // explore border edges
  while exist an unvisited border edge uv {
    mark uv as visited
    if (P(uv) > 0) {
      with probability r conquer vertex v and go to v (return)
    } else { // apply pressure
      if (AreaSize = -P(uv) - 1) {
        P(uv) ← 0
      } else {
        P(uv) ← AreaSize
      }
    }
  }
  while exist an unvisited non-border edge uv { // explore non-border edges
    mark uv as visited
    if (v is unvisited) { // go to new vertex
      go to v (return)
      }
    }
  if (there is a backtrack edge) {
    backtrack (return)
  } else {
    start new BDFS (return)
  }
}
```

Fig. 2: Balloon DFS Algorithm.

We will say that G_u is *stable* in a time period $[T_0, T_1]$ if for every two time cycles $t_0, t_1 \in [T_0, T_1]$, $G_u(t_0) = G_u(t_1)$. A configuration is *stable* in a time period if all the subgraphs are *stable* in that time period. It is convenient to define $\Delta u \triangleq 2|G_u| - 1$ which is agent u's route length. The following lemma describes the agent behavior when patrolling his domain.

Lemma 5.1. *When agent u patrols G_u which is* stable *in $[T_0, T_1]$: (1) Agent u will start a new DFS search at time $t_0 \in [T_0, T_0 + \Delta u]$. (2) In $[t_0, T_1]$, agent u will perform sequential DFS searches using exactly the same route according to the MLDFS algorithm.*

Proof. Assume on the contrary that in the time period $[T_0, T_0 + \Delta u]$ agent u does not start a new DFS search. However, the current DFS search can take at most Δu time steps - a contradiction (proving 1). From the beginning of this DFS search G_u is stable, according to the algorithm the agent will perform exactly the same route in every search as long as G_u is stable (proving 2). □

In order to calculate the transition probability between configurations, we derive the conquest probability over a border edge. We will abbreviate the sentence "agent u applies pressure on all border edges emanating from vertex u" to "u-press". According to the algorithm, after a u-press $P(uv) \geq 0$ and after a v-press $P(uv) \leq 0$. A conquest is possible when upon pressing a *border edge* the pressure is strictly positive. Consider a border edge uv. Upon a u-press, if the last press of edge uv was a v-press then $P(uv) \leq 0$ and agent u can not conquer vertex v. Denoting the event that agent u presses an edge uv and the last press of the edge was a u-press by "double u press". So a double press is a necessary condition for a conquest. However it is not sufficient as the next lemma shows.

Lemma 5.2. *Given a system in configuration c which is stable in* $[T_0, T_1]$ *and a border edge uv where* $|G_u^c| \leq |G_v^c|$ *then the probabilities of a conquest in any time period of length* $\Delta u \Delta v$ *in* $[T_0 + \Delta v, T_1]$ *are:*

$$p\,(\textit{agent } u \textit{ conquers vertex } v) = \begin{cases} (\Delta v - \Delta u)\, r & |G_u| \leq |G_v| - 2 \\ 0 & \textit{else} \end{cases} \tag{1}$$

$$p\,(\textit{agent } v \textit{ conquers vertex } u) = 0 \tag{2}$$

Proof. Observe any time period T' of length $\Delta u \Delta v$ i.e. $T' = [T, T + \Delta u \Delta v] \subseteq [T + \Delta v, T_1]$. Consider any vertex w, the agent who controls vertex w applies pressure over all border edges emanating from w every first visit to w in each DFS search. G_u and G_v are *stable* in T' so according to Lemma 5.1, there will be Δv u-presses and Δu v-presses in T'. Hence there are $\Delta v - \Delta u$ double u-presses and there are not any double v-presses in T'. Since a double v-press is a necessary condition for a conquest by agent v, equation 2 is proved.

In order to prove equation 1 three cases are considered:

(1) If $|G_u| = |G_v|$ there are no double presses so the conquest probability is zero.
(2) If $|G_u| = |G_v| - 1$ there are $\Delta v - \Delta u$ double u-presses. Consider a double u-press composed of the first and second u-presses. The press on edge uv preceding the first u-press is a v-press. So right before the first u-press $P(uv) = -|G_v|$ and as a result $|G_u| = -P(uv) - 1$. According to the algorithm, agent u will set $P(uv) \leftarrow 0$ so at the second u-press the pressure is not strictly positive and the conquest probability is zero.
(3) If $|G_u| \leq |G_v| - 2$ there are $\Delta v - \Delta u$ double u-presses. Consider a double u-press composed of the first and second u-presses. Whether the last press on

edge uv preceding the first u-press is a v-press or a u-press, right before the first u-press $|G_u| < -P(uv) - 1$ so agent u will set $P(uv) \leftarrow |G_u|$. According to the algorithm, in the second u-press $P(uv) > 0$ so the conquest probability is r. In the time period T' there are $\Delta v - \Delta u$ double u presses hence the total conquest probability in this time period is $(\Delta v - \Delta u)\, r$ for small r. □

The time period $T'_0 = [T_0, T_0 + \Delta v]$ is right after G_u or G_v have changed so it is possible that one of the agents has not yet found a stable DFS route. However, there is at most one v-press in that time period and if G_u and G_v are about the same size there are at most two u-presses. So the conquest probability in this time period can be bounded from above by:

$$\begin{aligned} p\,(\text{agent u conquers vertex v in } T'_0) &\leq \quad 2r \\ p\,(\text{agent v conquers vertex u in } T'_0) &\leq \quad\ \ r \end{aligned} \tag{3}$$

Since r is small we will neglect the probability of a conquest in this time period. We will later refer to a conquest in this time period as an unjustified conquest since it can happen whether $|G_u| \leq |G_v| - 2$ or not.

5.2. *Balanced configurations*

A balanced configuration is a configuration in which for all border edges uv, $||G_u| - |G_v|| \leq 1$ and all vertices are "controlled" by agents. The next lemma shows that the size of any subgraph in a balanced configuration is bounded.

Lemma 5.3. *For any balanced configuration b, the size of every subgraph G_i^b is bounded by $\frac{|G|}{k} - \frac{k}{2} \leq |G_i^b| \leq \frac{|G|}{k} + \frac{k}{2}$.*

Proof. We prove the upper bound by induction on k. The induction base is $k = 1$, $|G_1^b| = |G| \leq \frac{|G|}{1} + \frac{1}{2}$. Assuming the upper bound holds when the number of agents is at most k, we will show that it holds for $k+1$ agents. Consider a system composed of the graph G and $k+1$ agents patrolling it. Assume on the contrary that in a stable configuration b there is a subgraph with more than $\frac{|G|}{k+1} + \frac{k+1}{2}$ vertices. Without losing generality assume that $|G_1^b| > \frac{|G|}{k+1} + \frac{k+1}{2}$. The graph $G' \triangleq G - G_1^b$ is patrolled by k agents and is of size $|G'| = |G| - |G_1^b| < |G| - \frac{|G|}{k+1} - \frac{k+1}{2}$. Denote by g the component of G' for which the ratio between the size of the component to the number of agents patrolling it is the smallest. Let the number of agents patrolling g be k'. g is patrolled by $k' \leq k$ agents so by the induction hypothesis, in a stable configuration the size of each of the k' subgraphs of g is bounded by $|g_i| \leq \frac{|g|}{k'} + \frac{k'}{2}$. We have chosen g such as the ratio $|g|/k'$ is the smallest so we can write $|g_i| \leq \frac{|g|}{k'} + \frac{k'}{2} \leq \frac{|G'|}{k} + \frac{k}{2}$. Since G is connected there is a border edge connecting G_1^b to a subgraph of g. According to the stable configuration definition

$|g_i| \geq |G_1^b| - 1$. Using the above inequalities we get a contradiction.

$$\begin{aligned}|g_i| \geq |G_1^b| - 1 &> \frac{|G|}{k+1} + \frac{k+1}{2} - 1 \\ &= \frac{2|G|k + k^3 - k}{2k(k+1)} \geq \frac{2|G|k + k^3 - 2k - 1}{2k(k+1)} \\ &= \frac{|G| - \frac{|G|}{k+1} - \frac{k+1}{2}}{k} + \frac{k}{2} > \frac{|G'|}{k} + \frac{k}{2} \geq |g_i|\end{aligned}$$

The lower bound is proved using similar induction. □

After defining a balanced configuration we must ask "which graphs enable balanced configurations?"[a]. To the best of our knowledge this variation of the graph partition problem have never been asked. A work by Gyori [18] implies that every k-connected graph can be partitioned into k balanced subgraphs. Another sufficient condition is the existence of a Hamilton path i.e. every graph that contains a Hamilton path contains a balanced configuration for any k. This condition implies that grids contain a balanced configuration.

5.3. *The convergence conjecture*

We believe that if balanced configurations exist, the system will eventually converge to one of them. However, we could neither prove this nor find a counter example. One way to prove convergence is to show the following:

Conjecture 5.1.
If a balanced configuration exists, from every configuration there is a probability $p \geq \epsilon > 0$ that the system will reach a balanced configuration within a finite time.

Lemma 5.4. *When the system is in a balanced configuration there is a probability $p \geq \epsilon > 0$ that the system will remain in that configuration indefinitely.*

Proof. Consider a system in a balanced configuration at time $t = 0$. Let $\Delta \triangleq max_{u \in G}\{\Delta u\}$. There is a strictly positive probability that there will be no conquest in the time period $[0, \Delta]$ (see Equation 3). Assuming that in $[0, \Delta]$ no conquests occur, in the time period $[\Delta, \infty]$ all conquest probabilities are zero (by Lemma 5.2). So the system will remain in the balanced configuration indefinitely. □

We have proved Lemma 5.4. However, we could not prove Conjecture 5.1. Conjecture 5.1 is hard to prove because system transitions that lead to balanced configurations may be highly complex as shown in the next section.

---

[a]The exact question we ask is "which graphs can be partitioned into k connected subgraphs such as the size difference between any two adjacent subgraphs is at most 1?"

6. Discussion and Simulations

The quality of the partition can be measured using an "entropy" defined as:

$$E(c) = \begin{cases} -\sum |G_i^c| \cdot log|G_i^c| & \bigcup G_i^c = G \\ -\infty & else \end{cases}$$

The entropy is finite only if the subgraphs controlled by the agents cover the graph. In our case, a more balanced partition yields higher entropy.

Let uv be a border edge in configuration c where $u \in G_u^c$, $v \in G_v^c$, $\bigcup G_i^c = G$ and $|G_u^c| \leq |G_v^c| - 2$. Assuming agent u conquers vertex v, let c' be the resulting configuration with $G_u^{c'}$ and $G_v^{c'}$ as the resulting subgraphs. If v is not a cut vertex of G_v^c then the conquest yields a better partition and higher entropy. However, if v is a cut vertex of G_v^c then $G_v^{c'}$ is the component of $G_v^c \backslash v$ agent v happens to be in during the conquest. We call this event a *balloon explosion*. Right after a *balloon explosion* there are some vertices controlled by no one so $E(c') = -\infty$. These empty vertices will shortly be conquered and the entropy will be finite again. However, the resulting entropy might be lower then the entropy prior to the uv conquest. Note that *Balloon explosions* are due to the discrete space we work in and would not occur in the physical balloons model described in section 3.

Since any conquest which is a not a *balloon explosion* yields a better partition, *balloon explosions* might be considered destructive. It would be possible to change the algorithm so no *balloon explosions* will occur. However, *balloon explosions* are essential in order to escape deadlocks. An example for such a deadlock can be found in Figure 3. Several sequential configurations of a system in which three agents patrol a 4×4 grid are described in the figure. The different colors of vertices represent subgraphs controlled by different agents. In configuration c_1, the red and the black agents are controlling equal size subgraphs hence the conquest probability between them is zero. Any conquest of the yellow agent will result in a *balloon explosion*. In case we limit the agents to perform only conquests which does not cause *balloon explosions* the system is in a deadlock and will never reach an optimal partition. A possible scenario leading to a better partition, when *balloon explosions* are allowed, is described in the figure. Note that the entropy of configuration c_3, which is after the vertices left empty by the *balloon explosion* were recaptured, is lower than the entropy prior to the *balloon explosion*. So c_3 is less balanced than c_1, however c_3 is better configuration since its closer to a stable one as can be seen in c_4. So sometimes less balanced configurations are actually better than more balanced ones. That phenomena makes proving conjecture 5.1 hard since we could not "measure" which configurations are closer to being balanced than others.

The convergence process can be divided into two phases: The cover phase starts when the algorithm starts and ends when the graph is covered. The stabilization phase starts when the graph is covered and ends when a stable configuration is reached. Respectively, the cover time is the time needed to cover the graph and the convergence time is the time to reach a balanced configuration. The experiments

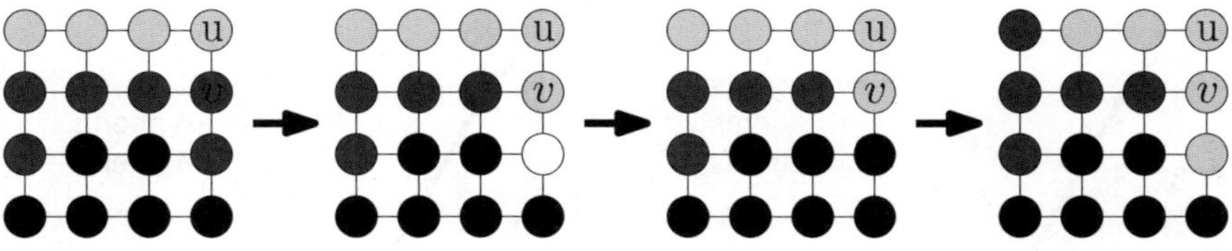

(c_1) E $\simeq -27$ (c_2) E $= -\infty$ (c_3) E $\simeq -27.2$ (c_4) E $\simeq -26.8$

Fig. 3: An example of a deadlock on a 4×4 grid resolved by a *balloon explosion.*

were performed on grids and random Hamiltonian graphs. As mentioned before, assuming Conjecture 5.1, the algorithm always converges to a balanced configuration on such graphs. A total of 42 grids were tested with size ranges between 4×5 to 15×16. A random Hamiltonian graph $H_n(p)$ is created by first constructing an n-cycle $v_1 v_2 \ldots v_n$ and then drawing additional edges at random with probability p for each possible edge. n is the number of vertices and the expected number of edges is $n + p\frac{n(n-3)}{2}$. The algorithm convergence time is influenced by the number of chords in the cycle. By setting $p \triangleq \alpha/n$ the number of chords is $\simeq \alpha n/2$ i.e. proportional to n and α. In order to avoid graphs where n/k is an integer, a primal number of vertices was chosen. The cover and convergence time of every experiment configuration was averaged over 500 runs. The algorithm converged in all our experiments - supporting the convergence conjecture.

Before performing simulations, the value of r (the conquest probability) must be set. r should be small enough to keep the probability of unjustified conquests low (see equation 3). On the other hand, choosing r too small will increase the convergence time because conquests will become rare. In our experiments $r = 0.01$ worked well.

Simulation results on random Hamiltonian graphs are presented in Figures 4 and 5. Observe Figure 4(a) for the average cover time as a function of the graph density. The solid lines represent the cover time and the dashed lines are the normalized cover time defined as the cover time multiplied by the number of agents. The cover time is decreasing with edge density. It is clear since the denser the graph there are more border edges and more conquest opportunities. For dense enough graphs the expansion rate of the area controlled by the agents is proportional to the number of agents. So we would expect the cover time to be proportional to $1/k$. The simulations revealed that for $\alpha > 10^{-0.8}$ our expectation is correct as the normalized cover times overlap. The very sparse graphs considered ($\alpha < 10^{-0.8}$) are rings with very few chords. In case of a single agent on a ring there are only two *border edges* no matter how big the subgraph controlled by the agent is. Furthermore, the path traveled by the agent when he controls a large subgraph is long hence the cover time is large. In case of k agents, there are about $2k$ *border edges* and the path traveled by each agent is relatively small. So using k agents instead of one improves the cover time with a factor larger than k. Therefore, experimentally, not only the absolute cover

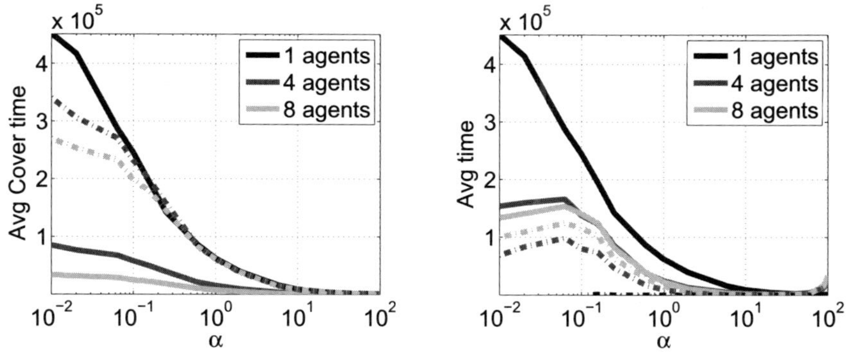

(a) The solid lines represent the cover time and the dashed lines are the normalized cover time.

(b) The solid lines represent the convergence time and the dashed lines are the stabilization time.

Fig. 4: Simulation results on random Hamiltonian graphs with varying edge-density and 101 vertices.

time improved when more agents were used but also the normalized cover time.

The average convergence and stabilization times as a function of the graph density are presented in Figure 4(b). We expected the convergence time to decrease when more edges are added to the graph since there will be more conquest opportunities and more balanced configurations. Simulations revealed that our expectation holds but for the two extremes of very sparse ($\alpha < 0.1$) and very dense ($\alpha > 50$) graphs. In the sparse extreme, the convergence time of graphs with $\alpha \simeq 0.1$ is larger than the convergence time of sparser graphs even though the cover time is smaller. The graphs with $\alpha < 0.1$ are rings with few chords so after they are covered there will be almost no *balloon explosions* delaying the stabilization. When α is increased more chords are added to the graph. The chords help to decrease the cover time but on the other hand they allow more *balloon explosions* during the stabilization phase resulting in longer stabilization phase and higher convergence time. On the other extreme, in very dense graphs there are many border edges so the probabilities of unjustified conquests can not be neglected. Hence unjustified conquests occur, and have a negative effect on the stabilization time.

Observe Figure 5(a) for the average convergence and cover times on random Hamiltonian graphs with $\alpha = 1$. The convergence and cover times are linear with the graph size. As expected, they are decreasing when more agents are added to the system. The convergence and cover time multiplied by the number of agents are presented in Figure 5(b). The normalized cover times overlap so the average cover time is $\Theta(n/k)$. However, the convergence time does not decrease linearly with k. When more agents are added the system, the stabilization phase is getting longer hence enlarging the convergence time. Formally, the convergence time is $\Theta(n/f(k))$ where f is a monotonic non-decreasing function.

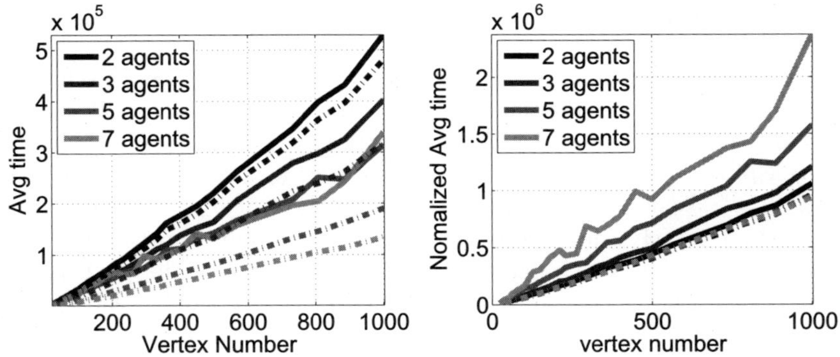

(a) (b)

Fig. 5: Simulation results on random Hamiltonian graphs with varying size. The solid lines represent the convergence time and the dashed lines are the cover time.

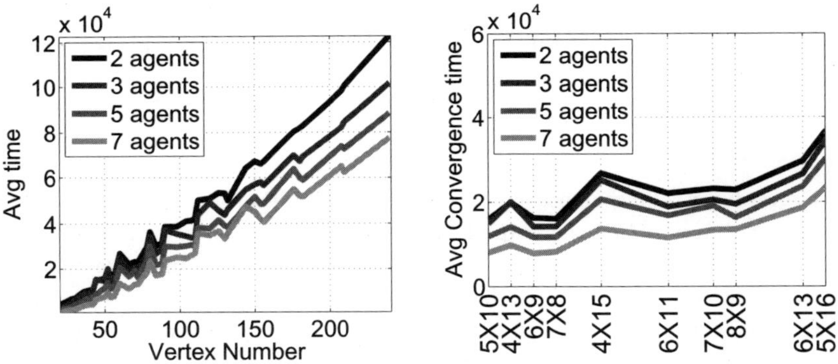

(a) (b) The labels are the grids dimensions.

Fig. 6: Average convergence time on Grids.

Observe Figure 6(a) for simulation results on grids. The convergence time was found to be linear with the number of vertices. However some grids are more challenging than others. Enlargement of the graph for grids of sizes $50 - 80$ vertices is presented in Figure 6(b). Note that the convergence time for a 4×15 grid is larger than for a 6×11 grid even though the second grid contains more vertices. The reason is that the second grid is more "square" so the algorithm is less limited by edge effects.

7. Weighted Graphs

In this section we will show how BDFS can be used to patrol weighted graphs. Let l_{max} (l_{min}) be the length of the longest (shortest) edge in the graph. The main idea, given a weighted graph G, is to convert it to an unweighted graph G' by prolonging all edges to l_{max}.

Prolonging all edges is easy when every agent knows l_{max}. It can be achieved in the following manner: Let the time required to traverse l_{max} be t_{max}. Every agent starts an internal clock whenever starting to traverse an edge. Upon reaching the other side of the edge, the agent will wait until time t_{max} and only then continue the algorithm. We do not address the problem of obtaining l_{max} and dispersing it between the agents. It can be easily done in preprocessing or "on the run".

Let G be a weighted graph containing n vertices and let G' be the graph obtained by prolonging all edges of G to l_{max}. The *worst idleness* of any k-agent patrolling strategy on G is at least $n \cdot l_{min}/k$ (when the graph contains a Hamilton cycle in which all edges length is l_{min}). The *worst idleness* of BDFS on G' is $l_{max} \cdot (2n/k + k)$. Assuming $n \gg k$,

$$\frac{worst\ idleness\,(BDFS)}{worst\ idleness\,(OPT)} \leq 2\frac{l_{max}}{l_{min}}$$

According to the equation above, the proposed expansion of BDFS to weighted graphs is efficient if the ratio l_{max}/l_{min} is low i.e. edges lengths do not vary too much. In case the graph is a result of a skeletonization process [19] this is frequently the case.

8. Conclusion

A novel graph patrolling algorithm was introduced. A group of k agents performing the algorithm partition the graph between them using simple local interactions. Graph partition into connected and size balanced components is an emergent behavior of the agent swarm. We have shown that in a balanced partition every subgraph is of size less than $\frac{|G|}{k} + \frac{k}{2}$. As a result, the time lag between two successive visits to any vertex is at most twice the optimal so the patrol quality is at least half the optimal. In case of a weighted graph we have shown that the algorithm achieves a competitive ratio of $2l_{max}/l_{min}$ when compared to the optimal patrol strategy where l_{max} (l_{min}) is the longest (shortest) edge in the graph. Simulations performed over grids and random Hamiltonian graphs showed that the average cover time is $\Theta\left(|G|/k\right)$. The average time to reach a balanced partition was experimentally found to be $\Theta\left(|G|/f(k)\right)$ where f is a monotonic non-decreasing function.

References

[1] A. Machado, G. Ramalho, J.-D. Zucker and A. Drogoul, Multi-agent patrolling: An empirical analysis of alternative architectures, in *MABS*, 2002.
[2] A. Almeida, G. Ramalho, H. Santana, P. A. Tedesco, T. Menezes, V. Corruble and Y. Chevaleyre, Recent advances on multi-agent patrolling, in *SBIA*, 2004.
[3] Y. Chevaleyre, F. Sempe and G. Ramalho, A theoretical analysis of multi-agent patrolling strategies, in *AAMAS*, (IEEE Computer Society, Washington, DC, USA, 2004).

[4] T. Sak, J. Wainer and S. K. Goldenstein, Probabilistic multiagent patrolling, in *SBIA*, 2008.
[5] Y. Elor and A. M. Bruckstein, Multi-agent graph patrolling and partitioning., in *IEEE/WIC/ACM Int. Joint Conf. on Web Intel. and Intel. Agent Technologies*, Sep. 2009.
[6] I. A. Wagner, M. Lindenbaum and A. M. Bruckstein, *Annals of Mathematics and Artificial Intelligence* **24**, 211 (1998).
[7] Y. Elor and A. M. Bruckstein., Autonomous multi-agent cycle based patrolling., in *Ants 2010: Seventh International Conference on Swarm Intelligence*, 2010.
[8] V. Yanovski, I. A. Wagner and A. M. Bruckstein, *Algorithmica* **37**, 165 (2003).
[9] U. Elsner, Graph partitioning - a survey (1997).
[10] M. Ahmadi and P. Stone, A multi-robot system for continuous area sweeping tasks *Robotics and Automation, 2006. ICRA 2006. Proceedings 2006 IEEE International Conference on* May 2006.
[11] F. Comellas and E. Sapena, A multiagent algorithm for graph partitioning., in *EvoWorkshops*, , Lecture Notes in Computer Science Vol. 3907 (Springer, 2006).
[12] T. Menezes, P. Tedesco and G. Ramalho, Negotiator agents for the patrolling task., in *IBERAMIA-SBIA*, eds. J. S. Sichman, H. Coelho and S. O. Rezende, Lecture Notes in Computer Science, Vol. 4140 (Springer, 2006).
[13] R. Diekmann, R. Preis, F. Schlimbach and C. Walshaw, *Parallel Computing* **26**, 1555 (2000), Graph Partitioning and Parallel Computing.
[14] H. Meyerhenke, B. Monien and T. Sauerwald, A new diffusion-based multilevel algorithm for computing graph partitions of very high quality *Parallel and Distributed Processing, 2008. IPDPS 2008. IEEE International Symposium on* April 2008.
[15] P. Paruchuri, J. P. Pearce, M. Tambe, F. Ordonez and S. Kraus, An efficient heuristic approach for security against multiple adversaries, in *AAMAS '07: Proceedings of the 6th international joint conference on Autonomous agents and multiagent systems*, (ACM, New York, NY, USA, 2007).
[16] N. Agmon, S. Kraus and G. Kaminka, Multi-robot perimeter patrol in adversarial settings *Robotics and Automation, 2008. ICRA 2008. IEEE International Conference on* May 2008.
[17] I. Wagner, M. Lindenbaum and A. Bruckstein, *Robotics and Automation, IEEE Transactions on* **15**, 918(Oct 1999).
[18] E. Gyori, On division of graphs to connected subgraphs., in *Combinatorics (Proc. Fifth Hungarian Colloq., Keszthely, 1976)*, 1976.
[19] B. W. Stout, *Game Developer* (October 1996).

THE ROLE OF NOISE IN BRAIN FUNCTION

S. ROY

Physics and Applied Mathematics Unit, Indian Statistical Institute
203 Barrackpore Trunk Road, Kolkata, 700108, India
E-mail: sisir@isical.ac.in

R. LLINÁS

New York University School of Medicine
530 First Avenue, NY, 10016 USA
E-mail: llinar01@med.nyu.edu

Noise plays a fundamental role in all living organisms from the earliest prokaryotes to advanced mammalian forms, such as ourselves. In the context of living organisms, the term 'noise' usually refers to the variance amongst measurements obtained from repeated identical experimental conditions, or from output signals from these systems. It is noteworthy that both these conditions are universally characterized by the presence of background fluctuations. In non-biological systems, such as electronics or in communications sciences, where the aim is to send error-free messages, noise was generally regarded as a problem. The discovery of Stochastic Resonances (SR) in non-linear dynamics brought a shift of perception where noise, rather than representing a problem, became fundamental to system function, especially so in biology. The question now is: to what extent is biological function dependent on random noise. Indeed, it seems feasible that noise also plays an important role in neuronal communication and oscillatory synchronization. Given this approach, it follows that determining Fisher information content could be relevant in neuronal communication. It also seems possible that the principle of least time, and that of the sum over histories, could be important basic principles in understanding the coherence dynamics responsible for action and perception. Ultimately, external noise cancellation combined with intrinsic noise signal embedding and, the use of the principle of least time may be considered an essential step in the organization of central nervous system (CNS) function.

Keywords: Stochastic Resonance, Synchronization, Principle of Least Time

1. Introduction

Noise has been generally regarded as a truly fundamental engineering problem particularly in electronics computation and communication sciences, where the aim has been reliability optimization. The discovery of Stochastic Resonances (SR) [1–6] in non-linear dynamics brought a shift to that perception, i.e. noise rather than representing a problem became a central parameter in system function, especially in biology. Indeed, noise plays a basic role in the development and maintenance of life as a system capable of evolution. The question at this point is, "to what extent is biological function dependent on random noise?". A crucial corollary to that question is: *is the significance of noise that depends on intrinsic system properties more meaningful than the noise brought in from the environment.* It is presently known, from both theoretical and experimental biological system research, that the addition of input noise improves detectability and transduction of signals in non-linear systems [7]. This effect is popularly known as Stochastic Resonance (SR). SR has been found to be an established phenomenon in sensory biology [8], but it is not presently determined to what extent SR is embedded in such systems. Thus, it is not clear, for instance, whether functionally significant SR can occur at the level of single ion channel in cell membranes or whether it is mostly an ensemble property of channel aggregates. Currently, both forms of SR seem to be present as it occurs based on external noise as well as in its absence, as exemplified by neuronal multimodal aperiodic spike firing in the absence of external input [6]. Both experimental findings and computer simulations of central neuron activity indicate that SR is enhanced when both noise and signals are simultaneously presented in neurons [9]. It is also known that spontaneous synchronization occurs if noise is added to coupled arrays of neurons. Indeed, coherence resonance has been observed in hippocampal neuron array [2]. In fact, it is widely recognized that Derksen and Verveen [10] changed the prevalent view of noise from a bothersome phenomenon to an essential component in biological systems. Recent developments in neuroscience further pointed out that the Central Nervous System (CNS) can utilize noise (which carries no signal information) to enhance the detection of signal through SR [9] thus emphasizing the fundamental role of noise in brain information processing. Moreover, the intrinsic noise associated with neuronal assemblies is known to produce synchronous oscillations utilizing the ISR or CR mechanism [10]. More fundamentally, the unicellular entities such as Diatom show quite evidently Brownian like noise following death, in contrast to the directed transport motion observed during its living state. Moreover, the ionic channels of Diatom remain active for the certain period of time following systemic death. In its living state, the ion channels of Diatom may produce a type of synchronous resonance that becomes asynchronous after death. However, a careful analysis of the noise is needed to understand this kind of phenomenology and its physical/biological significance. The analysis of the role of noise in Diatom may shed new light in the understanding of the transition between the living and diseased states.

The next question relates to noise sources. They can be broadly classified as follows :

(1) Basic Physics noise: Thermodynamics and quantum theory put physical limit to the efficiency of all information handling systems.
(2) Stimulus noise: Thermodynamics or quantum theory delineates the limit to the external stimuli and, thus, they are intrinsically noisy. During the process of perception, the stimulus energy is either converted directly to chemical energy(e.g. photoreception)or to mechanical energy, which is amplified and transformed to

electrical signals. The intrinsic noise in the external stimuli will be amplified and further amplification generated noise(transducer noise).
(3) Ionic channel noise: voltage, ligand and metabolic activated channel noise
(4) Cellular Contractile and Secretory noise: muscles and glands
(5) Macroscopic behavioral execution noise.

There exist two distinct sources behind execution noise. They are :

(1) Non-linear dynamics : where, in deterministic systems, the sensitivity to the initial conditions and to chaotic behavior engenders variability of initial conditions.
(2) Stochastic : where irregular fluctuations or stochasticity may be present intrinsicly or via the external world.

While these two sources generate noise from chaotic time series or via stochastic process, and while they do share some indistinguishable properties it is, nevertheless, possible to differentiate *noise* out of *chaos* from noise via *stochastic processes* [11]. The issue now is to define to what extent biological function is dependent on the presence of random noise i.e., whether noise can be considered as a useful property in biological systems [12]. The discovery of Stochastic Resonance in non-linear dynamics addresses the above questions directly.

2. Stochastic Resonance and Sensory Biology

Benzi et al [13] first used the term Stochastic Resonance(SR) for noise enhanced signal processing in climatology in 1980. Recently, Mark D. McDonnell1 and Derek Abbott published a comprehensive review on the subject devoted to biological systems [1]. Broadly speaking, SR is a phenomenon where in certain non-linear systems, subject to weak input signals, the presence of stochastic noise can enhance the coherence of the output instead of degrading its coherence. Initially SR was observed for periodic input signals, but more recently, it has been found in random aperiodic imput signals. In this case, mutual information or Fisher information is a useful measure of SR [14], rightly used by the phrase *noise benefits signal-processing systems* rather than noise-enhances signal processing and so we declare SR as a noise benefit. But, does this variability really have a useful function? The study of noise enhanced phase synchronization of coupled oscillator arrays [15] and nonidentical non-coupled noisy oscillators is a step ahead in understanding the benefits of noise in cell biology. Neiman et al [16] reported phase synchronization of nonidentical neuronal noisy oscillators both from experimental results as well as from numerical simulations in terms of stochastic synchronization. Such synchronization has been studied in various biological systems such as collective flashing fireflies, and in human cardiorespiratory synchronization. However, an issue not yet resolved concerns whether SR can occur at the level of single membrane bound ion channel or whether it is mostly an ensemble property of channel aggregates. This internal noise, if due to intrinsic channel noise, will, de facto, become ordered (even in the absence of external periodic signal) via a mechanism known as intrinsic coherence resonance. Indeed, McDonnell et al [1] have raised the question whether stochastic resonance is exploited by the nervous system and brain as part of the neural code. The answer is 'yes' and this has been recently demonstrated in the analysis of spontaneous oscillation of inferior olive cells following genetic mutation of particular ionic channel expression. In this case the absence of T or P type calcium channels results in a modification of coupled network oscillatory characteristics and in abnormal motor behavior [9]. More relevant to our present discussion, the inferior olive cells in these mutants fail to generate the chaotic phase synchronization characteristic of this nuclear ensemble. [17] This lack of phase reset is reflected both in the oscillatory properties of the individual neurons but also in their neuronal ensemble oscillation. The electrophysiological characterization of the subthreshold oscillation in these mice demonstrated, in addition to the lack of phase reset, an asymmetry in subthreshold membrane potential oscillation. Stein et al [18] discussed the neuronal variability and raised a very important issue as to whether this variability is neural noise or a part of the signal transmitted to other neurons. He argued that both temporal and rate coding are used in various parts of Central Nervous System(CNS) and both are useful to CNS to discriminate complex objects and produce movements. The noise in the ion channel is of the nature of *flicker noise*(FN) i.e. $\frac{1}{f^\alpha}$ where $\alpha > 0$. The mechanism of generation of FN in ion channel is not yet fully understood [19]. Recently, Faisal et al [20] investigated the effect of channel noise on the miniaturization of brain wiring. They consider that channel noise may put the limit to axon diameter given the limitations inherent by length and time constants of the nerve cable properties. The random or stochastic variation in gene expression under constant environmental conditions is considered as noise. Probably, this type of noise associated to gene expression put limit to the evolution of living systems. The field is still in its infancy as to whether their existance in noise induced enhancement of phase synchronization is possible for gene expression. In fact, it is not clear whether Stochastic Resonance(SR) occurs in vivo for neurons and brain function. Indeed even the occurence of SR indicate the existence of nonlinearity of the systems and it remains an open question whether and neurons use internally generated noise and SR effects. In the next section, we discuss the role of internal noise and that of the external world associated with the sensory stimuli. This is achieved through using the concept of principle of stationary phase, which is essentially related to the principle of least time.

3. Principle of Least Time and Sum Over Histories

The central issue in brain function is the internalization of the properties of the external world into an internal functional space [21]. By *internalization*, we mean the ability of the nervous system to fracture external reality into sets of sensory messages and to simulate such reality in brain reference frames. Over two decades ago, Pellionisz and Llinás [22, 23] proposed an integrated approach to address brain function. This was originally based on the assumption that the relationship between the brain and the external world is determined by the ability of the Central Nervous System (CNS) to construct an internal model of the world implemented through the interactive relationship between sensory and motor expression. In this model, the evolutionary realm provided the backbone for the development of an internal functional geometry space. Recently, it was proposed that if the CNS functional space is endowed with a stochastic metric tensor [24], such that a dynamic correspondence would exist between events in the external world and their specification in the internal space. One can think of this type of correspondence between incoming (sensory) covariant vectors and outgoing (motor) action i.e. contravariant vectors, as depending on internal states in a random fashion. We shall call this *dynamic geometry*. Here, dynamic geometry sets a state of readiness with respect to the external world, which is used in a predictive anticipation of motor action as well as serving as an ongoing contextual translator of covariant vectors into contravariant vectors. However, the motion of an object in the external world does not of itself engender simultaneity in space and time in its counterpart functional space in the brain, because conduction velocities are not constant amongst the various brain pathways. This is because, the time delay due to different conduction velocities lies within the 10 to 15 ms time resolution associated with gamma band (40 Hz) based cognitive resolution (10 - 15ms) [25] simultaneity may be considered, therefore, to have a granularity δt with in that temporal framework. We will refer to this δt as our operational definition of simultaneity. Here, the variation of functional state generated by the summed activation in all pathways and corresponding neurons, in response to external or intrinsic activity, within this δt will be referred to as a quantum of cognition. Instead of usual optimization for the axonal pathway with maximum conduction velocity, a weighted summation over all the pathways includes stationarity, which plays a preferential role in cognition and thus in our further understanding of the cognitive event. This is comparable to the so-called *sum over histories* considered by Feynman [26] concerning the reflection and refraction of light. He emphasized how quantum principles, through the superposition of multiplications of unitary complex factors along all possible photon follow paths, results in light travel between source to detector in the least possible time. Thus for light, a *sum of history* results in 'least time' displacement. By contrast, axonal paths with different conduction velocities can mimic light transmission in space. When viewed from a morpho-functional perspective, informational conduction in the CNS is associated with distinct and particular conduction pathways with specific conduction properties. Thus, in the brain, the set of different histories corresponding to different axonal paths summ in individual neurons to represent the external world. We suggest these conduction properties express a stochastic, a priori description of the world, consistent with a Bayesian perspective for brain function [27]. Anatomically, brain pathways connect functional spaces to one another;, it is here then that the *sum of history* principle and the resulting *least time* principle, due to the oscillatory grains, can support quantum cognition by synchronization. We wish to emphasize that the principle of least time is a fundamental principle of nature that may help to define the decision-making and behavioral control capacity of the brain. Note, however, that we are considering paths in a high dimensional functional space; it is not 'times' but 'actions' which are minimized. Before going into the details of sum over history approach, we will address the concept of simultaneity and 40Hz oscillations.

3.1. *40 Hz oscillations and the concept of simultaneity*

The concept of two types of simultaneity has been discussed in physics extensively, especially since the formulation of special theory of relativity by Einstein. As noted above, given differences in conduction time that are present in the Central Nervous System(CNS), a variable time delay for any given stimulus arising from the external world must be considered. Indeed, the system has two types of simultaneity problems to solve:

(1) simultaneity of two events in time related to the external events and the internally recognized counterpart, and
(2) the simultaneity of the same event at two different times due to the delays inherent in the differences of conduction velocities involved in relaying of information into the CNS.

Thus one has cognitive sense, when something happens and there is biophysical spread of conduction velocity along the different pathways. Indeed, in hitting one's thumb with a hammer, we feel the blow (fast conduction velocity) and know that the pain (slow conduction velocity) will *arrive* (be felt) shortly after. The question is : how does coherent perception emerge out of these ambiguities in the brain function? A possible solution is that the network which recognizes propagation differences (felling the blow and registering the resulting pain) act coherently (predict) even if conduction times are different. However, the first problem to be solved:
how to reduce the number of dimensions from a very large number to four.

The concept of Minkowski space as a four-dimensional space/time manifold was introduced in the special theory of relativity where the speed of light is assumed to be constant and maximum. By contrast, the conduction velocity in the CNS axons is 10 orders of magnitude slower, ranging from centimeter per second to hundreds of meter per second [28]. Thus, Minkowski space is not applicable as the description of functional geometry of the brain. By contrast, as the conduction velocity is low, events in the brain space have a far more distinct four dimensional quality necessitating the introduction of both types of simultaneity, not withstanding their fundamental differences. More generally, brain evolution favored inner space parameters that reduce sensory categories to those providing optimal information concerning external reality properties. This is similar to the construction of quotient space in mathematics. We will show later how this evolutionary pressure evolved the informational geometry characteristics of central nervous systems.

The two next fundamental issues concern:

(1) internal multiplicity, due to the presence of multiple path manifolds with different conduction velocities and
(2) the requirements of combinatorial annealing of the different varieties of sensory information required for a multisensory construction of objective reality.

These two sensory annealing requirements are essential to the generation of the appropriate motor command vectors supporting real-time interactions with the external world. We suggest that neurons, and the networks they weave, implement simultaneity in the same manner that the light follows the path of shortest time, by adding together a huge number of virtual path amplitudes, through oscillatory compensation.

Magnetic and electric recordings in the human brain have revealed the existence of coherent oscillatory activity near 40 Hz. Thus, magneto-encephalography (MEG) was used by Jolliot et al [25] to determine whether the 40 Hz oscillatory activities present during cognitive experience related to the temporal binding of sensory stimuli. The results indicate that the 40 Hz oscillations not only relate to primary sensory processing, but also reflect the temporal binding underlying cognition. Experimental results have shown that there is a time interval of $10 - 14$ ms which corresponds to the up-trajectory of 40 Hz oscillations. This establishes the minimum time required for the binding of sensory inputs into the cognition of any single event. This was proposed as the cognitive quanta of time and corresponds closely with the findings from psychophysical studies [29]. Because the delay in conduction speeds is found to lie within the range of $10 - 14$ ms, we can now propose an operational definition for simultaneity where the resolution of the instrument (here the quanta of time) makes *simultaneity* possible.

3.2. *Principle of least time*

Fermat introduced the principle of least time as the shortest time taken by light to travel along a path between a source and a receptor. Fermat was the first to investigate variational principles in general. In geometrical optics, every path of light ray can be predicted from this basic principle. The principle of least time proposes that the light explores all possible paths and played an important role in formulating path integral to quantum mechanics for microscopic particles by Feynman. Usually, the concept of all possible paths is closely related to two basic tenets concerning 1) the meaning of exploring all possible paths and 2) how the everyday objects (a stone) follow a particular path. Feynman made this issue tractable within his quantum mechanics formulation using his *sum over histories* and *path integration* concept. As Feynman noticed, the Maupertuis variational principle in Physics can be understood in quantum terms, by becoming a version of the mathematical *stationary phase principle*. More importantly, it became clear that such conjectures are not the exclusive property of Quantum Physics; they apply equally in ordinary geometrical optics.

3.3. *Sum over histories*

Let us remember that there is close correspondence between the sum over histories approach to quantum mechanics and the classical stochastic process. We can commence by discussing sum over histories in general terms. To describe the path of a particle in the physical world, $\gamma = \gamma(t)$, one needs to consider the path as a whole. For example, if we are to consider the Brownian motion as a stochastic process we must consider the space of all possible histories of $\gamma(t)$. In this framework, one can address dynamics by assigning the rule of probabilities to define certain class of paths. The issue lies in attaching a positive real number p to each history, and assigning to the class H a composite probability $p(H)$ as

$$p(H) = \sum_{\gamma \in H} p(\gamma).$$

Because the paths are unbounded in time, individual probability does not exist. Well defined individual probabilities can be obtained by truncating the histories at some final time; however, class probability is parameter of interest, and so, it is not necessary to have well defined individual probabilities. In the brain, the paths are inherently truncated in time when cognition occurs. This is similar to the quantum situation where one is interested in truncated paths or histories at time T known as collapse time. It is well known that the sum of excitatory and inhibitory synaptic currents and the intrinsic electrical properties of neuron are the fundamental to the CNS decision making process. This decision effect is closely related to the concept of collapse and is responsible for the recognition of any event. The rule of computation of probabilities is different in classical stochastic process and quantum paradigm.

3.3.1. *Phase-response properties of neurons*

Makarenko and Llinás et al [17] demonstrated that single neurons can behave like weakly chaotic oscillators. Intuitively speaking [30], the relative membrane potential oscillatory cycle in the timing of action potential or the responsiveness to excitatory and inhibitory synaptic inputs play an important role in defining the phase synchronization for rhythms that define global states. For instance, the presence of gamma band activity has been associated with wakefulness as well as in dream state [31]. The interactive Excitatory cells (E-cells) and Inhibitory cells(I-cells) bring the synchronous rhythm in neuronal networks. The mechanism of pyramidal-interneuronal gamma (PING) rhythms which is related to the activities of E-cells and I-cells is yet to be understood fully [32]. The effect of noise plays an important role in understanding PING. In sum over history approach, choice of the weightage given to the paths is an aspect of the formulation. One can construct several choices. Considering the oscillatory function like

$$w(\alpha) = \exp(i\beta S(\alpha)$$

where β is a large positive number and α the range of the conduction speed as the weightage function one can recover path integral similar to Feynman path integral and by exponential weighting function it gives Einstein-Smoluchovsky path integral. In case of oscillatory weighting function the small change of the trajectory may lead to large change in the sine and cosine function in the phase. Thus even if the trajectories are very close, the contributions from the trajectories will cancel each other due to rapid oscillations in sine and cosine. On the other hand, for a stationary phase trajectory as formulated by Fermat leads to small changes in the phase and the they produce constructive interference.

3.3.2. *Fisher information and noise in brain*

The idea developed by the authors(RS) [27] that external noise is adapted to the internal noise, fits well with Fisher's idea within the framework of parametric statistics because *noise* is defined here as a family of hypotheses about the possible forms of knowledge, and the covariant metric is the measure of the difficulty in determining the exact form of knowledge. Noise can decrease with further independent tests of the same parameter, as the case with individual neurons. Actions are planned, at least in part, according to a minimum variance principle [33],and their executions are conditioned by sensory information, at least in feedback. In some detection regimes (e.g. those involving many neurons, large time scales and facilitating synapses) the information gain in sensory areas follows the maximum of information metric [34].

It is worthmentioning that Fisher information plays two roles, serving as a measure of ability to estimate a parameter, and a measure of the degree of disorder. In the present context, it is more appropriate to consider Fisher information as a measure of ability to match a pattern, and a measure of the degree of uncertainty with respect to pattern matching. Matching is an act of perception associated with the covariant vector and an act of action associated with the contravariant vector; the two are connecting by the metric tensor. However, to date, no connection has been established between the representation of space-time and its oscillatory coding. We now propose such a link.

Let us call λ in the set Λ, the covariant space-time parameters of importance for action received through sensory areas and sent to the motor program, for example a spatial or temporal frequency. The maximum Fisher information required for its detection, expressed, for example, by the slope of a tuning curve, gives a metrical structure in λ. Thus matching is natural property of Fisher information, and is compatible with estimation in cases where repetition of independent measures is possible. Then, by generalized Fourier transform, the activity transmitted to motor neurons furnishes a metric in contra-variants for actions. However a paradox appears: a large frequency metric will result in the a priori generation of a small metric when expressed in position and time terms. Thus, the indeterminacy principle applies. The limit may be linked with the periodicity 40 Hz of oscillations. This type of situation is also seen in the Taylor series for movement plans as proposed by one of the authors [23]. In the case of the brain, the separation between the observer and the instrument does not exist and one can think of a continuum between the mind and brain.

4. Possible Implications

The above analysis addresses one of most pertinent questions related to noise: why do we study noise ? The limitation put by laws of physics like Thermodynamics and Quantum Theory force us to think of noise as and when we try design the electronic devices. This is essential to "measure" the physical world. On the other hand, the living systems, starting from unicellular object like Diatom -Bacteria to more complex object like the brain, utilize noise, as defined in physics, for its functioning. Recent developments in neuroscience further pointed out that the Central Nervous System (CNS) depends on noise (which carries no signal information) to enhance the detection of signal through Stochastic Resonance(SR) thus emphasizing the fundamental role of noise in information processing of brain. Moreover, the intrinsic noise associated to neuronal assemblies is known to produce synchronous oscillations utilizing the ISR or CR mechanism.

As it happens, then, it is via the dialog between reduction and systemics that the issue has finally taken form. As neuroscience addresses molecular biology in search for answers, and molecular biology reaches into biophysics and then into submolecular function, it is at that point that the issue arises in full force. We have become aware that the stochastic properties found in physico-chemistry are actually ruling what we had expected long ago to be deterministic. Once it is clear that the uncertainty principle applies in biology as well as in fundamental physics, a new game is on. And so the analysis of biological noise is born and, more to the point, the differences between the biological and the non-biological realms begin to merge. Two clear implications seems evident:

(1) biology is a branch of physics and
(2) physics, defined as what physicists do, a branch of biology.

References

[1] Mark D. McDonnell, Derek Abbott,*PLoS Comput. Biol.* **5**,1 (2009).

[2] D.M. Durand and W.C. Stacey,*Research Report* Accession Number: ADA409889 (2001)

[3] Priscilla E. Greenwood, Lawrence M. Ward, David F. Russell, Alexander Neiman, and Frank Moss.*Phys.Rev.Letters***84**,4773 (2000).

[4] Gerhard Schmid, Igor Goychuk, Peter Hanggi,*Physica A***325**,165 (2003)

[5] Gerhard Schmid, Igor Goychuk, and Peter Hanggi,*Europhys. Lett.* **56**,22 (2001)

[6] D. T. W. Chik,Yuqing Wang and Z. D. Wang, *Phys.Rev.Lett.***64**,021913 (2001).

[7] P. Hanggi,*Chemphyschem* **3**,285 (2002).

[8] Igor Goychuk, Peter Hañggi, *Phys.Rev.E* **61**,4272 (2000).

[9] Soonwook Choi et al,*J.Physol.* **16**, 3031 (2010).

[10] H.E. Derksen and A.A.Verveen ,*Science* **151**, 1388 (1966).

[11] O. A. Rosso, H. A. Larrondo, M. T. Martin A. Plastino and M. A.Fuentes, *Phys.Rev.Lett.***99**, 154102 (2007).

[12] Stephen F.Traynelis and Fernan Jaramillo,*TINS* **21**, 137 (1998).

[13] R.Benzi, A.Sutera A, A.Vulpiani,*Journal of Physics A: Mathematical and General* **14**,L453 (1981).

[14] B.Kosko,*Noise* (Viking, 2006).

[15] Changsong Zhou, Jurgen Kurths, Istvan Z.Kiss and John L. Hudson,*Phys.Rev.Letts.* **89**, 014101 (2002).

[16] Alexander B.Neiman and David F. Russell,*Phys.Rev.Letts.* **88**, 138103 (2002).

[17] V.Makarenko and R.Llinaás. *Proc. Natl. Acad.Sci.U.S.A.***95**, 15747 (1998).

[18] Richard B.Stein, E.R.Gossen and Kelvin E.Jones,*Nature Review Neuroscience***6**, 389 (2005).

[19] Z.Siwy and A.Fulinski, *Phys.Rev.Letts.***89**, 158101 (2002).

[20] A. Aldo Faisal, John A. White and Simon B. Laughlin, *Current Biology* **15**,1143 (2005).

[21] Rodolfo Llinás ,*I of the Vortex* (The MIT Press, Cambridge Massachusetts, London,England ,2002).

[22] A.Pellionisz and R.Llinás,*Neuroscience* **7**,2949 (1982).

[23] A.Pellionisz and R.Llinaás, *Neuroscience* **16**,245 (1985).

[24] S. Roy and R.Llinás,*Progress of Brain Research***168**,133 (2008)

[25] M.Joliot, U.Ribary and R.Llinaás, *Proc. Nat.Acad. Sc(USA)* **91**,11748 (1994).

[26] R.P.Feynman,*QED, the strange theory of light and matter* (Princeton University Press, Princeton and Oxford, 1968).

[27] R.Llinaás and S.Roy, *Phil Trans. R.Soc. B* **364**,1301 (2009).

[28] J.Erlanger and H.S.Gasser,*Electrical Signs of Nervous Activity* (University of Pennsylvania Press, Philadelphia, 1937).

[29] A.B.Kristoffwerson,*Ann NY Acad. Sc.***423**, 3 (1984).

[30] XIAO-JING WANG, *Physiol. Rev.* **90**, 1195 (2010).

[31] R.Llinaás,and D. Pare .*Neuroscience* **44**,521 (1991).

[32] B.Christoph and Nancy Kopell, *Neural Computation* **17**,557 (2005).

[33] C.M.Harris, D.M.Wolpert,*Nature* **394**, 780 (1998).

[34] N.Brunel and J.P.Nadal ,*Neural Computation* **10**, 1731 (1998).

F-GRANULATION, GENERALIZED ROUGH ENTROPY AND IMAGE ANALYSIS

SANKAR K. PAL

Center for Soft Computing Research, Indian Statistical Institute
203 Barrackpore Trunk Road, Kolkata, 700108, India
E-mail: sankar@isical.ac.in

Role of f-granulation in handling uncertainty in machine intelligence and its significance as a facet of natural computation are discussed. Its modeling through judicious integration of rough and fuzzy sets is explained. Several tasks like case generation, clustering, classification and segmentation are described in rough-fuzzy framework demonstrating the role of f-granulation and the resulting merits. A definition of generalized rough sets is given considering both the set and granules as crisp and/or fuzzy. Based on that new entropy measures are defined using exponential and logarithmic gain functions. The superiority of rough-fuzzy integration, in terms of performance and computation time, is illustrated for the tasks of case mining in large data sets, segmenting brain MR images, and classifying remotely sensed images as examples. Neighbourhood rough sets are capable of better dimensionality reduction. The effect of fuzzy-granules and generalizations in rough sets is demonstrated for image segmentation. The article includes some of the results published elsewhere.

Keywords: soft computing, fuzzy granulation, rough-fuzzy entropy, image processing, case based reasoning, natural computing.

1. Introduction

Rough set theory [1] provides an effective means for analysis of data by synthesizing or constructing approximations (upper and lower) of set concepts from the acquired data. The key notions here are those of "information granule" and "reducts". Information granule formalizes the concept of finite precision representation of objects in real life situation, and reducts represent the core of an information system (both in terms of objects and features) in a granular universe. The theory provides a popular mathematical framework for granular computing. Its focus is on the ambiguity caused by limited discernibility of objects in the domain of discourse. Granules are formed as objects and are drawn together by the limited discernibility among them. Rough set based techniques have been used in the fields of pattern recognition, image processing, data mining and knowledge discovery [2, 3] from large data sets. They were also found to have extensive application in dimensionality reduction and knowledge encoding particularly when the uncertainty is due to granularity in the domain of discourse.

Rough-fuzzy or fuzzy-rough techniques [4] are efficient hybrid methods based on judicious integration of the principles of rough sets and fuzzy sets. While the membership functions of fuzzy sets enable efficient handling of overlapping classes, the concept of lower and upper approximations of rough sets deals with uncertainty, vagueness, and incompleteness in class definition using the notion of granularity. This integration provides a stronger paradigm for uncertainty modeling in an ambiguous situation.

It may be mentioned that the concept of rough-fuzzy computing has a significant role in modeling the fuzzy-granulation (f-granulation) characteristics of *Computational theory of perceptions* (CTP) [5–7] which is inspired by the remarkable human capability to perform a wide variety of physical and mental tasks, including recognition, without any measurements and computations. Reflecting the finite ability of the sensory organs and (finally the brain) to resolve details, perceptions are inherently fuzzy-granular (f-granular). That is, boundaries of perceived classes are unsharp and the values of the attributes they can take are granulated (a clump of indistinguishable points or objects). One may note that perception and granulation are two constituting technologies of natural computation [8] that are abstracted from natural phenomena for problem solving.

The present article deals with the significance of f-granulation in rough-fuzzy computing for uncertainty analysis, in general, and pattern recognition and image processing in particular. The organization of the article is as follows. Section 2 presents, in brief, the role of f-granulation in granular computation. Section 3 explains the application of rough-fuzzy granulation for the problem of case generation. Section 4 describes a classification method demonstrating the power of neighborhood rough sets in feature selection from f-granulated space. Section 5 demonstrates the concept of rough-fuzzy clustering. The problem of segmenting brain MR images is considered, as an example. Section 6 deals with generalized rough sets, entropy and image ambiguity measures. Its application to image segmentation, as an example, is shown. Concluding remarks are given in Section 7.

2. f-granules and Granular Computing

In many practical situations, when a problem involves incomplete, uncertain and vague information, it may be difficult to differentiate distinct elements and one is forced to consider granules. On the other hand, in some situations though detailed information is available, it may be sufficient to use granules in order to have an efficient and practical solution. Depending on the nature of problems and data, granules could be fuzzy rather than crisp; thereby leading to f-granulation. Granulation is an important step in the human cognition process. This is also a computational paradigm, among others such as self-organization, self-replication, evolution, perception, functioning of brain, group behavior, cell membranes, and morphogenesis, which are abstracted from natural phenomena.

Granular computing (GrC) refers to that where computation and operations are performed on information granules (clump of similar objects or points). Since it deals with compressed information (granules) rather than individual data points, it leads to have both data compression and gain in computation time, and finds wide applications such as in data mining and knowledge discovery. There are two aspects of granular computing, one deals with formation, representation and interpretation of granules (algorithmic aspect) while the other deals with utilization of granules for problem solving (semantic aspect).

Several approaches for granular computing have been suggested in literature. Paradigms considered include fuzzy set theory, rough set theory, power algebras and interval analysis. The rough set theoretic approach [1] is based on the principles of set approximation and provides an attractive framework for data mining and knowledge discovery. The key notions here are those of "information granule" and "reducts". Information granule formalizes the concept of finite precision representation of objects in real life situation, and reducts represent the core of an information system (both in terms of objects and features) in a granular universe. An important use of rough set theory and granular computing in data mining has been in generating logical rules for classification and association. These rules correspond to different important regions of the feature space, and represent various data clusters.

3. Rough-Fuzzy Granulation and Case Generation

In this section we present briefly the concept of case generation in rough-fuzzy framework since it forms the basic principles of f-granulation as applicable in several other applications. A case may be defined as a contextualized piece of knowledge representing evidence that teaches a lesson fundamental to achieving goals of the system. Case based reasoning (CBR) [9] is a novel Artificial Intelligence (AI) problem-solving paradigm. It involves adaptation of old solutions to meet new demands, explanation of new situations using old instances (called cases), and performance of reasoning from precedence to interpret new problems. The significance of soft computing to CBR problems has been adequately explained by Pal et al [10] and Pal and Shiu [11]. In this section we provide an example [12, 13] of using the concept of f-granulation for performing the task of *case generation* in large scale CBR systems. While case selection deals with selecting informative prototypes from the data with no dimensionality reduction, case generation concerns with construction of 'cases' that need not necessarily include any of the given data points and has, in addition, the possibility of reducing the dimension varying with cases.

For generating cases, linguistic representation of patterns is used to obtain a fuzzy granulation of the feature space. Rough set theory is used to generate dependency rules corresponding to informative regions or clusters in the granulated feature space. The fuzzy membership functions appearing in the rules corresponding to the informative regions are stored as cases. Figure 1 shows an example of such case generation for a two dimensional data having two classes. The granulated feature space has $3^2 = 9$ granules. These granules of different sizes are characterized by three membership functions along each axis, and have ill-defined (overlapping) boundaries. Two dependency rules: $class_1 \leftarrow L_1 \wedge H_2$ and $class_2 \leftarrow H_1 \wedge L_2$ are obtained using rough set theory. The parameters (e.g., center point and bandwidth) of the fuzzy membership functions corresponding to sets *low* along feature F_1 and *high* along feature F_2, marked bold, are stored as cases for class 1. Similarly, the center points and bandwidths of the fuzzy membership functions corresponding to sets *high* along feature F_1 and *low* along feature F_2, marked bold, are stored as cases for class 2.

Unlike the conventional case selection methods, the cases here are cluster granules and not sample points. Also, since all the original features may not be required to express the dependency rules, each case involves a reduced number of features and the number of such reduced features may vary from case to case. The methodology is therefore suitable for mining data sets, large both in dimension and size, due to its low time requirement in case generation as well as retrieval.

The aforesaid characteristics are demonstrated in Figure 2 [12, 13] on a real life forest cover-type GIS data with number of features 10 and number of samples 586012. Its superiority over IB3, IB4 [9] and random case selection algorithms, in terms of classification accuracy (with one nearest neighbor rule), case generation (t_{gen}) and retrieval (t_{ret}) times, and average storage requirement (average feature) per case, is evident. The number of cases considered for comparison is 545. Based on the similar concept, Li et al reported a CBR based classification system combining efficient feature reduction and case selection [14].

Note that here the granules considered are class independent where the number of fuzzy granules generated in an n-dimentional space, using three linguistic terms *small, medium* and *high*, is 3^n. In the next section we describe a classification method where the granules are class dependent. The number of class dependent fuzzy granules required for modeling c classes in an n-dimentional space is c^n.

4. Rough-Fuzzy Classification

The rough-fuzzy classification model has three steps of operation as in Fig. 3.

The first step generates the class-dependent (CD) fuzzy granulated feature space of input pattern vector. For fuzzy granulation of a feature space containing **L** number of classes, we used **L** number of fuzzy sets (p-type membership function) to characterize the feature values of each pattern vector. Each feature is thus represented by **L** [0,1]-valued membership functions (MFs) representing **L** fuzzy sets or characterizing **L** fuzzy granules along the axis. The p-type MF explores the degree of belonging of a pattern into different classes based on individual features and the granules thus provide an improved class-wise representation of input patterns. The granules preserve the interrelated class information to build an informative granular space which is potentially useful for improved classification for the data sets with overlapping classes. In the granulation process, each feature value is represented with more than one membership values and thus the feature dimension

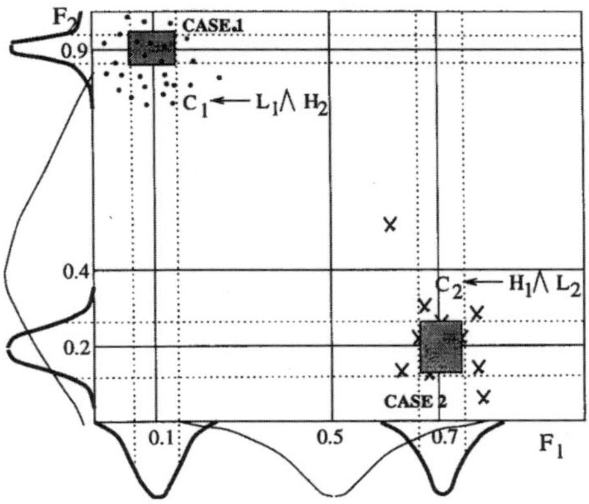

Fig. 1: Rough-fuzzy case generation for a two dimensional data.

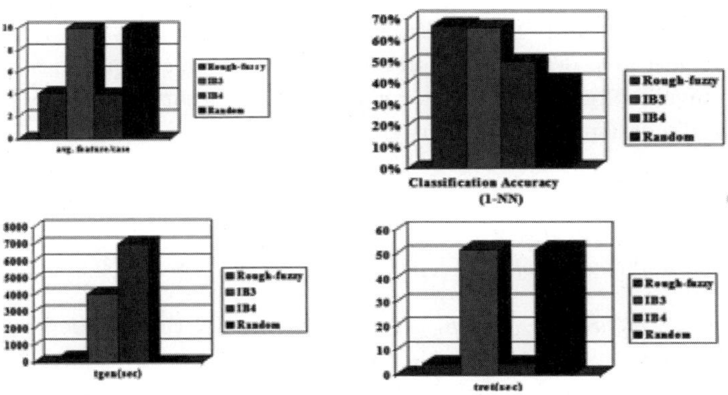

Fig. 2: Performance of different case generation schemes for the forest cover-type GIS data set with 7 classes, 10 features and 586012 samples.

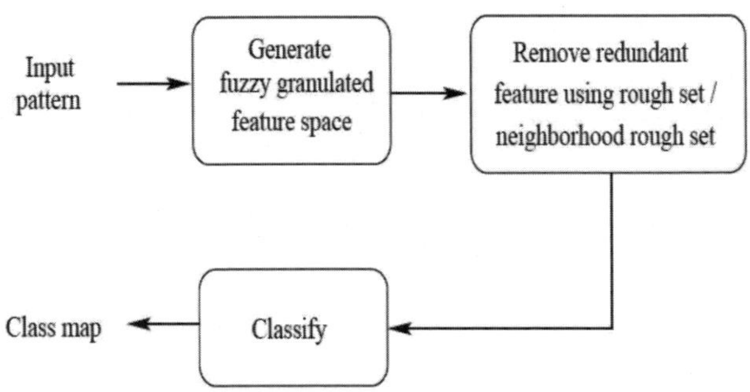

Fig. 3: Schematic diagram for rough-fuzzy pattern classification.

increases. The increased dimension brings great difficulty in solving many tasks of pattern recognition, machine learning and data mining. This motivates for selecting a subset of relevant and non-redundant features. In this regard, we have used the neighborhood rough set [1, 4, 15] (NRS) based feature selection method in the second step (Fig. 3). The advantage in the use of NRS is that it can deal with both numerical and categorical data. NRS does not require any discretisation of numerical data and is suitable for the proposed fuzzy granulation of features. Further, the neighboring concept facilitates to gather the possible local information through neighbor granules that provide better class discrimination information. Thus the combination of these two steps of operations can be a better framework for the classification of patterns in overlapping class environment. The integrated model thus takes the advantage of both class-dependent fuzzy granulation and NRS feature selection methods. After the features are selected, we use a classifier as in the third step of Fig. 3 to classify the input pattern based on the selected features. The aforesaid steps are now discussed in brief.

Let a pattern (object) **F** be represented by n numeric features and be expressed as: $\mathbf{F} = [F1,\ F2,...,\ Fn]$. Thus **F** is visualized as a point in n-dimensional vector space. Each feature is described in terms of its fuzzy membership values corresponding to **L** linguistic fuzzy sets. Thus, an n-dimensional pattern vector is expressed as (n x **L**)-dimensional vector and is given by

$$\begin{aligned}\mathbf{F} = \ & [\mu_1^1(F_1), \mu_2^1(F_1), \ldots, \mu_c^1(F_1), \ldots, \mu_{\mathbf{L}}^1(F_1); \\ & \mu_1^2(F_2), \mu_2^2(F_2), \ldots, \mu_c^2(F_2), \ldots, \mu_{\mathbf{L}}^2(F_2); \\ & \mu_1^n(F_n), \mu_2^n(F_n), \ldots, \mu_c^n(F_n), \ldots, \mu_{\mathbf{L}}^n(F_n)];\ (c = 1, 2, \ldots, \mathbf{L})\end{aligned}$$

where $\mu_1^n\,(F_n)$, $\mu_2^n\,(F_n)$,..., $\mu_n^n\,(F_n)$,..., $\mu_L^n\,(F_n)$ signify the membership values of F_n to **L** number of fuzzy sets along n^{th} feature axis and $\mu(F_n) \in [0,1]$. That means each feature F_n is expressed separately by **L** number of membership functions. In other words, each feature F_n characterizes **L** number of fuzzy granules along each axis and thus comprises $\mathbf{L}^n$ fuzzy granules in an n-dimensional feature space. Fig. 4 shows a crisp visualization of 16 ($= 4^2$) such class dependent granules using 0.5-cut when the number of classes is four in the F_1-F_2 feature space. The shape and size of the granules are dependent on the nature of overlapping of the classes and class-wise feature distribution.

For implementation of the concept of neighbourhood rough sets in feature selection, let us assume an information system denoted by $I = (U,\ A)$ where U (the universal set) is a non-empty and finite set of samples $\{x_1,\ x_2,\ ...,\ x_n\}$; $A = \{C\ \cup\ D\}$, where A is the finite set of features $\{a_1, a_2, ...,\ a_m\}$, C is the set of conditional features and D is the set of decision features. Given an arbitrary $x_i\ \in\ U$ and $B \subseteq C$, the neighborhood $\phi_B\ (x_i)$ of x_i for the feature set $\beta$is [15]

$$\phi_B\,(x_i) = \{x_i\,|x_j \in U,\ \Delta^{\mathrm{B}}\,(x_i, x_j) \leq \phi\} \tag{1}$$

where Δ is a distance function.

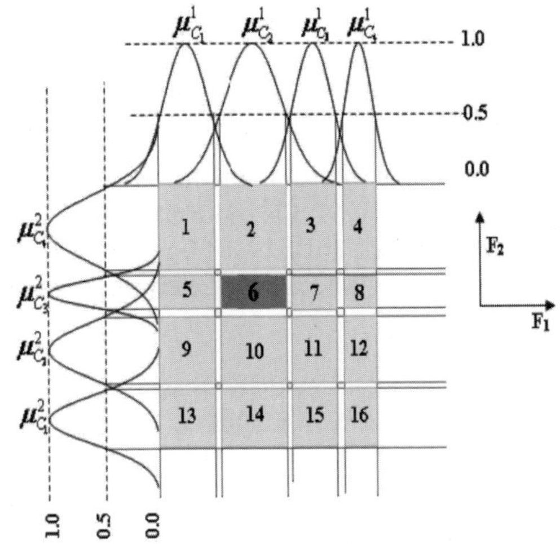

Fig. 4: Sixteen class dependent crisp granules for L = 4 in F_1-F_2 space.

$\phi_B(x_i)$ in Eqn. (1) represents the neighborhood information granule centered at sample x_i. That is, each sample x_i generates granules with a neighbourhood relation. For a metric space (U, Δ), the set of neighbourhood granules $\{\phi\ (x_i)|x_i \in\ U\}$ forms an elemental granule system, that covers the universal space rather than partitions it as done by Pawlak's rough set (PaRS). A neighbourhood granule degrades to an equivalence class when $\phi = 0$. In this case, samples in the same neighbourhood granules are equivalent to each other and the neighbourhood model degenerates to Pawlak's rough set. Thus NRS can be viewed as a generalization of PaRS.

Generation of neighborhood depends on both distance function Δ and parameter ϕ. The first one determines the shape and the second controls the size of neighborhood granule. For example, with Euclidean distance the parameter ϕ acts as the radius of the circle region developed by Δ function. Both these factors play important roles in neighbourhood rough sets (NRS) and control the granularity of data analysis. The significance of features varies with the level of granularity. Accordingly, the NRS based algorithm selects different subsets of feature as the Δ and ϕ are changed.

Performance of rough-fuzzy feature selection is demonstrated with a k-NN classifier, as an example, on various data sets. Here we present some results for k = 1 on remotely sensed images where the different regions are highly overlapping and the number of available training samples is small. Table 1 shows the comparative performance of various models in terms of β value [16] and Davies-Bouldin (DB) value [17] on IRS-1A and SPOT images with partially labelled samples. Partially labelled means, the classifiers are initially trained with labelled data of six land cover types and then the said trained classifiers are applied on the unlabeled image data to partition into six regions.

Five different models considered are [18]:

(1) Model 1: 1-NN classifier,
(2) Model 2: CI fuzzy granulation + PaRS based feature selection + 1-NN classifier,
(3) Model 3: CI fuzzy granulation + NRS based feature selection + 1-NN classifier,
(4) Model 4: CD fuzzy granulation + PaRS based feature selection + 1-NN classifier,
(5) Model 5: CD fuzzy granulation + NRS based feature selection + 1-NN classifier.

Table 1: Comparative performance of models using 1-NN classifier with partially labelled data sets (for $\phi = 0.45$ and Δ = Euclidean distance

Model	β value		DB value	
	IRS-1A	SPOT	IRS-1A	SPOT
Training samples	9.4212	9.3343	0.5571	1.4893
1	6.8602	6.8745	0.9546	3.5146
2	7.1343	7.2301	0.9126	3.3413
3	7.3559	7.3407	0.8731	3.2078
4	8.1372	8.2166	0.779	2.8897
5	8.4162	8.4715	0.7345	2.7338

As expected, the β value is the highest and DB value is the lowest for the training set (Table 1). It is also seen that model 5 yielded superior performance in terms of both the indexes, compared to other four models. The gradation of performance of five models can be established with the following β relation:

$$\beta training > \beta model5 > \beta model4 > \beta model3 > \beta model2 > \beta model1 \qquad (2)$$

Similar gradation of performance was also observed with DB value supporting the superiority of model 5.

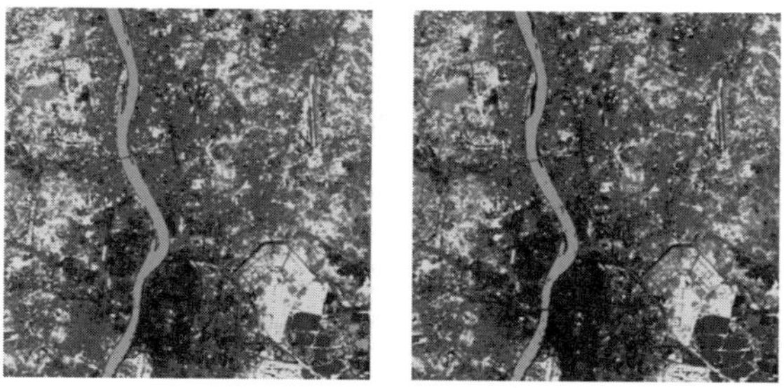

Fig. 5: Classified IRS-1A images with (a) model 1 and (b) model 5.

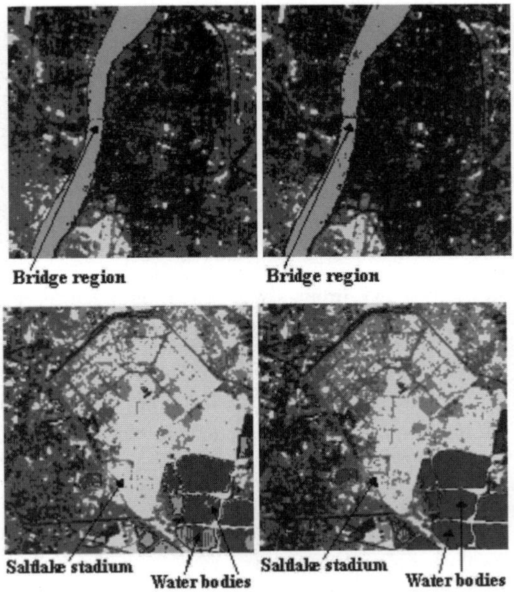

Fig. 6: (Zoomed) Two selected regions of classified IRS-1A image with (a and c) model 1, and (b and d) model 5.

In order to demonstrate the significance of granular computing visually, let us consider Figs.5a and 5b depicting the output corresponding to models 1 (without granulation) and 5 (with granulation), say for IRS-1A. It is clear from the figures that model 5 performed well in segregating different areas by properly classifying the land covers. For example, the *Howrah bridge* over the south part of the *river Hooghly* is more prominent in Fig.5b, whereas it is not so in Fig.5a. A zoomed version of the said *bridge* region is shown in Figs. 6a and 6b to have an improved visualization. Similarly, the regions such as *Saltlake stadium* and *water bodies* are more distinct and well shaped with model 5 as shown in Fig.6d (zoomed version). Similar is the case for SPOT image where Fig.7b is superior to Fig.7a in the sense that the different structures (e.g., *roads* and *canals*) are more prominent. These observations also justify the significance of the indexes β and DB in reflecting the performance of the models automatically without visual intervention.

5. Rough-Fuzzy Clustering

In the classification method, described in Section 4 f-granulation was used to model the overlapping characteristics of classes, whereas rough sets were considered for dimensionality reduction. In this section we describe a rough-fuzzy c-means algorithm (RFCM) which adds the concept of membership of fuzzy sets, and lower and upper

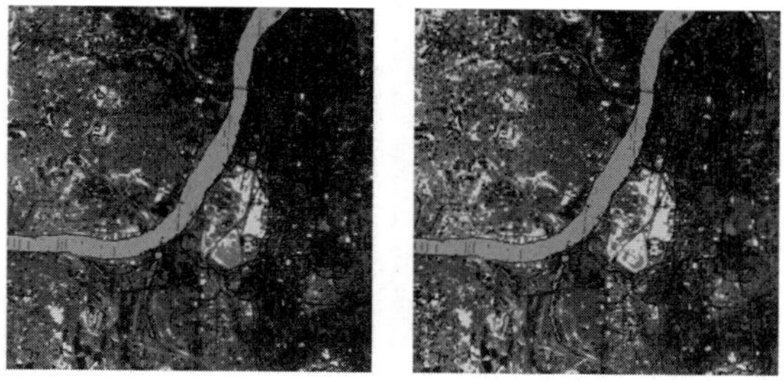

Fig. 7: Classified SPOT images with (a) model 1 and (b) model 5.

approximations of rough sets into hard c-means algorithm. While the membership of fuzzy sets enables efficient handling of overlapping partitions, the rough sets deal with uncertainty, vagueness, and incompleteness in class definition [19, 20].

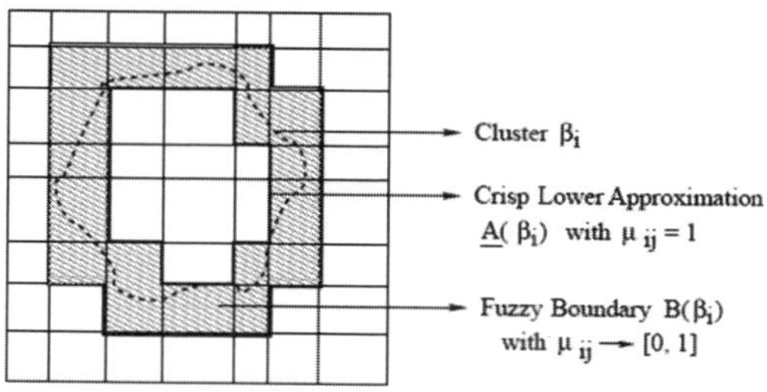

Fig. 8: Rough-fuzzy c-means: each cluster is represented by crisp lower approximations and fuzzy boundary.

In RFCM, each cluster is represented by a centroid, a crisp lower approximation, and a fuzzy boundary (Fig. 8). According to the definitions of lower approximations and boundary of rough sets, if an object belongs to lower approximation region (LAR) of a cluster, then the object does not belong to any other clusters. That is, the object is contained in that cluster definitely. Thus, the weights of the objects in LAR of a cluster should be independent of other centroids and clusters, and should not be coupled with their similarity with respect to other centroids. Also, the objects in LAR of a cluster should have equal influence on its centroid. Whereas, if an object belongs to the boundary of a cluster, then the object possibly belongs to that cluster and potentially belongs to another cluster(s). Hence, the objects in boundary regions should have different influence on the centroids and clusters. So, in RFCM, the membership values of objects in LAR are made equal to 1 (crisp), while those in boundary region are considered to be fuzzy as in fuzzy c-means algorithms. The new centroid is calculated based on the weighting average of the crisp LAR and fuzzy boundary region. *In essence, rough-fuzzy clustering compromises between restrictive (hard clustering) and descriptive (fuzzy clustering) partitions.*

The effectiveness of RFCM algorithm is shown, as an example, for segmentation of brain MR images. For this purpose, 100 MR images with different sizes and 16 bit gray levels were tested. These MR images were collected from Advanced Medicare and Research Institute (AMRI), Kolkata, India. The comparative performance of different c-means is shown in Fig. 9 with respect to β index [16].

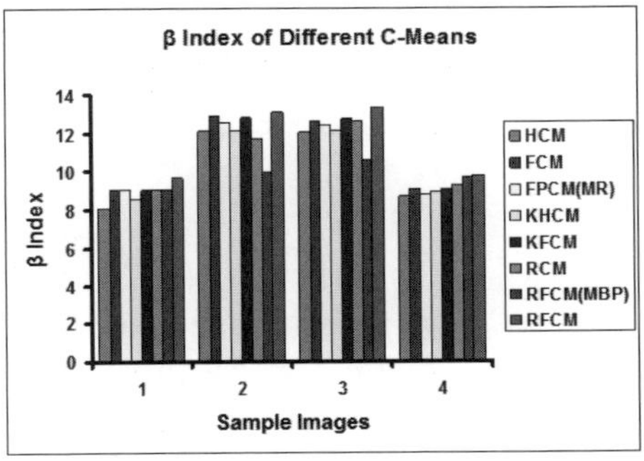

Fig. 9: Comparison of β index of HCM, FCM [17], RCM [21], RFCM^{MBP} [22], and proposed RFCM.

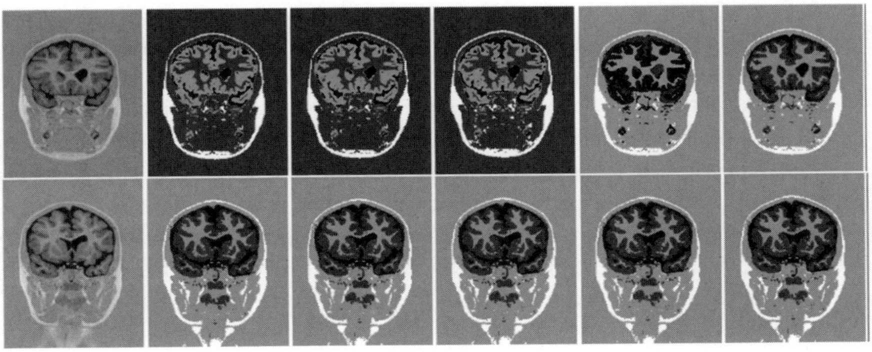

Fig. 10: An Original and its segmented images by HCM, FCM [17], RCM [21], RFCM^{MBP} [22], and proposed RFCM.

An example original image and its segmented versions with different c-means are shown in Fig. 10. The results in Figs. 9 and 10 confirm that the proposed RFCM algorithm produces segmented images which are more promising than do the other conventional methods, both visually and quantitatively.

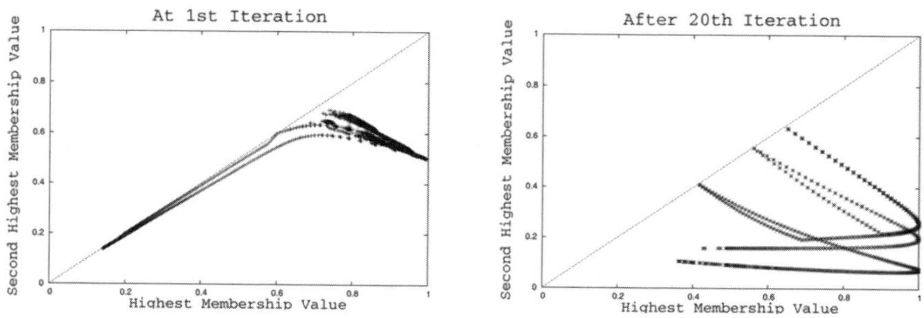

Fig. 11: Scatter plots of two highest membership values of all the objects in image data.

Figure 11 shows the scatter plots of the highest and second highest membership of all the objects in the data set of image at first and final iterations respectively, considering w=0.95, ($\acute{m}_1 = 2.0,$) and $c = 4$. The diagonal line represents the zone where two highest memberships of objects are equal. It is observed that though the average differences δ between two highest memberships of the objects is very low $(= 0.145)$ at first iteration it becomes ultimately very high $(= 0.652)$ at the final iteration.

6. Generalized Rough-fuzzy Sets: Entropy and Image Ambiguity Measures

In previous examples we have demonstrated the role of granules in modeling overlapping classes, representing linguistic rules and in defining class exactness. Here we demonstrate how fuzzy boundaries of image regions, rough resemblance between nearby gray levels and rough resemblance between nearby pixels *give rise to ambiguity in images,* where the significance of granules in determining *roughly resemblance in* gray levels and pixels is evident [23]. First, we describe the concept of generalized rough sets where different cases considering both the sets and granules as crisp and/or fuzzy are explained. This is followed by entropy measures and its applications to image processing.

Generalized Rough Sets - Lower & Upper Approximation:

Lower and upper approximations of a set Xdepend on the type of relation R and whether $X$is a crisp or a fuzzy set. Here we describe their expressions for different cases, i.e., when R denotes an equivalence or a fuzzy equivalence relation and X is a crisp or a fuzzy set.

Case 1: When R denotes an equivalence relation and X is a crisp set, the expressions for the lower and upper approximations of the set X are:

$$\underline{R}X = \{u | u \in U : [u]_R \subseteq X\}$$

$$\overline{R}X = \{u | u \in U : [u]_R \cap X \neq \emptyset\} \tag{3}$$

where $[u]_R$ denotes the granule to which the element u belongs. Here, the pair of sets $< \underline{R}X, \overline{R}X >$ is referred to as the *rough set of* X and $< U, R >$ is a crisp equivalence approximation space.

Case 2: When R denotes an equivalence relation and X is a fuzzy set, the expressions for the lower and upper approximations of the set X are:

$$\underline{R}X = \{(u, \inf_{z \in [u]_R} \mu_X(z)) | u \in U\}$$

$$\overline{R}X = \{(u, \sup_{z \in [u]_R} \mu_X(z)) | u \in U\} \tag{4}$$

where μ_X is the membership function associated with$X$. Here, the pair of fuzzy sets $< \underline{R}X, \overline{R}X >$ is referred to as the *rough-fuzzy set of* X and $< U, R >$ is a crisp equivalence approximation space.

Case 3: Consider the case when R refers to a fuzzy equivalence relation, that is, when the belongingness of every element (u) in the universe (U) to a granule $Y \in U/R$ is specified by a membership function, say, m_Y, that takes values in the interval $[0, 1]$ such that $\sum_Y m_Y(u) = 1$. In such a case, when X is a crisp set, the expressions for the lower and upper approximations of the set X are:

$$\underline{R}X = \{(u, \sum_{Y \in U/R} m_Y(u) \times \inf_{\varphi \in U} \max(1 - m_Y(\varphi), C)) | u \in U\}$$

$$\overline{R}X = \{(u, \sum_{Y \in U/R} m_Y(u) \times \sup_{\varphi \in U} \min(m_Y(\varphi), C)) | u \in U\} \tag{5}$$

where

$$C = \begin{cases} 1, \varphi \in X \\ 0, \varphi \notin X \end{cases}$$

Here, the symbols $\sum$ (sum) and $\times$ (product) respectively represent specific fuzzy union and intersection operations. Note that, one may consider any fuzzy union and intersection operation instead of the sum and product operations by judging their suitability with respect to the underlying application. The pair of fuzzy sets $< \underline{R}X, \overline{R}X >$ is referred to as the *fuzzy rough set of* X in this case, and $< U, R >$ is a fuzzy equivalence approximation space.

Case 4: In *Case 3* with R referring to a fuzzy equivalence relation, when X is a fuzzy set, the expressions for the lower and upper approximations of the set X are:

$$\underline{R}X_{-} = \{(u, \sum_{Y \in U/R} m_Y(u) \times \inf_{\varphi \in U} \max(1 - m_Y(\varphi), \mu_X(\varphi)))|u \in U\}$$

$$\overline{R}X_{-} = \{(u, \sum_{Y \in U/R} m_Y(u) \times \sup_{\varphi \in U} \min(m_Y(\varphi), \mu_X(\varphi)))|u \in U\} \tag{6}$$

The pair of fuzzy sets $< \underline{R}X, \overline{R}X >$ is referred as the *fuzzy rough-fuzzy set of* X, and $< U, R >$ is again a fuzzy equivalence approximation space. From the above explanation, it is obvious that the sets of expressions in *Cases 1-3* are special cases of the set of expressions for the lower and upper approximations given in *Case 4*. Pictorial diagram of lower and upper approximations for *Case 4* is shown in Fig 12.

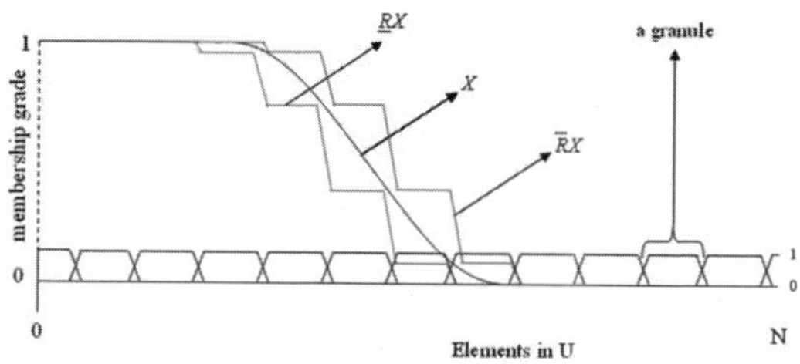

Fig. 12: The pair $< \underline{R}X, \overline{R}X >$is referred to as the *fuzzy rough-fuzzy set of X*.

Entropy Measures:

Let us now provide two classes of entropy measures based on roughness measures of a set and its complement in order to quantify the incompleteness of knowledge about a universe. One of them is based on logarithmic gain function, defined as [23]

$$H_R^L(X) = -\frac{1}{2}[\rho_R(X)\log_\beta(\frac{\rho_R(X)}{\beta}) + \rho_R(X^c)\log_\beta(\frac{\rho_R(X^c)}{\beta})] \tag{7}$$

where β denotes the base of the logarithmic function used and $X \subseteq U$ stands for the complement of the set X in the universe. The various entropy measures of this class can be obtained by calculating the roughness values $\rho_R(X)$ and $\rho_R(X^c)$ considering the different ways of obtaining the lower and upper approximations of the vaguely definable set$X$. Note that, the 'gain in incompleteness' term is taken as $-\log_\beta(\frac{\rho_R}{\beta})$ in (7) and for $\beta > 1$ it takes a value in the interval$[1, \infty]$. The other class of entropy measures, as obtained by considering an exponential function to measure the 'gain in incompleteness', is:

$$H_R^E(X) = \frac{1}{2}[\rho_R(X)\beta^{(1-\rho_R(X))} + \rho_R(X^c)\beta^{(1-\rho_R(X^c))}] \tag{8}$$

where β denotes the base of the exponential function used. Similar to the class of entropy measures H_R^L, the various entropy measures of this class can be obtained by using the different ways of obtaining the lower and upper approximations of X in order to calculate $\rho_R(X)$ and $\rho_R(X^c)$. The 'gain in incompleteness' term is taken as $\beta^{(1-\rho_R)}$ in (8) and for $\beta > 1$ it takes a value in the finite interval $[1, \beta]$.

In Figure 13, the values of H_R^L and H_R^E are shown for all possible values of the roughness measures A and B considering $\beta = e$. Figure 14 shows the plots of the proposed entropies for different values of β, when$A = B$.

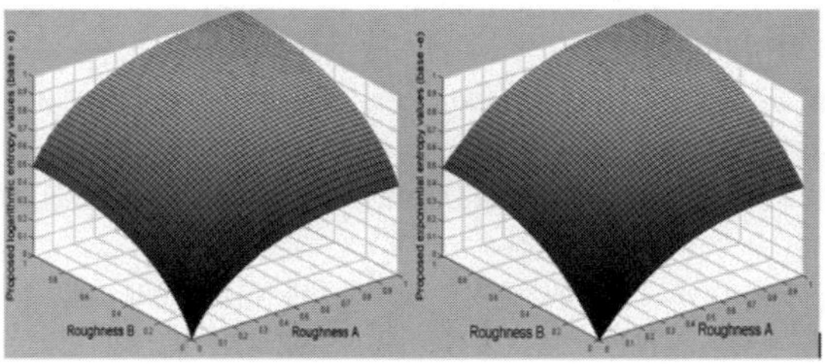

Fig. 13: Plots of the proposed classes of entropy measures for various roughness values A=$\rho_R(X)$ and B=$\rho_R(X^c)$.

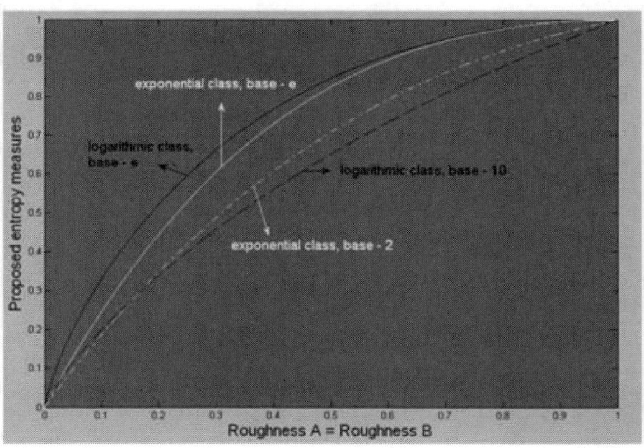

Fig. 14: Proposed entropy measures for a few different β values, when $\rho_R(X)$=$\rho_R(X^c)$($A = B$).

Image Ambiguity measures and Segmentation:

Using the aforesaid entropy definitions, we compute grayness and spatial ambiguity measures of an image. Grayness ambiguity refers to indefiniteness associated with deciding whether a pixel or a clump of pixels (granule) is white or black. That is, it concerns with the indefiniteness due to fuzziness as well as granularity in gray values. Spatial Ambiguity, on the other hand, refers to indefiniteness in shape and geometry of various regions where indefiniteness is concerned with both intensity and spatial location of individual pixel or group of pixels. These ambiguity measures can be minimized by changing the cross-over point of the membership function in order to find a set of minima corresponding to different thresholds of an image.

Fig 15 shows the segmentation results of three images, as an example, using grayness ambiguity measures based on rough-fuzzy entropy (proposed) and fuzzy entropy [24]. In the former case, membership of a pixel is dependent on the granule (defined over one-dimensional gray scale) to which it belongs, and it is independent of its spatial location. In the latter case, the membership of a pixel is entirely dependent only on its own gray value, and it is independent of its spatial location. Therefore the improvement in segmentation results by rough-fuzzy entropy as compared to fuzzy entropy in Fig. 15 is due to the effect of granules. The same is quantitatively demonstrated in Fig. 16 for 45 other images where β-index for segmentation is seen in almost all cases to be higher for outputs corresponding to rough-fuzzy entropy.

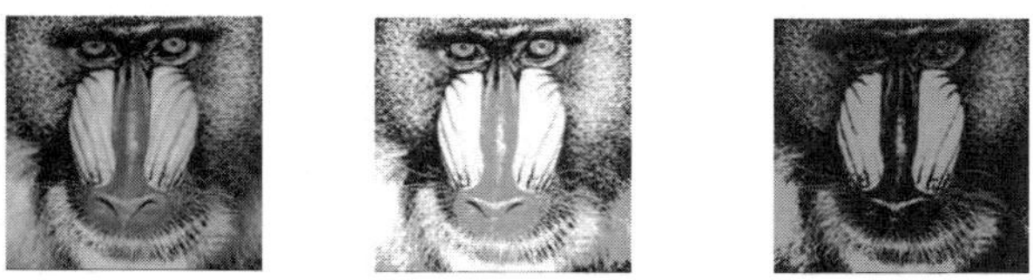

(a) The baboon image (b) Using proposed (c) Using fuzzy entropy

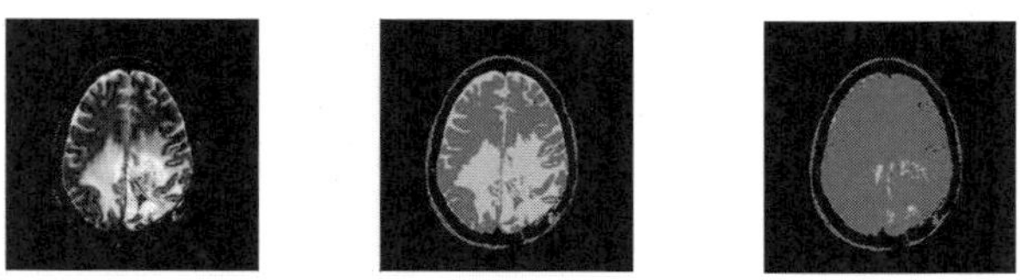

(a) A brain MR image (b) Using proposed (c) Using fuzzy entropy

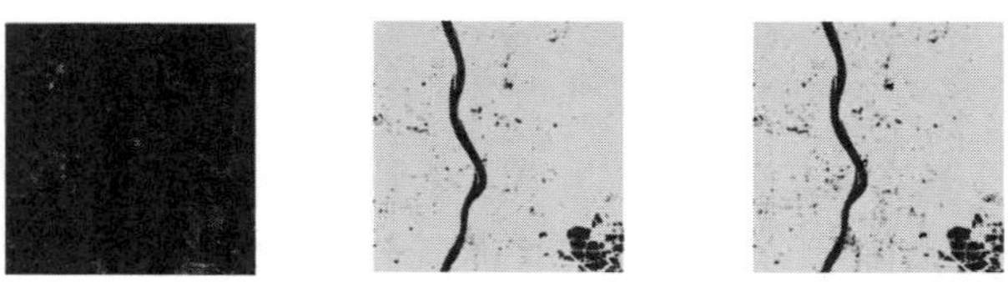

(a) A remote sensing image (b) Using proposed (c) Using fuzzy entropy

Fig. 15: Comparative segmentation results.

Figs. 17 and 18 show the effect of generalization in rough sets (*Cases 1- 4*) on

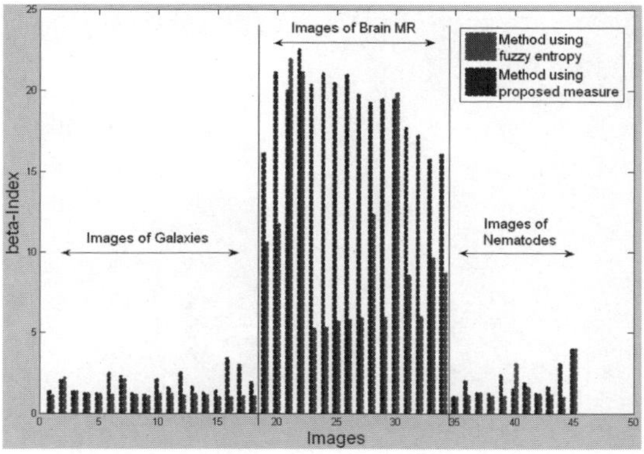

Fig. 16: β-index for segmentation results on 45 images (Significance of adding the concept of granules to gray level fuzziness is evident).

object extraction. As seen from the figures, *Case 4* (when both the sets and granules are fuzzy) results in highest β-index for both the images.

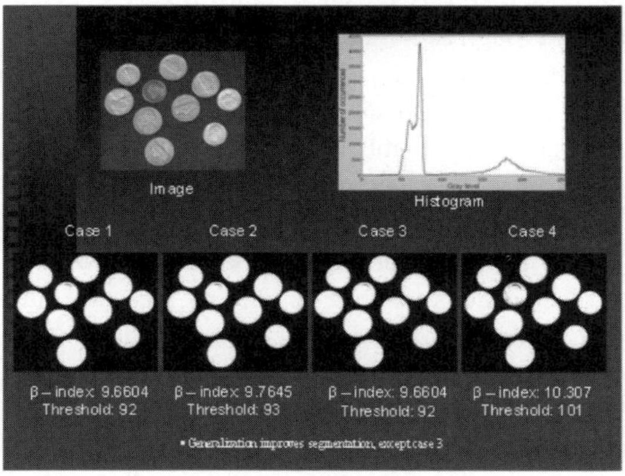

Fig. 17: Effect of generalization on object extraction: input, histogram and extracted objects with β-index for four cases.

7. Conclusions

Problem of integrating rough and fuzzy sets for modeling uncertainty is considered for different tasks of pattern recognition and knowledge discovery. The concept of knowledge encoding using rough sets and the role of f-granulation to make it more efficient are illustrated. Different tasks like case generation, classification, clustering and image segmentation are considered. Various rough-fuzzy entropy measures

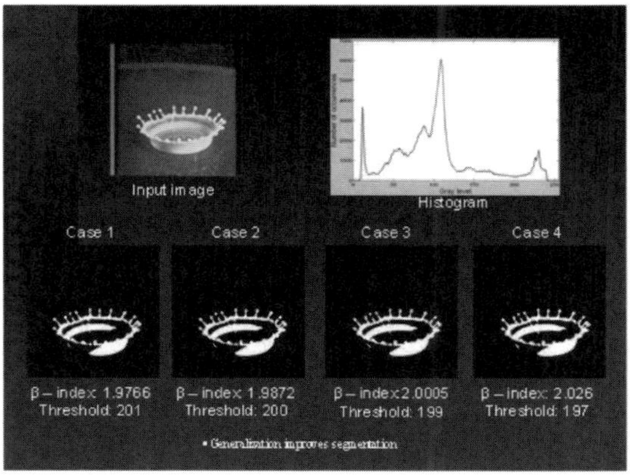

Fig. 18: Effect of generalization on object extraction: input, histogram and extracted objects with β-index for four cases.

with their characteristics are described. Cases generated are efficient for dimensionality reduction, classification and retrieval. Segmentation results of brain MR images are superior both qualitatively and quantitatively. Class dependent granulation with neighborhood rough sets is seen to have better class discrimination ability than class independent granulation with Pawlak's rough sets. The algorithm is very useful even when the number of available training samples is small. The effect of granularity, considering rough resemblance in nearby gray values and pixels, in improving the quality of image segmentation *vis-à-vis* considering only the fuzziness in gray values is adequately established. Incorporating fuzziness in both sets and granules provides the best performance. The same characteristic features can be demonstrated for other image processing operations. Recently, an application of the rough-fuzzy entropy for defining association error for image segmentation problem is reported [25]. Although the significance of the rough-fuzzy methods is demonstrated on some typical examples, they can be treated as basic modules to be made applicable to other real life recognition or decision-making problems. The model of rough-fuzzy granular space has recently been extended to create a rough-wavelet granular space which is applied for classification of multispectral remote sensing image [26].

The concept of aforesaid rough-fuzzy computing has a significant role in modeling the computational theory of perception (CTP). The work embodying f-granulation also enriches the literature in natural computational framework.

Acknowledgement

The article is dedicated to the memory of Professor Vito di Gesu, our great friend.

The author acknowledges J.C. Bose Fellowship of the Govt. of India, and his colleagues/students Drs. Pabitra Mitra, Pradipta Maji, Saroj Meher and Debashis

Sen for their contributions.

References

[1] Z. Pawlak, *Rough Sets: Theoretical Aspects of Reasoning about Data* (Kluwer, Dordrecht, 1991).

[2] S. K. Pal and J. F. Peters, *Rough-fuzzy Image Analysis: Foundations and Methodologies* (Chapman & Hall/CRC, Boca Raton, Florida, 2010).

[3] S. K. Pal and A. Skowron (eds.), *Pattern Recognition Letters, Special issue on Rough sets, pattern recognition and data mining* **24** (2003).

[4] S. K. Pal and A. Skowron (eds.), *Rough-Fuzzy Hybridization: A New Trend in Decision Making*, 1st edn. (Springer-Verlag New York, Inc., Secaucus, NJ, USA, 1999).

[5] L. A. Zadeh, *AI Magazine* **22**, 73 (2001).

[6] S. K. Pal and S. Mitra, *Neuro-fuzzy pattern recognition: methods in soft computing*, 1st edn. (John Wiley & Sons, Inc., 1999).

[7] S. K. Pal, Computational theory perception (ctp), rough-fuzzy uncertainty analysis and mining in bioinformatics and web intelligence: A unified framework, in *Transactions on Rough Sets XI*, eds. J. Peters and A. Skowron, Lecture Notes in Computer Science, Vol. 5946 (Springer Berlin / Heidelberg, 2010) pp. 106–129.

[8] H. Nguyen, S. Pal and A. Skowron (eds.), *Theoretical Computer Science, Part C, Special issue on Rough Sets and Fuzzy Sets in Natural Computing, (to appear)* .

[9] J. Kolodner, *Case-Based Reasoning* (Morgan Kaufmann, San Mateo, CA, 1993).

[10] S. K. Pal, T. S. Dillon and D. S. Yeung (eds.), *Soft computing in case based reasoning* (Springer-Verlag, London, UK, 2001).

[11] S. Shiu and S. K. Pal, *Foundations of Soft Case-Based Reasoning* (John Wiley & Sons, 2004).

[12] S. K. Pal and P. Mitra, *IEEE Trans. Knowledge and Data Engineering* **16**, 292 (2004).

[13] S. K. Pal and P. Mitra, *Pattern Recognition Algorithms for Data Mining* (Chapman & Hall/CRC, Boca Raton, Florida, 2004).

[14] Y. Li, S. Shiu and S. K. Pal, *IEEE Trans. Knowledge and Data Engineering* **18**, 415 (2006).

[15] Q. Hu, D. Yu, J. Liu and C. Wu, *Inform. Sciences* **178**, 3577 (2008).

[16] S. K. Pal, A. Ghosh and B. Sankar, *Int. Journal of Remote Sensing* **2**, 2269 (2000).

[17] J. Bezdek, *Pattern Recognition with Fuzzy Objective Function Algorithms* (Plenum, New York, 1981).

[18] S. K. Pal, S. Meher and S. Dutta, Pattern classification using class-dependent rough-fuzzy granular space, in *Proceedings of Fifth Int. Conf. on Rough Set and Knowledge Technology*, eds. J. Yu, S. Greco, P. Lingras, G. Wang and A. Skowron, Lecture Notes in Artificial Intelligence, Vol. 6401 (Springer Berlin / Heidelberg, Beijing, China, 24-30 June 2010) pp. 54–61.

[19] P. Maji and S. K. Pal, *IEEE Trans. on System, Man and Cybernetics, Part B* **37**, 1529 (2007).

[20] P. Maji and S. Pal, *IEEE Trans. Knowledge and Data Engineering* **19**, 859 (2007).

[21] P. Lingras and C. West, *Journal of Intelligent Information Systems* **23**, 5 (2004).

[22] S. Mitra, H. Banka and W. Pedrycz, *IEEE Trans. on Systems, Man, and Cybernetics - Part B: Cybernetics* **36**, 795 (2006).

[23] D. Sen and S. K. Pal, *IEEE Trans. on Systems, Man, and Cybernetics - Part B: Cybernetics* **39**, 117 (2009).

[24] S. K. Pal, R. King and A. Hashim, *Pattern Recognition Letters* **1**, 141 (1983).

[25] D. Sen and S. K. Pal, *IEEE Trans. Image Processing* **18**, 879 (2009).

[26] S. Meher and S. Pal, *Applied Soft Computing, (to appear)* .

FAST REDSHIFT CLUSTERING WITH THE BAIRE (ULTRA) METRIC

F. MURTAGH*[1] and P. CONTRERAS*[2]

** Department of Computer Science, Royal Holloway University of London, Egham, Surrey TW20 0EX, UK*
**E-mail: fionn@cs.rhul.ac.uk, pedro@cs.rhul.ac.uk*
www.cs.rhul.ac.uk/~fionn, www.cs.rhul.ac.uk/~pedro

1 Wilton Park House, Wilton Place
Dublin 2, Ireland

2 Thinking SAFE Ltd.
Egham, Surrey TW20 0EX, UK

The Baire metric induces an ultrametric on a dataset and is of linear computational complexity, contrasted with the standard quadratic time agglomerative hierarchical clustering algorithm. In this work we evaluate empirically this new approach to hierarchical clustering. We compare hierarchical clustering based on the Baire metric with (i) agglomerative hierarchical clustering, in terms of algorithm properties; (ii) generalized ultrametrics, in terms of definition. We apply the Baire distance to spectrometric and photometric redshifts from the Sloan Digital Sky Survey using, in this work, about half a million astronomical objects. We want to know how well the (more costly to determine) spectrometric redshifts can predict the (more easily obtained) photometric redshifts, i.e. we seek to regress the spectrometric on the photometric redshifts, and we use clusterwise regression for this.

Keywords: Clustering; Redshift; Ultrametric; Agglomerative hierarchical clustering; Baire distance.

1. Introduction: Agglomerative Hierarchical Clustering Algorithms

A metric space (X, d) consists of a set X on which is defined a distance function d which assigns to each pair of points of X a distance between them, and satisfies the following four axioms for any triplet of points x, y, z:

A1: $\forall x, y \in X, d(x, y) \geq 0$ (positiveness)
A2: $\forall x, y \in X, d(x, y) = 0$ iff $x = y$ (reflexivity)
A3: $\forall x, y \in X, d(x, y) = d(y, x)$ (symmetry)
A4: $\forall x, y, z \in X, d(x, z) \leq d(x, y) + d(y, z)$ (triangle inequality)

When considering an ultrametric space we need to consider the strong triangular inequality or ultrametric inequality defined as:

A5: $d(x, z) \leq max\ \{d(x, y),\ d(y, z)\}$ (ultrametric inequality)

and this in addition to the positivity, reflexivity and symmetry properties (properties A1, A2, A3) for any triple of point $x, y, z \in X$.

If X is endowed with a metric, then this metric can be mapped onto an ultrametric. In practice, endowing X with a metric can be relaxed to a dissimilarity. An often used mapping from metric to ultrametric is by means of an agglomerative hierarchical clustering algorithm. A succession of $n - 1$ pairwise merge steps take place by making use of the closest pair of singletons and/or clusters at each step. Here n is the number of observations, i.e. the cardinality of set X. Closeness between singletons is furnished by whatever distance or dissimilarity is in use. For closeness between singleton or non-singleton clusters, we need to define an inter-cluster distance or dissimilarity. This can be defined with reference to the cluster compactness or other property that we wish to optimize at each step of the algorithm. In terms of advising a user or client, such a cluster criterion, motivating the inter-cluster dissimilarity, is best motivated in turn by the data analysis application or domain.

Since agglomerative hierarchical clustering requires consideration of pairwise dissimilarities at each stage it can be shown that even in the case of the most efficient algorithms, e.g. those based on reciprocal nearest neighbors and nearest neighbor chains [1], $O(n^2)$ or quadratic computational time is required. The innovation in the work we present here is that we carry out hierarchical clustering in a different way such that $O(n)$ or linear computational time is needed. As always in computational theory, these are worst case times.

A hierarchy, H, is defined as a binary, rooted, node-ranked tree, also termed a dendrogram [1–4]. A hierarchy defines a set of embedded subsets of a given set of objects X, indexed by the set I. These subsets are totally ordered by an index function ν, which is a stronger condition than the partial order required by the subset relation. A bijection exists between a hierarchy and an ultrametric space.

Let us show these equivalences between embedded subsets, hierarchy, and binary tree, through the constructive approach of inducing H on a set I.

Hierarchical agglomeration on n observation vectors with indices $i \in I$ involves a series of $1, 2, \ldots, n-1$ pairwise agglomerations of observations or clusters, with the following properties. A hierarchy $H = \{q | q \in 2^I\}$ such that (i) $I \in H$, (ii) $i \in H\ \forall i$, and (iii) for each $q \in H, q' \in H : q \cap q' \neq \emptyset \implies q \subset q'$ or $q' \subset q$. Here we have denoted the power set of set I by 2^I. An indexed hierarchy is the pair (H, ν) where the positive function defined on H, i.e., $\nu : H \to \mathbb{R}^+$, satisfies: $\nu(i) = 0$ if $i \in H$ is a singleton; and (ii) $q \subset q' \implies \nu(q) < \nu(q')$. Here we have denoted the positive reals, including 0, by $\mathbb{R}^+$. Function ν is the agglomeration level. Take $q \subset q'$, let $q \subset q''$ and $q' \subset q''$, and let q'' be the lowest level cluster for which this is true. Then if we define $D(q, q') = \nu(q'')$, D is an ultrametric.

In practice, we start with a Euclidean or alternative dissimilarity, use some criterion such as minimizing the change in variance resulting from the agglomerations, and then define $\nu(q)$ as the dissimilarity associated with the agglomeration carried out.

2. Baire or Longest Common Prefix Distance

Agglomerative hierarchical clustering algorithms are constructive hierarchy-constructing algorithms. Such algorithms have the aim of mapping data into an ultrametric space, or searching for an ultrametric embedding, or ultrametrization [5].

Now, inherent ultrametricity leads to an identical result with most commonly used agglomerative criteria [1]. Furthermore, data coding can help greatly finding how inherently ultrametric data is [6]. In certain respects the hierarchy determined by the Baire distance can be viewed as a particular coding of the data because it seeks longest common prefixes in pairs of (possibly numerical) strings. We could claim that determining the longest common prefix is a form of data compression because we can partially express one string in terms of another.

2.1. *Ultrametric Baire Space*

A Baire space consists of countably infinite sequences with a metric defined in terms of the longest common prefix: the longer the common prefix, the closer a pair of sequences. What is of interest to us here is this longest common prefix metric, which we call the Baire distance [7, 8].

Consider real-valued or floating point data (expressed as a string of digits rather than some other form, e.g. using exponent notation). The longest common prefixes at issue are those of precision of any value. For example, let us consider two such values, x_i and y_j, with i and j ranging over numeric digits. When the context easily allows it, we will call these x and y.

We take x and y to be bounded by 0 and 1. Each are of some precision, and we take the integer $|K|$ to be the maximum precision.

Thus we consider ordered sets x_k and y_k for $k \in K$. In line with our notation, we can write x_k and y_k for these numbers, with the set K now ordered. So, $k = 1$ is the first decimal place of precision; $k = 2$ is the second decimal place; $\ldots$; $k = |K|$ is the $|K|th$ decimal place. The cardinality of the set K is the precision with which a number, x_k, is measured.

Take as examples $x_k = 0.478$; and $y_k = 0.472$. In these cases, $|K| = 3$. Start from the first decimal position. For $k = 1$, we find $x_k = y_k = 4$. For $k = 2$, $x_k = y_k$. But for $k = 3$, $x_k \neq y_k$.

We now introduce the following distance (case of vectors x and y, with 1 attribute, hence unidimensional):

$$d_{\mathcal{B}}(x_K, y_K) = \begin{cases} 1 & \text{if } x_1 \neq y_1 \\ \inf\ 2^{-k} & \quad x_k = y_k \quad 1 \leq k \leq |K| \end{cases} \tag{1}$$

We call this $d_{\mathcal{B}}$ value Baire distance, which is an ultrametric [6, 7, 9–11] distance.

Note that the base 2 is given for convenience. When dealing with binary data x, y, then 2 is the chosen base. When working with real numbers the base can be redefined to 10 if needed.

2.2. *Constructive Hierarchical Clustering Algorithm versus Hierarchical Encoding of Data*

The Baire distance was introduced and described by Bradley [12] in the context of inducing a hierarchy on strings over finite alphabets. This work further pursued the goal of embedding a dendrogram in a p-adic Bruhat-Tits tree, informally characterized as a "universal dendrogram".

By convention we denote a prime by p, and a more general, prime or non-prime, positive integer by m.

A geometric foundation for ultrametric structures is presented in Bradley [13]. Starting from the point of view that a dendrogram, or ranked or unranked, binary or more general m-way, tree, is an object in a p-adic geometry, it is noted that: "The consequence of using p-adic methods is the shift of focus from imposing a hierarchic structure on data to finding a p-adic encoding which reveals the inherent hierarchies."

This summarizes well our aim in this work. We seek hierarchy and rather than using an agglomerative hierarchical clustering algorithm which is of quadratic computational time (i.e., for n individuals or observation vectors, $O(n^2)$ computational time is required) we instead seek to read off a p-adic or m-adic tree. In terms of a tree, p-adic or m-adic mean p-way or m-way, respectively, or that each node in the tree has at most p or m, respectively, sub-nodes.

Furthermore, by "reading off" we are targeting a linear time, or $O(n)$ algorithm involving one scan over the dataset, and we are imposing thereby an encoding of the data. (We recall that n is the number of observations, or cardinality of the observation set X.)

In practice we will be more interested in this work in the hierarchy, and the encoding algorithm used is a means towards this end. For a focus on the encoding task, see [14].

2.3. *A Baire-Based Hierarchical Clustering Algorithm*

A straightforward algorithm for hierarchical clustering based on the Baire distance, as described in section 2.1 is as follows. Because of working with real numbers in our case study below, we define the base in relation 1 as 10 rather than 2.

For the first digit of precision, $k = 1$, consider 10 "bins" corresponding to the digits $0, 1, \ldots, 9$. For each of the nodes corresponding to these bins, consider 10 subnode bins corresponding to the second digit of precision, $k = 2$, associated with $0, 1, \ldots, 9$ at this second level. We can continue for a third and further levels. In practice we will neither permit not wish for a very deep (i.e., with many levels) storage tree. For the base 10 case, level one (corresponding to $k = 1$) gives rise to up to 10 clusters. For level two (corresponding to $k = 2$) we have up to 100 clusters. We see that in practice a small number of levels will suffice. In one pass over the data we map each observation (recall that it is univariate but we are using its ordered set of digits, i.e. ordered set K) to its bin or cluster at each level. For ℓ levels, the computation required is $n \cdot \ell$ operations. For a given value of ℓ we therefore have $O(n)$ computation – and furthermore with a very small constant of proportionality since we are just reading off the relevant digit and, presumably, updating a node or cluster membership list and cardinality.

3. Inducing a Hierarchy on the SDSS Data using the Baire Ultrametric

The Sloan Digital Sky Survey (SDSS) [15] is systematically mapping the sky producing a detailed image of it and determining the positions and absolute brightnesses of more than 100 million celestial objects. It is also measuring the distance to a million of the nearest galaxies and to a hundred thousand quasars. See [16] for a description of the data available in this catalog.

The aim here is to build a mapping from $z_{spec} \rightarrow z_{phot}$ to help calibrating the redshifts, based on the z_{spec} observed values. Traditionally we could map $f : z_{phot} \longrightarrow z_{spec}$ based on trained data. The mapping f could be linear (e.g. linear regression) or non-linear (e.g. multilayer perceptron) as used by D'Abrusco [17]. These techniques are global. Here our interest is to develop a locally adaptive approach based on numerical precision. That is the direct benefit of the (very fast, hierarchical) clustering based on the Baire distance.

We look specifically into four parameters: right ascension (RA), declination (DEC), spectrometric (z_{spec}) and photometric (z_{phot}) redshift. The spectrometric technique uses the spectrum of electromagnetic radiation (including visible light) which radiates from stars and other celestial objects. The photometric technique uses a faster and economical way of measuring the redshifts.

3.1. *Clustering SDSS Data*

In order to perform the clustering process introduced in section 2.1 and further described in 2.3, we compare every z_{spec} and z_{phot} data point searching for common prefixes based on the longest common prefix (see section 2.1). Thereafter, the data points that have digit coincidences are grouped together to form clusters.

Data characterization is presented in Figure 1. The left panel shows the z_{spec} and z_{phot} sky coordinates of the data currently used by us to cluster redshifts. This section of the sky presents approximately 0.5 million object coordinate points with the current data. As can be observed, various sections of the sky are represented in the data. We find this useful since preliminary data exploration has shown that correlation between z_{spec} and z_{phot} is consistent in different parts of the sky. For example, when taking correspondences between z_{spec} and z_{phot} as shown in Figures 3 and 4, and plotting them in RA and DEC space (i.e. astronomical coordinate space) we obtain the same shape as presented in Figure 1.

This leads us to conclude that digit coincidences of the redshift measures are distributed approximately uniformly in the sky and are not concentrated spatially. The same occurs for all the other clusters. We will concentrate on the very near astronomical objects, represented by redshifts between 0 and 0.6. When we plot z_{spec} versus z_{phot} we obtain a highly correlated signal as shown in Figure 1, right panel. The number of observations that we therefore analyze is 443,014.

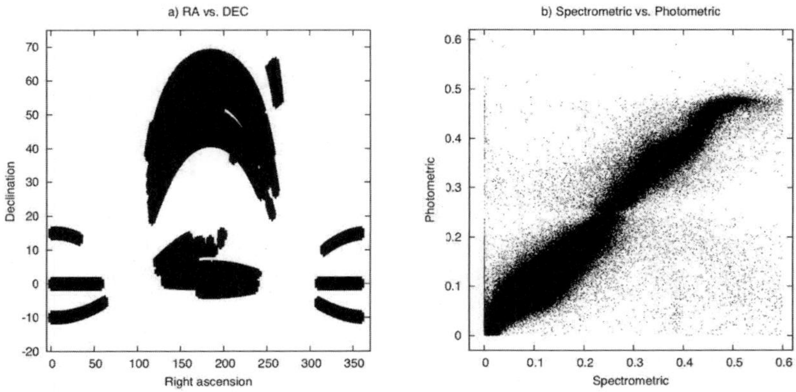

Fig. 1: Left: right ascension (RA) versus declination (DEC); Right: z_{spec} versus z_{phot}. SDSS data selection used for redshift analysis.

Looking at Figure 2 it can be seen clearly that most data points fall in the range between 0 and 0.2. Here the histogram on the top shows the z_{phot} data points distribution, and the histogram on the right the z_{spec} data points distribution. The heat plot also highlights the area where data points are concentrated, where the yellow color (white region in monochrome print) shows the major density.

Consequently, now we know that most cluster data points will fall within this range (0 and 0.2) if common prefixes of digits in the redshift values, taken as strings, are found.

Figures 3 and 4 show graphically how z_{spec} and z_{phot} correspondences look at different levels of decimal precision. On one hand we find that values of z_{spec} and z_{phot} that have equal precision up to the 3rd decimal digit are highly correlated. On the other hand when z_{spec} and z_{phot} have only the first digit in common, correlation is weak. For example, let us consider the following situations for plots 3 and 4:

- Figure 3 left: let us take the values of $z_{spec} = 0.437$ and $z_{phot} = 0.437$. We have that they share the first digit, the first decimal digit, the second decimal digit, and the third decimal digit. Thus, we have a highly correlated signal of the data points that share only up to the third decimal digit.

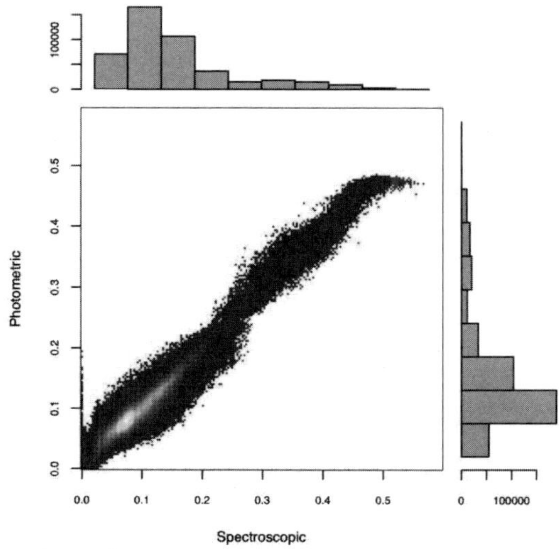

Fig. 2: Heat plot and histogram for z_{spec} versus z_{phot}. Histogram at the top shows the z_{spec} frequencies, histogram at the bottom shows z_{phot} frequencies.

- Figure 3 right: let us take the values of $z_{spec} = 0.437$ and $z_{phot} = 0.439$. We have that they share the first digit, the first decimal digit, and the second decimal digit. Therefore, the plot shows data points that share only up to the second decimal digit.
- Figure 4 left: let us take the values of $z_{spec} = 0.437$ and $z_{phot} = 0.474$. We have that they share the first digit, and the first decimal digit. Thus, the plot shows data points that share only up to the first decimal digit.
- Figure 4 right: let us take the values of $z_{spec} = 0.437$ and $z_{phot} = 0.571$. We have that they share only the integer part of the value, and that alone. Furthermore, this implies redshifts that do not match in succession of decimal digits. For example, if we take the values 0.437 and 0.577, the fact that the third digit is 7 in each case is not of use.

Table 1 (see also Figure 5) shows the clusters found for all different levels of precision. In other words this table allows us to define empirically the confidence levels for mapping of z_{phot} and z_{spec}. For example, we can expect that 82.8% of values for z_{spec} and z_{phot} have at least two common prefix digits. This percentage of confidence is derived as follows: the data points that share six, five, four, three, two, and one decimal digit (i.e., $4 + 90 + 912 + 8,982 + 85,999 + 270,920 = 366,907$ data points. Therefore 82.8% of the data).

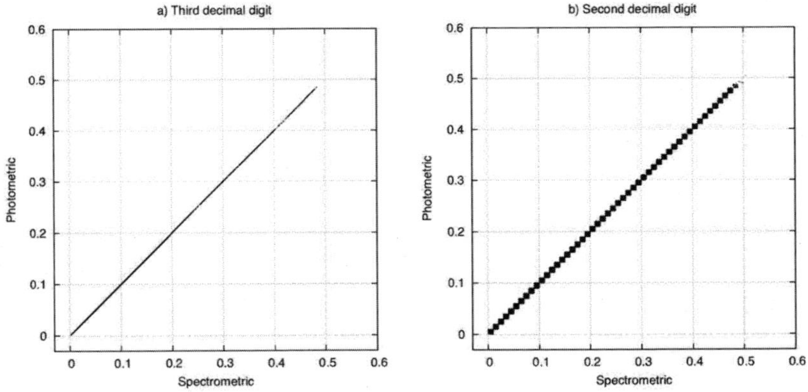

Fig. 3: Prefix-wise clustering frequencies depicting 3rd decimal digit coincidences (left panel), and two decimal digit coincidences (right panel).

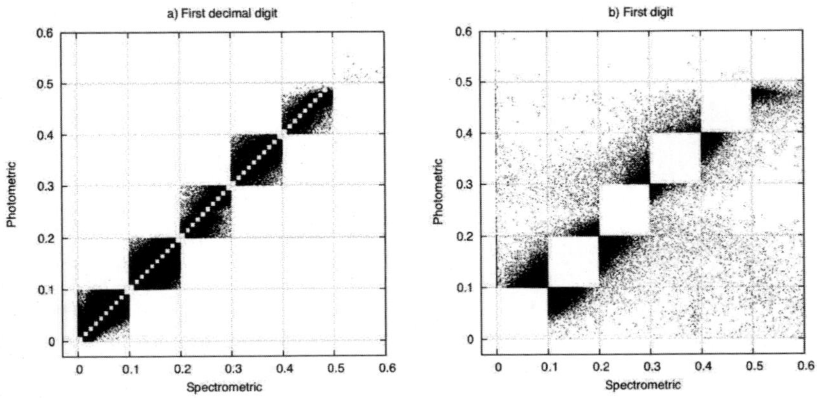

Fig. 4: Prefix-wise clustering frequencies depicting the 1st decimal digit coincidences (left panel), and first digit coincidences (right panel).

Additionally we observe that around a fifth of the observations share at least 3 digits in common. Namely, $4 + 90 + 912 + 8,982 + 85,999 = 95,987$ data points, which equals 21.7% of the data. In a previous work [18] we have shown that the clusters generated by means of the Baire ultrametric are similar to the clusters generated by k-means, also see [19].

Digit	No.	%
1	76,187	17.19
Decimal digit	No.	%
1	270,920	61.14
2	85,999	19.40
3	8,982	2.07
4	912	0.20
5	90	0.02
6	4	—
	443,094	100

Table 1: Data points based on the longest common prefix for different levels of precision.

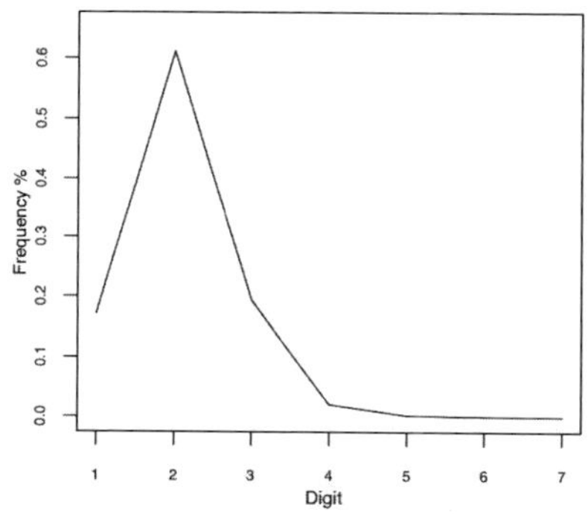

Fig. 5: Frequency distribution for Table 1. The abscissa shows the digit positions, where 1 is the first digit, 2 the first decimal digit, 3 the second decimal digit and so on.

4. Spectrometric and Photometric Digit Distribution

We have seen that the Baire ultrametric produces a strict hierarchical classification. In the case of z_{spec} and z_{phot} this can be seen as follows. Let us take any observed measurement of either case of $z_{spec} = z_{phot}$. Let us say $z_{spec} = z_{phot} = 0.1257$. Here we have that for $|K| = 4$, $z_{spec} = z_{phot}$. Hierarchically speaking we have that the root node is 0, for the first level where there potentially exist 6 nodes (i.e. 0,1,...,5); for the second level potentially there are 60 nodes; and so on until $k = |K| = 4$, and $z_{spec} = z_{phot}$, where potentially there are $6 \cdot 10 \cdot 10 \cdot 10 = 6,000$ nodes.

Of course not all nodes will be populated. In fact we can expect that a large number of these potential nodes will be empty if the number of observations n is lower than the potential number of nodes for a certain precision $|K|$ (i.e. $n \leq |K|^{10}$). Note that this points to a big storage cost, but in practice the tree is very sparsely populated and $|K|$ small.

A particular interpretation can be given in the case of an observed data point. Following up the above example if we take $z_{spec} = z_{phot} = 0.1257$, a tree can be produced to store all observed data that falls within this node. Doing this has many advantages from the viewpoint of storing. Access and retrieval, for example, is very fast and it is easy to retrieve all the observations that fall within a given node and its children.

With this tree it is a trivial task to build bins for data distribution. Figure 6 depicts the frequency distribution for a given digit and precision. There are 100 data points that have been convolved with a Gaussian kernel to produce surface planes in order to assemble three-dimensional plots.

This helps to build a cluster-wise mapping of the data. Following the Figure 6 top panel we observe that for the first decimal digit most data observations are concentrated in the digits 0, 1, 2, and 3. Then the rest of decimal precision data is uniformly distributed, gradually going towards zero when the level of precision increases. There is the exception of two peaks, for precision equal to 8. This turns out to be useful because when comparing the z_{spec} and z_{phot} digit distribution we do not find the same peaks in z_{phot}. This is very useful because now we can discriminate which observations are more reliable in z_{phot} through different characteristics of the data associated with the peaks.

5. Conclusions

In this work the distance called the Baire distance is presented. This distance has been very recently introduced into data analysis. We show how this distance can be used to generate clusters in a way that is computationally inexpensive when compared with more traditional techniques. The distance is also an ultrametric and therefore furnishes an ultrametric too. Hence the new Baire distance lends itself very well to the new hierarchical clustering method that we have introduced here.

In the astronomy case clusters generated with the Baire distance can be useful when calibrating redshifts. In general, applying the Baire method to cases where digit precision is important can be of relevance, specifically to highlight data "bins" and some of their properties.

Note that when two numbers share 3 prefix digits, and base 10 is used, we have a Baire distance of $d_{\mathcal{B}} = 10^{-3}$. We may not need to define the actual (ultra)metric values. It may be, in fact, more convenient to work on the hierarchy, with its different levels.

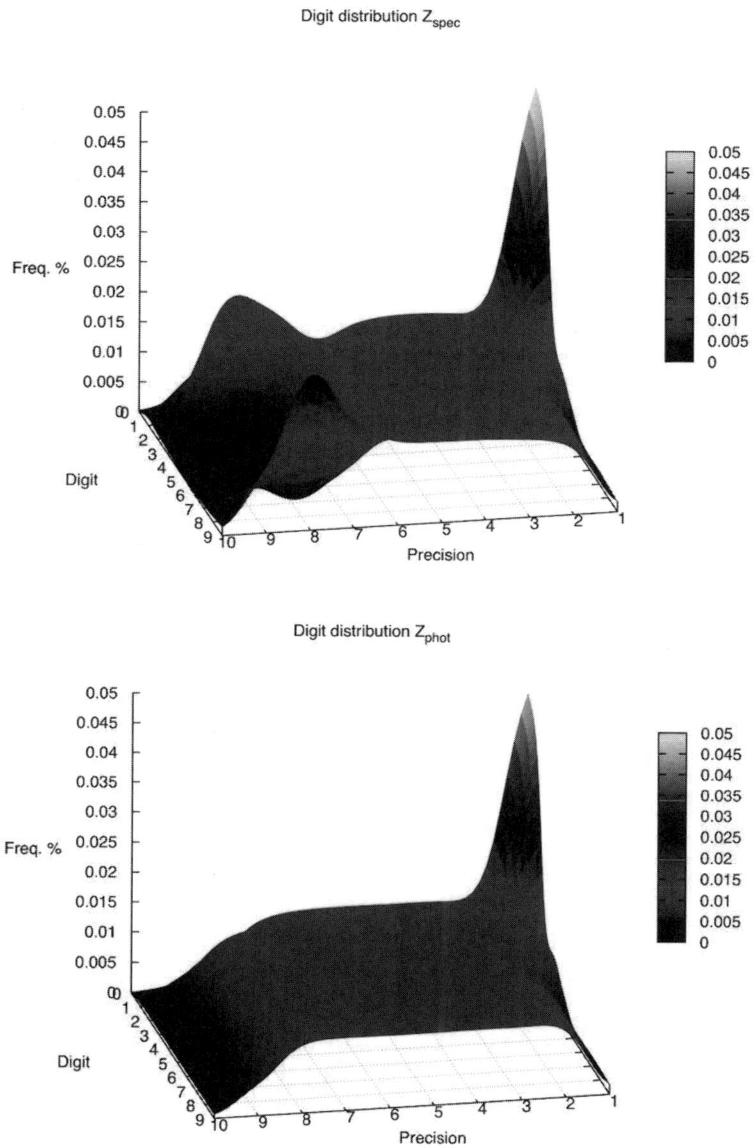

Fig. 6: Digit distribution for z_{spec} and z_{phot}; Top: Spectrometric digit distribution; Bottom: Photometric digit distribution. Note that digit distribution for z_{spec} has three peaks, but z_{phot} has only one.

In section 3.1 we showed how we could derive that 82.8% of values for z_{spec} and z_{phot} have at least two common prefix digits. This is a powerful result in practice when we recall that we can find very efficiently where these 82.8% of the astronomical objects are.

Using the Baire distance we showed in section 4 that z_{spec} and z_{phot} signals can be stored in a tree like structure. This is advantageous when measuring the digit distribution for each signal. When comparing these distributions, it can easily be seen where the differences arise.

The Baire distance has proved very useful in a number of cases, for instance in [7] this distance is used in conjunction with random projection [20] as the basis for clustering a large dataset of chemical compounds achieving results comparable to k-means but with better performance due to the lower computational complexity of the Baire-based clustering method.

Other application areas include text mining and semantic preservation [21]. For more details refer to [19] where a number of examples are discussed.

References

[1] F. Murtagh, *Multidimensional Clustering Algorithms* (Physica-Verlag, 1985).
[2] J.-P. Benzécri, *La Taxinomie*, 2nd edn. (Dunod, Paris, 1979).
[3] S. Johnson, *Psychometrika* **32**, 241 (1967).
[4] I. C. Lerman, *Classification et Analyse Ordinale des Données* (Dunod, Paris, 1981).
[5] A. van Rooij, *Non-Archimedean Functional Analysis* (Marcel Dekker, 1978).
[6] F. Murtagh, *Journal of Classification* **21**, 167(September 2004).
[7] F. Murtagh, G. Downs and P. Contreras, *SIAM Journal on Scientific Computing* **30**, 707(February 2008).
[8] P. Contreras and F. Murtagh, Evaluation of hierarchies based on the longest common prefix, or Baire, metric (2007), Classification Society of North America (CSNA) meeting, University of Illinois. Urbana-Champaign. IL, USA.
[9] F. Murtagh, Quantifying ultrametricity, in *COMPSTAT 2004 – Proceedings in Computational Statistics*, ed. J. Antoch (Springer, Prague, Czech Republic, 2004).
[10] F. Murtagh, Thinking ultrametrically, in *Classification, Clustering, and Data Mining Applications. Proceedings of the Meeting of the International Federation of Classification Societies (IFCS)*, eds. D. Banks, L. House, F. R. McMorris, P. Arabie and W. Gaul (Springer, Illinois Institute of Technology, Chicago, July 2004).
[11] F. Murtagh, *The European Physical Journal B* **43**, 573(February 2005).
[12] P. Bradley, *Computer Journal* **53**, 393 (2010).
[13] P. Bradley, *Journal of Classification* **25**, 27 (2008).
[14] F. Murtagh, *Proceedings of Steklov Institute of Mathematics* **265**, 177 (2009).
[15] SDSS, Sloan Digital Sky Survey (2008), `http://www.sdss.org`.
[16] J. K. Adelman-McCarthy, *The Astrophysical Journal Supplement Series* **172**, 634 (2007).
[17] R. D'Abrusco, A. Staiano, G. Longo, M. Paolillo and E. D. Filippis, Steps toward a classifier for the virtual observatory. I. Classifying the SDSS photometric archive 1st Workshop of Astronomy and Astrophysics for Students- Naples(April, 2006), `http://arxiv.org/abs/0706.4424`.

[18] P. Contreras and F. Murtagh, Fast hierarchical clustering from the Baire distance, in *Classification as a Tool for Research*, eds. H. Locarek-Junge and C. Weihs (Springer Berlin Heidelberg, Dresden, Germany, 2009).

[19] P. Contreras, *Search and Retrieval in Massive Data Collections* (PhD thesis. Royal Holloway, University of London, 2010).

[20] S. S. Vempala, *The Random Projection Method. DIMACS: Series in Discrete Mathematics and Theoretical Computer Science* (American Mathematical Society, 2004).

[21] J. Pereira, F. Schmidt, P. Contreras, F. Murtagh and H. Astudillo, Clustering and semantics preservation in cultural heritage information spaces, in *RIAO'2010, 9th International Conference on Adaptivity, Personalization and Fusion of Heterogeneous Information*, (Paris, France, 2010).

INTERACTIVE CLASSIFICATION ORIENTED SUPERRESOLUTION OF MULTISPECTRAL IMAGES

P. RUIZ

Department of Computer Science and Artificial Intelligence, Universidad de Granada
Granada, 18071, España
E-mail: mataran@decsai.ugr.es
decsai.ugr.es/vip

J.V. TALENTS

Image Processing Laboratory (IPL), Universitat de Valencia
Valencia, 46071 España
E-mail: juatano@.uv.es

J. MATEOS and R. MOLINA

Department of Computer Science and Artificial Intelligence, Universidad de Granada
Granada, 18071, España
E-mail: {rms, jmd}@decsai.ugr.es

A.K. KATSAGGELOS

Department of Electrical Engineering and Computer Science, Northwestern University
Evanston, IL, 60208-3118 USA
E-mail: aggk@eecs.northwestern.edu

Classification techniques are routinely utilized on satellite images. Pansharpening techniques can be used to provide super resolved multispectral images that can improve the performance of classification methods. So far, these pansharpening methods have been explored only as a preprocessing step. In this work we address the problem of adaptively modifying the pansharpening method in order to improve the precision and recall figures of merit of the classification of a given class without significantly deteriorating the performance of the classifier over the other classes. The validity of the proposed technique is demonstrated using a real Quickbird image.

Keywords: Pansharpening, super-resolution, classification, LDA, SVM.

1. Introduction

Satellite images are of great interest due to the numerous applications they can be utilized. Drawing maps, delimitation of parcels, studies on hydrology, forest or agriculture are just a few examples where these images are used.

Due to physical and technological constraints, satellites usually have sensors that capture two types of images. One sensor captures a multispectral (MS) image composed of several spectral bands with low spatial resolution (LR). The other sensor captures a high spatial resolution (HR) image, named panchromatic (PAN) image, with a low spectral resolution. While the first image allows to distinguish features spectrally but not spatially, the second allows to distinguish features spatially but not spectrally.

Pansharpening is an image fusion approach that combines the LR MS and PAN images to obtain an image with the spectral resolution of the MS image and the spatial resolution of the PAN image. Many techniques have been proposed in the literature to carry out the pansharpening procedure (see Ref. 1 for a complete review of pansharpening methods).

Many satellite image applications involve the classification of pixels in an image into a number of classes. In supervised classification, starting from a small set of samples previously labeled by the user, classification is carried out automatically by the classifiers. Bruzzone et al. [2] showed that the use of pansharpening methods that do not introduce significant spectral distortion helps the classifier to obtain higher accuracy, especially for pixels at the borders of objects.

While in the past pansharpening techniques have only been used as a preprocessing step, in this work we address the problem of adaptively modifying the pansharpening method in order to improve the precision and recall figures of merit of the classification of a given class without deteriorating the performance of the classifier over the other classes.

The rest of paper is organized as follows: In section 2 we describe the pansharpening technique we use. The used classifiers are briefly explained in section 3. The proposed method to estimate the pansharpening parameters to improve the performance of the classifier on a given class is described in section 4. Section 5 presents experimental results on real data. Finally, section 6 concludes the paper.

2. Pansharpening Algorithm

In this paper we use the pansharpening method proposed by Amro et al. [3] and the parameter estimation procedure described in Ref. 1. This method makes use of the non-subsampled contourlet transform [4] (NSCT) to decompose the details of the PAN and each band of the MS image into different scales and different directions. Then, the hierarchical Bayesian framework is used to model those observations and their relations with the original high resolution multispectral image and Bayesian inference is applied to estimate the HR MS image and the model parameters. Let us now explain in detail the used pansharpening method.

The used contourlet based pansharpening algorithm takes as input the PAN image, x, of size $p = m \times n$, and the observed LR MS image, Y, with B bands, $Y_b, b = 1, \ldots, B$, each of size $P = M \times N$ pixels with $M < m$ and $N < n$. Initially, each band of the LR MS image Y is upsampled to the size of the PAN image by bicubic interpolation. We will denote by s_b each band b of the $p = m \times n$ upsampled image.

Then, using the NTSC transform we can write the PAN and the upsampled MS images as:

$$x = x^r + \sum_{l=1}^{L}\sum_{d=1}^{D} x^{ld}, \quad s_b = s_b^r + \sum_{l=1}^{L}\sum_{d=1}^{D} s_b^{ld}, \quad b = 1, \ldots, B \tag{1}$$

where the superscript r denotes the residual (low pass filtered version) NSCT coefficient band and the superscript ld refers to the detail bands, with $l = 1, \ldots, L$, representing the scale and $d = 1, \ldots, D$, representing the direction for each coefficient band. The pansharpening goal is to estimate the HR MS image coefficients y_b^{ld} from the observed x^{ld} and s_b^{ld} coefficients. Finally, each band of the pansharpened HR MS image will be obtained by the inverse NSCT from the corresponding residual band of the upsampled MS image s_b^r and the estimated detail bands y_b^{ld}.

We will model the coefficient bands using the hierarchical Bayesian framework. This framework has two stages. In the first stage, knowledge about the structural form of the noise in the coefficients bands and the structural behavior of the HR MS image coefficients is used in forming $p(s_b^{ld}, x^{ld}|y_b^{ld}, \Omega_b^{ld})$ and $p(y_b^{ld}|\Omega_b^{ld})$, respectively. These noise and image models depend on the unknown parameters Ω_b^{ld} that need to be estimated. In the second stage a hyperprior on the parameters is defined, thus allowing the incorporation of information about these hyperparameters into the process. Let us define the probability distribution involved in each stage.

Following Refs. 3, 5, we chose a prior model based on the Total Variation (TV) for the HR MS image coefficient bands, y_b^{ld}, given by

$$p(y_b^{ld}|\alpha_b^{ld}) \propto (\alpha_b^{ld})^{p/2} \exp\left\{-\alpha_b^{ld} TV(y_b^{ld})\right\}, \tag{2}$$

with $TV(y_b^{ld}) = \sum_{i=1}^{p} \sqrt{(\Delta_i^h(y_b^{ld}))^2 + (\Delta_i^v(y_b^{ld}))^2}$ where $\Delta_i^h(y_b^{ld})$ and $\Delta_i^v(y_b^{ld})$ represent the horizontal and vertical first order differences at pixel i, respectively, and α_b^{ld} is the model parameter of the MS band b coefficients at level l and direction d. The idea behind this model is to consider the coefficient bands as a set of relatively smooth regions separated by strong edges, such as the coefficients of the NSCT.

Since the MS bands coefficients and the PAN image coefficients are independent given the HR MS image coefficients, we define $p(s_b^{ld}, x^{ld}|y_b^{ld}, \Omega_b^{ld}) = p(s_b^{ld}|y_b^{ld}, \Omega_b^{ld}) \times p(x^{ld}|y_b^{ld}, \Omega_b^{ld})$. The conditional distribution of the upsampled MS coefficients given the HR MS coefficients is defined as [3]

$$p(s_b^{ld}|y_b^{ld}, \beta_b^{ld}) \propto (\beta_b^{ld})^{p/2} \exp\left\{-\frac{1}{2}\beta_b^{ld} \left\|s_b^{ld} - y_b^{ld}\right\|^2\right\}, \tag{3}$$

where β_b^{ld} is the inverse of the unknown noise variance of the detail band at level l and direction d of the MS band b. The relationship between the HRMS band coefficients and the PAN image is modeled by the conditional probability distribution

$$p(x^{ld}|y_b^{ld}, \gamma_b^{ld}) \propto (\gamma_b^{ld})^{p/2} \exp\left\{-\frac{1}{2}\gamma_b^{ld} \left\|x^{ld} - y_b^{ld}\right\|^2\right\}. \tag{4}$$

where γ_b^{ld} is the inverse of the unknown noise variance at each NSCT decomposition level, l, and direction, d, of PAN image. Note that, with this modeling, we have decoupled each one of the bands of the contourlet transform and, since they are uncorrelated, we can do the estimation of each band independently of the other bands.

In the second stage of the hierarchical Bayesian framework we define the distribution on the parameters by using a gamma distribution

$$p(w|a_w, c_w) = \Gamma(w|a_w, c_w), \tag{5}$$

where $w > 0, w \in \Omega_b^{ld} = (\alpha_b^{ld}, \beta_b^{ld}, \gamma_b^{ld})$ denotes a hyperparameter, and $a_w > 0$ and $c_w > 0$ are, respectively, the shape and the inverse scale parameters of the distribution.

Finally, combining the first and second stages of the problem modeling, and defining $\Omega_b^{ld} = \{\alpha_b^{ld}, \beta_b^{ld}, \gamma_b^{ld}\}$, we have the global distribution

$$\begin{aligned} p(\Omega_b^{ld}, y_b^{ld}, x^{ld}, s_b^{ld}) =& p(\alpha_b^{ld})\, p(\beta_b^{ld})\, p(\gamma_b^{ld})\, p(y_b^{ld}|\alpha_b^{ld}) \\ &\times p(s_b^{ld}|y_b^{ld}, \beta_b^{ld})\, p(x^{ld}|y_b^{ld}, \gamma_b^{ld}), \end{aligned} \tag{6}$$

where $p(y_b^{ld}|\alpha_b^{ld})$, $p(s_b^{ld}|y_b^{ld}, \beta_b^{ld})$ and $p(x^{ld}|y_b^{ld}, \gamma_b^{ld})$ are given in Eqs. (2), (3), and (4), respectively.

The Bayesian paradigm dictates that inference on the parameters and the image, $(\Omega_b^{ld}, y_b^{ld})$, should be based on $p(\Omega_b^{ld}, y_b^{ld}|s_b^{ld}, x^{ld}) = p(\Omega_b^{ld}, y_b^{ld}, s_b^{ld}, x^{ld})/p(s_b^{ld}, x^{ld})$. Since $p(s_b^{ld}, x^{ld})$ cannot be calculated analytically, then $p(\Omega_b^{ld}, y_b^{ld}|s_b^{ld}, x^{ld})$ can not be found in closed form. We apply the variational methodology to approximate the posterior distribution by another distribution, $q(\Omega_b^{ld}, y_b^{ld})$, that minimizes the Kullback-Leibler(KL) divergence. We choose to approximate the posterior distribution $p(\Omega_b^{ld}, y_b^{ld}|s_b^{ld}, x^{ld})$ by the distribution $q(\Omega_b^{ld}, y_b^{ld}) = q(\Omega_b^{ld})q(y_b^{ld})$, where $q(y_b^{ld})$ and $q(\Omega_b^{ld})$ denote distributions on y_b^{ld} and Ω_b^{ld}, respectively.

The estimation of the parameters and the image is done iteratively. First, an estimation of each parameter $w \in \Omega_b^{ld}$ is selected as the mean of the posterior gamma distribution $q(w)$ and then the estimation of the Gaussian distribution of the HR MS coefficients, $q(y_b^{ld})$, is performed.

3. Classification

Once the pansharpened image has been obtained, its classification is carried out. The approach we follow (which will be described later) to improve the classification rate of one class will be tested on two classification methods which, for completeness, are briefly described now: linear discriminant analysis (LDA) and support vector machines (SVM).

LDA is an effective subspace technique that optimizes Fisher's score [6]. Subspace methods are a particular class of algorithms focused on finding projections of the original hyperdimensional space to a lower dimensional space where class separation is maximized. In addition, LDA does not require the tuning of free parameters. These good attributes have resulted in its extensive use and practical exploitation in remote sensing applications mainly focused on image classification and band selection. LDA is related to Fisher's linear discriminant and, roughly speaking, both aim at finding a linear combination of features that characterize or separate two or more classes.

SVM is one of the most successful examples of kernel methods, being a linear classifier that implements maximum margin separation between classes in a high dimensional Hilbert space $\mathcal{H}$. Kernel methods embed the data observed in the input space $\mathcal{X}$ into a higher dimensional space, the feature space $\mathcal{H}$, where the data are more likely to be linearly separable. Therefore, it is possible to build an efficient linear classifier in $\mathcal{H}$, that translates into a nonlinear classifier in the input space. The mapping function to perform such an embedding is denoted as $\mathbf{\Phi} : \mathcal{X} \to \mathcal{H}$. Computing the explicit mappings $\mathbf{\Phi}(\mathbf{x})$ of all the observed data points can be computational demanding, especially if the dimensionality of $\mathcal{H}$ is high. To avoid this problem and build efficient algorithms, kernel methods compute the similarity between training samples $\{\mathbf{x}_i\}_{i=1}^n$ using inner products between mapped samples instead of computing the dot product in the higher dimensional space explicitly. The so-called kernel matrix $\mathbf{K}_{ij} = K(\mathbf{x}_i, \mathbf{x}_j) = \langle \mathbf{\Phi}(\mathbf{x}_i), \mathbf{\Phi}(\mathbf{x}_j) \rangle$ contains all necessary information to perform many classical linear algorithms in the feature space, which are non-linear in the input space [7].

It is important to note that, both for training and using the SVM for testing, one only needs to work with a valid kernel function, which should accurately reflect the similarity between samples. Valid kernels are functions representing a dot product in $\mathcal{H}$. The radial basis function (RBF), $K(\mathbf{x}, \mathbf{z}) = \exp\left(-\|\mathbf{x} - \mathbf{z}\|^2/2\sigma^2\right)$, $\sigma \in \mathbb{R}^+$ was the kernel function selected in this work. To implement SVM for multiclass problems we used the one-versus-all strategy given the particular characteristics of the proposed scheme.

4. Improving the Classification Performance for a Single Class

Once the pansharpening method described in section 2 has been used on a LR MS image and one of the classification methods described above has been applied, the user may be interested in boosting the performance of the classifier on a given class. In this section we propose to recalculate the parameters of the pansharpening method in order to obtain a new pansharpened image with an improved classification rate for the class of interest.

By examining the HR classified image, both visually and numerically (using for instance the confusion matrix), the user selects a class to improve its classification figures of merits. A new estimation of the image and parameters is performed. Utilizing the already estimated pansharpened image, the parameters for the new reconstruction are estimated utilizing only the pixels belonging to the class of interest in this image. Using those parameters a new pansharpened image is obtained. No iteration between parameter and image estimates is required.

This result in an estimation of the image whose spectral and spatial characteristics are more tailored to the pixels in the class of interest and, hence, will hopefully increase the classification performance for the elements of the class. Note however that this may imply, as we will see in the experimental section, that the classification performance on the other classes may decrease.

5. Experimental Results

Experiments were run on a Quickbird image. The MS image, depicted in real color in Figure 1a, has a spatial resolution of 256×256 pixels with each pixel covering a square area with a side of 2.4 m and four spectral bands: blue (450-520 nm), green (520-600 nm), red (630-690 nm), near-IR (760-900 nm). The PAN image (see Fig. 1b) has a resolution of 1024×1024 pixels with a size of 0.6 m covering the whole spectral interval (405-1053 nm). The result of the pansharpening process, with the parameters automatically estimated using all the MS and PAN images, is shown in Fig. 1d.

Using the MS and PAN images, a small number of pixels were classified into ten different classes (cars, water, forest, ...). This set of pixels, depicted in Fig. 1c, is considered our ground truth. We randomly chose 20% of the samples of each class to train the LDA and SVM classifiers and the rest was used for testing. In order to incorporate both spectral and spatial characteristics for each pixel into the classification process, we used a descriptor composed of the value of each pixel under consideration and its four nearest neighbors. Since each pixel has associated five values, four corresponding to the MS bands and another one for the panchromatic, the descriptor for each pixel has 25 components.

The classification quality is measured using the precision and recall values on a given class defined as

$$recall = \frac{TP}{TP + FN}, \quad precission = \frac{TP}{TP + FP} \tag{7}$$

where TP is the number of pixels in class correctly classified, FN is the number of pixels in the class incorrectly classified and FP is number of pixels not belonging to a class incorrectly classified as belonging to the class. Table 1 presents the figures of merit for each classifier on the pansharpened image in Fig. 1d. This image presents a very high level of spatial detail with no chromatic distortion. The classification figures show that SVM outperforms LDA although both classifiers perform well for all the classes except classes 10 and 8 where they perform poorly, especially the LDA classifier.

A class is now selected to improve its classification figures. In this case class 10 (isolated tree) was selected although similar results were obtained when selecting the other classes. Using only the pixels of the MS and PAN image belonging to the selected class, the parameters were estimated using the procedure described in section 4 and a new pansharpened image, depicted in Fig. 1e, was obtained.

Using this image, the classifiers were trained and a new classification step was performed obtaining the results presented in Table 2. Although the reconstructed images using the parameters estimated from the whole image (Fig. 1d) and the parameters estimated using only the pixels of the class 10 (Fig. 1e) are very similar

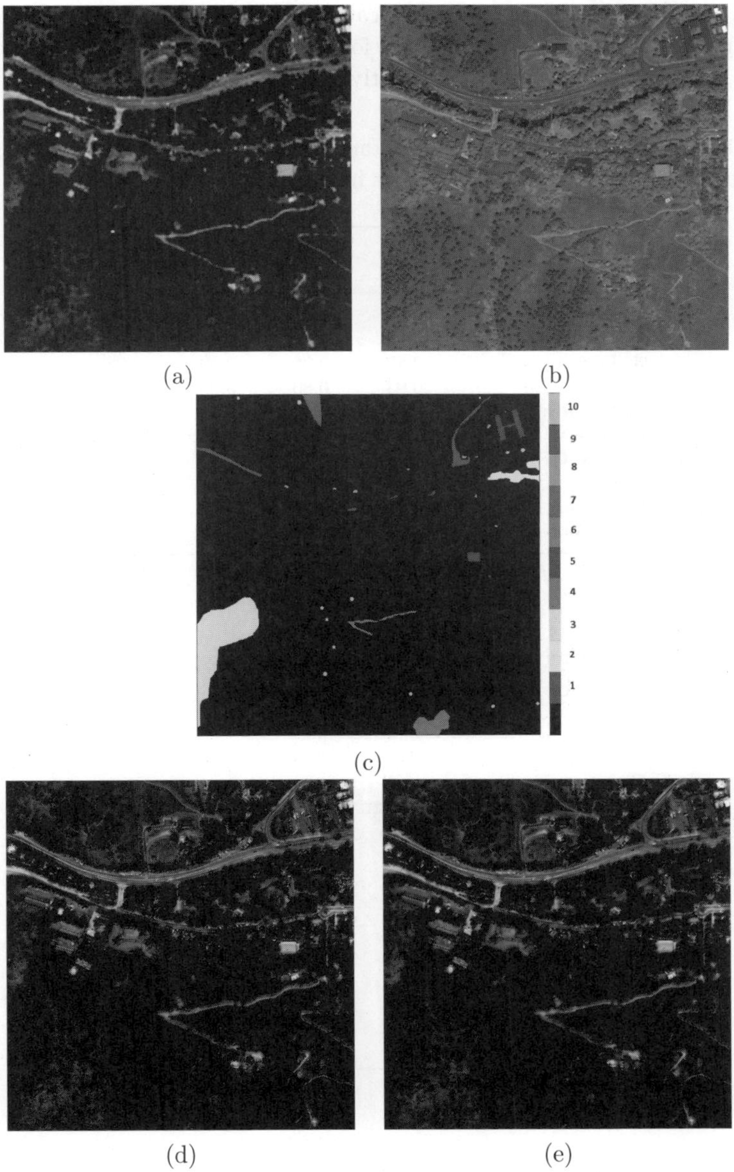

Fig. 1: (a) MS and (b) PAN images. (c) Ground truth. (d) Pansharpened image using the super resolution method described in section 2 with the parameters estimated using the whole image. (e) Pansharpened image utilizing only the pixels of the training set in class 10 to estimate the model parameters.

from a visual point of view, the classification figures show a higher precision and recall for the selected class 10 and, also, for many others. Note however, that some classes, like classes 6 or 8, perform slightly worse with those parameters.

Table 1: Recall and Precision values obtained using the pansharpened image with parameters estimated from all the pixels in the image.

Class	LDA recall	LDA precision	SVM recall	SVM precision
1. Asphalt	0.89	0.91	0.99	0.98
2. Dense Forest	0.75	0.72	0.91	0.93
3. Forest	0.87	0.98	0.99	0.98
4. Bare Soil	0.93	0.86	0.99	0.99
5. Building	0.84	0.93	0.99	0.98
6. Grass	0.82	0.82	0.96	0.94
7. Dry Grass	0.99	0.77	0.99	0.99
8. Car	0.63	0.66	0.81	0.98
9. Water	0.93	0.74	0.97	0.98
10. Isolated Tree	0.89	0.29	0.82	0.89

Table 2: Recall and Precision values obtained using the pansharpened image with parameters estimated only from pixels of the class 10.

Class	LDA recall	LDA precision	SVM recall	SVM precision
1. Asphalt	0.89	0.90	0.99	0.97
2. Dense Forest	0.72	0.73	0.92	0.93
3. Forest	0.88	0.99	0.99	0.99
4. Bare Soil	0.94	0.87	0.99	0.99
5. Building	0.82	0.93	0.99	0.98
6. Grass	0.83	0.80	0.96	0.95
7. Dry Grass	0.99	0.80	0.99	0.99
8. Car	0.54	0.56	0.75	0.94
9. Water	0.94	0.72	0.99	0.99
10. Isolated Tree	0.89	0.30	0.88	0.92

6. Conclusions

In this paper we have shown that pansharpening techniques can be used to increase the performance of classification methods when are applied to MS images. We have addressed the problem of adaptively modifying a pansharpening method in order to improve the precision and recall figures of merit of the classification on a given class without deteriorating the performance of the classifier over the other classes. The validity of the proposed technique has been demonstrated using a real Quickbird image. Work is being currently carried out to theoretically justify the used approach.

Acknowledgements

This work has been supported by the Consejería de Innovación, Ciencia y Empresa of the Junta de Andalucía under contract P07-FQM-02701, by the Comisión Nacional de Ciencia y Tecnología under contract TIN2010-15137, and the Spanish research program Consolider Ingenio 2010: MIPRCV (CSD2007-00018).

References

[1] I. Amro, Multispectral Image fusion using Multiscale and Super-resolution methods, PhD thesis, Dept. of Computer Science and Artificial Intelligence, Universidad de Granada (2011).

[2] L. Bruzzone, L. Carlin, L. Alparone, S. Baronti, A. Garzelli and F. Nencini, *Image and Signal Processing for Remote Sensing XII*, **6365** (2006).

[3] I. Amro, J. Mateos, M. Vega, General contourlet pansharpening method using Bayesian inference, in *2010 European Signal Processing Conference (EUSIPCO-2010)* (2010).

[4] A. L. da Cunha, J. Zhou and, M. N. Do, *IEEE Trans. Image Proc.* **15**, 3089 (2006).

[5] M. Vega, J. Mateos, R. Molina and A.K.. Katsaggelos, Super resolution of multispectral images using TV image models, in *in Porc of 2th Int. Conf. on Knowledge-Based and Intelligent Information & Eng. Sys.*, (2008).

[6] R. Duda and P. Hart, *Pattern classification and scene analysis* (Wiley, New York, USA, 1973).

[7] J. Shawe-Taylor and N. Cristianini, *Kernel Methods for Pattern Analysis* (Cambridge University Press, 2004).

BLIND PROCESSING IN ASTROPHYSICAL DATA ANALYSIS

E. SALERNO and L. BEDINI
National Research Council of Italy - CNR,
Institute of Information Science and Technologies - ISTI,
Via Moruzzi 1, 56124 Pisa, Italy
E-mail: name.surname@isti.cnr.it

Some blind approaches to a typical problem in astrophysical data analysis are briefly recalled. The problem is to disentangle different radiation maps from a multichannel map where signals coming from different astrophysical sources are mixed. Some separation methods assume a perfect knowledge of the mixing operator. However, if the model does not explain the data, this approach can lead to very poor results. Conversely, a blind approach, that is, one that does not enforce a rigid data model, is able to match the data while preserving all the desirable properties of the result.

Keywords: Blind source separation; Astrophysical imaging; Cosmic microwave background.

1. Introduction

The radiation intensity measured by an instrument looking at the sky is the sum of the individual radiations of all the astrophysical sources distributed along the line of sight. A radiotelescope placed somewhere in our galaxy measures a mix of the diffuse emissions from the interstellar medium, the radiation of the extragalactic compact and pointlike sources, and the cosmic background, which is the farthest (and oldest) source of light in the universe. Each source has its own emission spectrum, thus the mixing proportions of the different signals vary with frequency. Given a set of observed maps on different channels, our problem is to separate the astrophysical sources, that is, to obtain distinct maps for all the different sources, taking into account that the instrumental output is always affected by errors. One error, always present, is the noise produced by any electronic sensor. Another error is related to the angular resolution of the telescope: each data point does not measure the radiation from a unique line of sight, but is the integral of the radiation as a function of the bearing, weighted by the telescope's beam profile, whose size and shape often depend on frequency. These errors can often be treated easily, since they depend on known features of the system. Another serious problem, however, complicates the estimation of the sources: the model error. Indeed, the mixing proportions depend on the physics of the sources, which is often unknown. The relationships between the source intensities are thus unknown functions of frequency.

This is a very common problem in astrophysical data analysis. The source separation problem arising from ESA's cosmological mission *Planck Surveyor Satellite* [1], is taken here as a case study. *Planck* has been launched on 14th May 2009, and its data analysis goals include the extraction of the cosmic microwave background (CMB) from the all-sky, high-resolution, and high-accuracy maps that will be available at the end of the observation period. The center-frequencies of *Planck*'s nine observation channels are distributed between 30 GHz and 857 GHz. The dominant radiations in that range, besides CMB, come from the thermal emission of the galactic dust, the galactic synchrotron and free-free emissions, and extragalactic sources such as individual galaxies and clusters of galaxies. Details on features, operation and performances of the *Planck* instruments at the time of launch, as well as the scientific program of the mission, are reported in Ref. 2.

The mixing in this case can safely be assumed linear. By assuming that the mixing coefficients are known, the radiation sources can be separated by solving a linear system with a noisy data vector. Several algorithms are long available to solve this problem, which proved to be very effective on simulated *Planck* data (see, for example, Refs. 3–6).

However, since the individual emission spectra are only known roughly, the mixing coefficients are known with a coarse approximation. Relying on incomplete physical information to build a rigid data model normally leads to poor results, especially if, as in this case, the data are affected by a strong noise component. This is why the methods mentioned above were modified to cope with the more and more sophisticated and realistic simulations that were being performed in the preparatory phase of the *Planck* mission [7–9]. To try a different approach, we first proposed to use totally blind separation methods, that is, we just assumed a linear mixing model, with no further assumption on the mixing coefficients [10, 11]. To compensate this loss of information, we made assumptions on the source functions. As a first attempt, we proposed to assume that all the sources are mutually independent, and to solve the problem by some independent component analysis (ICA) strategy [12]. The first results obtained by ICA were so encouraging that many research teams, in the *Planck* collaboration and beyond, started to explore blind strategies to solve this separation problem in observational cosmology.

Leaving the data model unconstrained is apparently an advantage, in that no unphysical assumption can hamper the solution. On the other hand, exploiting the significant amount of information thus neglected could make the solution more accurate and robust. Furthermore, the mutual independence assumption is not strictly justified for all the sources. Relaxing the independence assumption and exploiting the available information is thus necessary. With this aim, in parallel to the development of totally blind techniques, many researchers have been trying to include as much prior information as possible in the problem and to drop the independence assumption.

Besides standard ICA, there are other strategies that can be adopted, which use different assumptions, such as spatial structure or multidimensional separation. If spatial structure is taken into account, then an ICA separation can be obtained by simply constraining the second-order statistics, rather than maximizing independence [13]. Moreover, if some of the sources are not independent, an ICA separation can be achieved between the individual independent sources and the subspaces spanned by the subsets of dependent sources [14]. Second-order statistics can also be used to separate dependent sources, provided that their spatial correlations are sufficiently rich and some information to further constrain the model parameters is available. This is the case with the *Planck* data, since the emission spectra of the astrophysical sources can be parametrized, thus reducing the total number of unknowns. The result is that even dependent sources can be separated, and their covariance matrices can be estimated [15, 16]. This strategy has proved very effective and accurate in estimating the mixing operator [17], thus allowing the reconstruction of the sources to be performed by standard techniques.

Theoretically, however, the best way to include prior information is the Bayesian approach. It allows all the unknown parameters to vary within the ranges permitted by their prior probabilities, and the fit to the data is enforced through the likelihood probability density. By composing all this information, a posterior density can be built, through which all the relevant parameters can be estimated statistically. The problems with this approach are that finding appropriate priors can be difficult, and the posterior can be very complicated to be treated analytically or numerically. The unknown parameters could thus need to be estimated by random sampling. Although fully Bayesian methods are still much slower than any kind of non-Bayesian or approximate-Bayesian methods, new sampling strategies that are speeding up convergence significantly are being developed. Details on many of these methods can be found in Ref. 18. In principle, a truly Bayesian and well-constrained solution can be very accurate and naturally equipped with sound error estimates, which are normally much more reliable than the ones obtained by analytical methods. In fact, to produce accurate error estimates many researhers have been resorting to Monte Carlo strategies even on non-Bayesian solutions [19].

This paper is organized as follows. In Section 2, a data model for astrophysical source separation is formalized. Section 3 describes explicitly two strategies that are being used to separate the *Planck* sources, and Section 4 shows an illustrative example. Section 5 concludes the paper.

2. A data model for radioastronomical images

A telescope with known, frequency-dependent, beam profile $\tilde{h}(\xi, \eta; \nu)$ looks at the sky with bearing (ξ, η) and feeds a multichannel sensor that measures the total radiation received on N channels centered at frequencies $\{\nu_i\}, i = 1, \ldots N$, and adds a measurement error $n_i(\xi, \eta)$ to the i-th channel. The total radiation coming to the telescope for each bearing and frequency is the sum of the radiations from M astrophysical sources $\tilde{s}_j(\xi, \eta; \nu), j = 1, \ldots M$. Let us now assume that $\tilde{s}_j(\xi, \eta; \nu_i) =$

$s_j(\xi,\eta)\cdot f_j(\nu_i)$ and that $\tilde{h}(\xi,\eta;\nu_i)=h_i(\xi,\eta)$ does not depend on frequency within the i-th channel. The total radiation x_i measured at bearing (ξ,η) and channel i is the convolution between the sum of the individual radiations and the telescope's beam profile at frequency ν_i, plus noise:

$$x_i(\xi,\eta)=h_i(\xi,\eta)*\sum_{j=1}^{M}a_{ij}s_j(\xi,\eta)+n_i(\xi,\eta) \tag{1}$$

where the asterisk means convolution, and the *mixing coefficients* a_{ij} depend on the (unknown) emission spectra $f_j(\nu_i)$ (see Ref. 19 for details). Equation (1) can be written in vector form by letting $\mathbf{x}=[x_i]_{i=1,\ldots N}$, $\mathbf{s}=[s_j]_{j=1,\ldots M}$, $\mathbf{H}=\text{diag}[h_i]_{i=1,\ldots N}$, and $\mathbf{A}=[a_{ij}]_{i=1,\ldots N;j=1,\ldots M}$:

$$\mathbf{x}(\xi,\eta)=[\mathbf{H}*\mathbf{A}\cdot\mathbf{s}](\xi,\eta)+\mathbf{n}(\xi,\eta) \tag{2}$$

The separation problem consists in estimating the source angular templates, $\mathbf{s}(\xi,\eta)$ and the mixing coefficients, a_{ij}. Matrix $\mathbf{H}$ is perfectly known, as it only depends on the features of the telescope, and the noise vector $\mathbf{n}$ can be characterized statistically, as it only depends on the sky scanning strategy and on the features of the sensors. A particular case of data model (2) occurs when the beam profiles do not depend on the channel, that is, $h_i(\xi,\eta)=h(\xi,\eta)\ \ \forall i=1,\ldots N$. In this case, the model reduces to

$$\mathbf{x}(\xi,\eta)=\mathbf{A}\cdot\hat{\mathbf{s}}(\xi,\eta)+\mathbf{n}(\xi,\eta) \tag{3}$$

where $\hat{s}_j=[\hat{\mathbf{s}}]_j=h*s_j$ are the source templates at the common resolution of all the channels. Problem (3) is no more convolutional, and can easily be treated in the angular space [15]. Problem (2), conversely, is easy to treat in the Fourier space [16]. Even if the resolution is not independent of the channel, model (2) can be reduced to model (3) by smoothing the channel data maps so to downgrade their resolution to the worst available [11]. It is to note,however, that obtaining a channel-independent resolution in this way entails modifying the spatial structure of the noise. This must be taken into account when designing a solution algorithm or evaluating the estimation error.

3. Blind separation of astrophysical sources

In this section, two strategies to solve either problem (2) or (3) are briefly described. The first assumes mutual independence among sources, the second allows the sources to be correlated, and exploits the autocorrelation of each source and a physically sound parametrization of matrix $\mathbf{A}$ to solve the problem through second-order statistics.

Under mild additional assumptions [12] the ICA principle ensures that a number of independent random processes can be disentangled from a linear mixture thereof through a linear process that maximizes independence. An independence criterion and an optimization algorithm for that criterion must thus be chosen. Choosing the criterion establishes statistical properties of the result, such as consistency and robustness. Other features, such as convergence and stability, are influenced by the optimization algorithm.

The FASTICA algorithm [12] takes nongaussianity as the independence criterion, and a fast fixed-point optimization algorithm. A version that takes into account a white and Gaussian noise component, with known covariance, has been used in Ref. 11 to process simulated *Planck* data. The data are first *quasi whitened* through a spectral analysis of the empirical data covariance matrix $\mathbf{C}_x$ subtracted of the known noise covariance, $\mathbf{C}_n$. Then, the independent components are computed by multiplying the quasi-whitened data by a *separating matrix* $\mathbf{W}$, whose generic row is evaluated through the following iteration

$$\mathbf{w}_{\text{new}} = E\{\hat{\mathbf{x}}g(\mathbf{w}^T\hat{\mathbf{x}})\} - (\mathbf{I} + \hat{\Sigma})\mathbf{w}E\{g'(\mathbf{w}^T\hat{\mathbf{x}})\} \tag{4}$$

where $\hat{\mathbf{x}}$ are the quasi-whitened data, g is a suitable nonlinear function, $\hat{\Sigma}$ is a modified noise covariance matrix, and $E\{\cdot\}$ means expectation. Since $\mathbf{W}$ must have unit norm and mutually orthogonal rows, before continuing the iteration, the current estimate $\mathbf{w}_{\text{new}}$ is orthonormalized to the other rows. So far, this strategy is the fastest experimented on *Planck* data [9].

However, mutual independence does not hold true everywhere among the *Planck* sources. Indeed, whereas the CMB is independent of all the foregrounds, the galactic sources are normally correlated. Attempting a separation where the foregrounds are strong can thus produce unsatisfactory results. Fortunately, matrix $\mathbf{A}$ can be parametrized through a few *spectral indices* [17] that we denote collectively as a vector Γ. The number of parameters that define the mixing operator can thus be reduced from dozens to two or three. The source functions, in turn, are spatially correlated, and some information about the structure of the covariance matrix is available. Exploiting the known relationship between the source covariance and the data covariance matrices at several shift pairs (τ, ψ), we proposed an algorithm to estimate the spectral indices Γ and the source covariances $\mathbf{C}_s(\tau, \psi)$ by minimizing [15]

$$L(\Gamma, \mathbf{C}_s(:,:)) = \sum_{\tau,\psi} ||\mathbf{A}(\Gamma)\mathbf{C}_s(\tau,\psi)\mathbf{A}^T(\Gamma) - \mathbf{C}_x(\tau,\psi) + \mathbf{C}_n(\tau,\psi)||^2 \tag{5}$$

where the data covariances $\mathbf{C}_x$ can be estimated directly from the data. This minimization is well posed if there are a sufficient number of shift pairs for which the data covariance matrix is nonzero, and if some prior information on the source covariances $\mathbf{C}_s$ is avaiable (*e.g.*, if some source pairs are uncorrelated, their cross-covariances are zero). This minimization solves the separation problem modeled as in (3). A Fourier domain version of this strategy [16] solves the problem in the form (2). Once estimates for Γ and $\mathbf{C}_s(\tau, \psi)$ (or the source cross-spectra, in the case of Ref. 16) are available, the source functions $\mathbf{s}(\xi, \eta)$ can be estimated by standard techniques. In Ref. 19, after generating a space-varying map for spectral indices, a generalized least squares solution is implemented to estimate the sources, and an

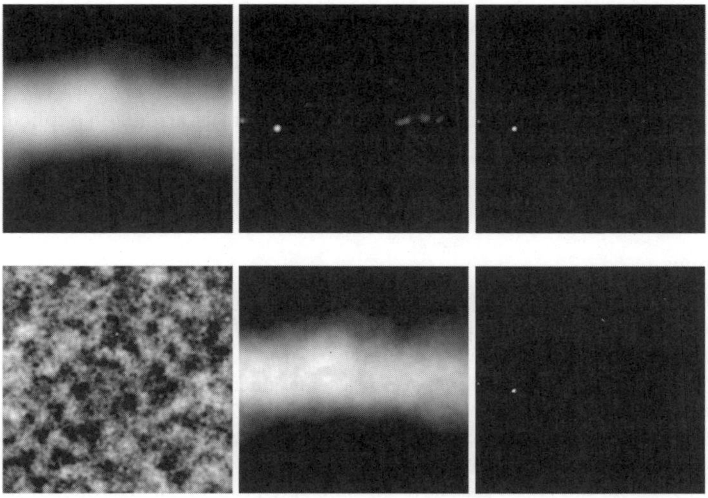

Fig. 1: Top row, left to right: *Planck* channel maps at 30, 100 and 857 GHz, in galactic coordinates, of a square sky patch centered at 0° latitude. Bottom row: reconstructed source maps at 100 GHz; left to right, CMB, galactic synchrotron, thermal dust radiation.

effective error evaluation strategy is envisaged. This technique has proved to be very effective even with *Planck* polarization data, which are known to be much more difficult to treat than the total intensity data.

4. Example solution

An example solution from simulated *Planck* data is shown in Fig. 1 [16]. The top row shows three channel maps (at 30, 100, and 857 GHz) of a $15° \times 15°$ sky patch cut across our galactic plane. The angular resolutions of these three maps are significantly different since the telescope beamwidth varies strongly with frequency (33, 9.2, and 5 arcminutes, respectively). The bottom row shows the reconstructed source functions at 100 GHz. Note that the galactic plane is the region where we have the strongest interference from the foreground radiations, and that the CMB is completely dominated by the synchrotron radiation at 30 GHz and by the dust radiation at 857 GHz. In this case, from the estimated vector Γ and the source cross-spectra, the source maps in the selected sky patch are obtained by multifrequency Wiener filtering [6].

5. Conclusions

This paper is a short outline of blind processing applied to a typical problem in astrophysical data analysis. For lack of space, not all the strategies attempted to solve this problem are mentioned and not all the details are made explicit. A complete information, however, can be found in the bibliography, especially in Refs. 9 and 18.

References

[1] ESA *Planck* website, http://www.esa.int/SPECIALS/Planck/index.html
[2] J. A. Tauber *et al.* *Planck* pre-launch status: The *Planck* mission, *Astr. & Astroph.* **520**, A1 (2010).
[3] M. Tegmark, How to make maps from cosmic microwave background data without losing information, *Astroph. J. Lett.* **480**, L87 (1997).
[4] M. P. Hobson *et al.* Foreground separation methods for satellite observations of the cosmic microwave background, *Mon. Not. Roy. Astr. Soc.* **300**, 1 (1998).
[5] M. P. Hobson and A. N. Lasenby, The entropic prior for distributions with positive and negative values, *Mon. Not. Roy. Astr. Soc.* **298**, 905 (1998).
[6] F. R. Bouchet *et al.* Multifrequency Wiener filtering of cosmic microwave background data with polarization, *Mon. Not. Roy. Astr. Soc.* **302**, 663 (1999).
[7] V. Stolyarov *et al.* All-sky component separation for the *Planck* mission, *Mon. Not. Roy. Astr. Soc.* **336**, 97 (2002).
[8] V. Stolyarov *et al.* All-sky component separation in the presence of anisotropic noise and dust temperature variations, *Mon. Not. Roy. Astr. Soc.* **357**, 145 (2005).
[9] S. M. Leach *et al.* Component separation methods for the *Planck* mission, *Astr. & Astroph.* **491**, 597 (2008).
[10] C. Baccigalupi *et al.* Neural networks and the separation of cosmic microwave background and astrophysical signals in sky maps, *Mon. Not. Roy. Astr. Soc.* **318**, 769 (2000).
[11] D. Maino *et al.* All-sky astrophysical component separation with Fast Independent Component Analysis (FASTICA), *Mon. Not. Roy. Astr. Soc.* **334**, 53 (2002).
[12] A. Hyvärinen *et al.* *Independent Component Analysis* (Wiley, New York, 2001).
[13] J. Delabrouille *et al.* Multidetector multicomponent spectral matching and applications for cosmic microwave background data analysis, *Mon. Not. R. Astr. Soc.* **346**, 1089 (2003).
[14] J. F. Cardoso, Multidimensional independent component analysis, in *Proc. ICASSP 1998*, 1941 (IEEE, Piscataway, 1998).
[15] L. Bedini *et al.* Separation of correlated astrophysical sources using multiple-lag data covariance matrices, *Eurasip J. on Appl. Sig. Proc.* **2005**, 2400 (2005).
[16] L. Bedini and E. Salerno, Extracting astrophysical sources from channel-dependent convolutional mixtures by correlated component analysis in the frequency domain, *Lect. N. Art. Int.* **4694**, 9 (2007).
[17] A. Bonaldi *et al.* Estimating the spectral indices of correlated astrophysical foregrounds by a second-order statistical approach, *Mon. Not. R. Astr. Soc.* **373**, 271 (2006).
[18] E. E. Kuruoğlu, Bayesian source separation for cosmology, *IEEE Sig. Proc. Mag.* **27**, 43 (2010).
[19] S. Ricciardi *et al.*, Correlated component analysis for diffuse component separation with error estimation on simulated data, *Mon. Not. R. Astr. Soc.* **406**, 1644 (2010).

THE EXTINCTION MAP OF THE ORION MOLECULAR CLOUD

G. SCANDARIATO*, I. PAGANO
Dipartimento di Fisica e Astronomia, Università di Catania, Catania 95123, Italy
**E-mail: gas@oact.inaf.it*

M. ROBBERTO
Space Telescope Science Institute, Baltimore, MD 21218, USA

The Orion Nebula and its associated young stellar cluster are located at the front-side of the optically thick OMC-1 molecular cloud. In order to disentangle the cluster members from background contamination, it is important to know the extinction provided by the OMC-1, which is poorly known, the available measurements yielding contradictory results. We analyze the most recent near-infrared catalog of stars to study the distribution of reddening across a 0.3 deg^2 area covering the Orion Nebula Cluster. On the basis of the observed photometry, we establish a statistical criterion for disentangling contaminants from bona-fide cluster members. For contaminant stars, interstellar reddenings are estimated by comparison with a synthetic galactic model. A statistical analysis is then performed to consistently account for local extinction, reddening and star-counts analysis. By means of this statistical approach, we derive the extinction map of the OMC-1 with angular resolution $< 3'$, deriving that the extinction provided by the OMC-1 is highly variable on spatial scales of a few arcminutes, while showing a general increase from the outskirts ($A_V \sim 6$) to the direction of the Trapezium asterism ($A_V \gtrsim 30$).

Keywords: Interstellar medium and extinction; Open clusters: Orion Nebula Cluster; Statistical analysis

1. Introduction

The closest event of massive star formation is occurring at present in the direction of the Galactic anticenter ($l = 209°, b = -19°$), in the Orion constellation. The Orion Nebula (ON) represents the most spectacular signature of star formation activity in this region. The ON is a blister HII region carved into the OMC-1 giant molecular cloud by the UV flux emitted by a handful of OB stars, the so-called Orion "Trapezium" [1, 2]. The Trapezium stars are the most massive members of a rich ($n \simeq 2000$ members [3]) cluster of young (1-3 Myr old [4]) Pre-Main-Sequence objects (Orion Nebula Cluster, ONC). Given its youth, vicinity, and low foreground extinction, the observed luminosity function of the ONC can be converted into a true initial mass function with relatively modest assumptions [3]. It is largely for this reason that the Orion Nebula and its associated cluster are regarded as a critical benchmark for our understanding of the star formation process.

In order to build a more reliable luminosity function, especially for the substellar regime (brown dwarfs and planetary mass objects), or to explore its spatial variations with the distance from the cluster center, it is necessary to remove the contribution of non-cluster sources. In principle, this would require the acquisition of thousand of spectra of faint sources distributed over the bright nebular background. On the other hand, the number of contaminant sources, both galactic and extragalactic, can be estimated using the most recent models for stellar and galaxy counts at various wavelengths. The main complication in this case is due to the presence of the OMC-1, which provides a backdrop to the ONC of high and non-uniform extinction.

In this paper we present a statistical reconstruction of the OMC-1 extinction map based on the analysis of the near-IR source catalog of Ref. 5. In Sec. 2 we illustrate our statistical method to disentangle contaminants from the cluster population. We also describe how we measure the interstellar extinction affecting each contaminant star, and how we combine our measurements to derive the OMC-1 extinction map. The main features of the map are discussed in Sec. 3, where we also compare our map to previous studies.

2. The OMC-1 extinction map

To compute a new extinction map of the OMC-1 region, we use the near-infrared photometric catalog assembled by Ref. 5 (R10 hereafter). This catalog contains JHK_S photometry in the 2MASS system for 7759 sources, 6630 labeled as point-like sources, spread over an area of $\sim 30' \times 40'$ roughly centered on θ^1Ori-C (RA=$05^h35^m16.46^s$, DEC=$-05°23'23.2''$). A key aspect of the present study is that, while the catalog is slightly shallower than previous studies (e.g. Ref. 3 or Ref. 6), it covers a sky area ~50 times larger roughly centered on the Trapezium asterism.

In order to maximize the number of detected background sources, we exclude in our analysis the J measurements: we take into account just the H and K_S bands since they are the least affected by the selective extinction provided by OMC-1. The (H,H-K_S) color-magnitude diagram (CMD) of sources with measured H and K_S magnitudes (Fig. 1) shows a characteristic bimodal distribution: a first group of stars (the ONC) is clustered at H-K_S=0.5 and H=12, whereas a second group of fainter objects appears clustered around $H\sim 17$. This second group also has a peak, but this is a selection effect, as the number of faint sources drops at $H > 17.5$ in correspondence of the sensitivity limit of the R10 survey.

Figure 1 also shows a 2Myr isochrone appropriate for the ONC (solid line) and the density contour of the galactic stellar population along the line of sight of the ONC (solid line). The galactic population has been derived from the Besançon galactic model [7][a], computed at the galactic coordinates of the ONC over the R10 survey area of 0.329 deg^2, and for each star we computed synthetic 2MASS

[a]http://bison.obs-besancon.fr/modele/

photometry using the latest grid of atmosphere models provided by Ref. 8. Since the galactic model also provides for each star the distance and interstellar extinction, both parameters were accounted for in the synthetic photometry.

Of the 4868 sources with $H \leq 18$ returned by the model, 4616 ($\approx$95%) lie at distances larger than 420 pc. Their colors will generally be further reddened by the OMC-1. In Fig. 1 we also plot (dashed lines) the galactic model reddened by $A_V = 6$, an *ad-hoc* value chosen to shift the contours in the same region of the second peak in the (H,H-K_S) diagram at $H \sim 17$. As anticipated by R10, this confirms that the second peak in Fig. 1 is fully compatible with the reddened galactic population.

The peculiar morphology of the CMD, with the two well separated peaks associated to two well distinct stellar populations, the ONC and the reddened galactic background, allows us to assume that sources fainter than H=15 or redder than H-K_S=1.3 (respectively below and to the right of the dotted lines in Fig. 1) are non-cluster, contaminant field star, being either too faint or too red to belong to the ONC.

We are aware of the fact that there still may be brown dwarfs and/or planetary objects belonging to the cluster and improperly included in the bona-fide contaminant sample. Based on the latest results on the ONC initial mass function [1], the fraction of sources having substellar masses decreases with decreasing mass and therefore we argue, with reference to Fig. 1, that the contaminant sample is poorly affected by the presence of ONC member with very low masses.

While it is possible to assume that the intrinsic color of cluster stars is (neglecting disk excess emission) roughly the one corresponding to the 2 Myr isochrone, for the background population there is no isochrone. Adapting to our case the algorithm outlined by Ref. 6, we have taken a statistical approach to compute the extinction affecting each contaminant star. In the following subsections we will describe how we estimate the extinction toward each contaminant star (Sec. 2.1) and how these estimates are combined to consistently derive the extinction map of the OMC-1 (Sec. 2.2).

2.1. *The extinction affecting each background star*

The basic idea is illustrated in the left panel of Fig. 2. The star is represented in the CMD as a 2D probability density distribution (the ellipse at the bottom right of the figure), where the shape of the distribution is due to the correlation between the H and H-K_S measures with their uncertainties. This density distribution is regularly binned in a 20×20 grid, resulting in 400 grid points.

The next step projects each grid point backward in the reddening direction, tracing a stripe which intersects the galactic population model. For each grid point we have a certain number of galactic stars at different extinctions, without any restriction on the maximum extinction allowed. Each of these stars is a candidate to represent the dereddened photometry of the given grid point. The distribution of candidates along the stripe is given by the galactic counts model, and the num-

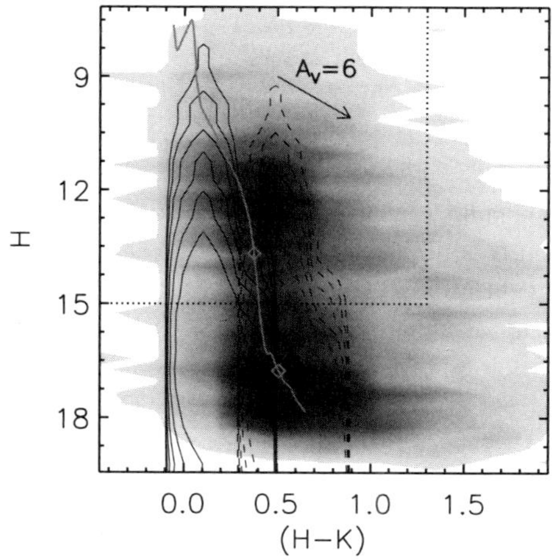

Fig. 1: Hess format of the (H,H-K_S) CMD of all point-like sources in the R10 catalog with measured HK_S photometry (color-filled areas), together with the 2 Myr isochrone (solid line). Two main populations are apparent: the ONC clustered at H=12 and the bona-fide contaminant stars at H∼17. The latter population is to be compared with the synthetic galactic population provided by Ref. 7 (solid contours) after accounting for the extinction provided by the OMC-1. The case of a constant A_V=6 (indicated by the arrow) across the cloud is shown in dashed contours, the highest ones roughly overlapping with the observed contaminant population. The dotted line represents our safe selection criterion: any source fainter than H=15 or redder than H-K_S=1.3 is considered a field star.

ber of candidates with extinctions less than A_V obviously increases with A_V. The cumulative distribution $F(A_V)$ of the dereddened candidates with extinction less than A_V, for each grid point, is given in Fig. 2, central panel.

Then, for each grid point, we randomly draw 100 A_V–values whose statistical distribution is taken from the cumulative distribution $F(A_V)$, and assign to the full set of 100 estimates of A_V a weight given by the 2D probability density distribution of the given grid point. Repeating this process for all grid points, we obtain a total of 400×100=40000 probability estimates, each one with its own weight: performing the weighted mean of this sample, we obtain the most likely estimate of the extinction affecting a given observed star. The right panel in Fig. 2 shows the weighted probability density function $f(A_V)$ of the 40000 A_V random estimates; to avoid data binning and for the sake of smoothness, this function is computed using a gaussian kernel density estimator [9]. This plot shows that the 40000 estimates are distributed following a distribution with a central peak value and decreasing wings.

This computation is repeated for all contaminant sources, providing the extinction estimate, and relative uncertainty, for each star in the contaminant sample.

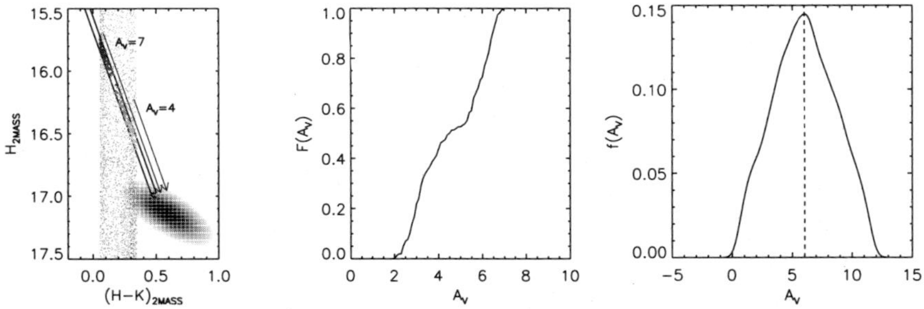

Fig. 2: *Left panel* – Example of the dereddening algorithm for the star with ID=578 in the R10 catalog. The star is represented in the CMD by a regularly gridded 2D density distribution, and to each grid element it corresponds a stripe of the galactic model (represented in dots) parallel to the reddening direction. Dereddened candidates for the outlined grid point are coded from light gray to dark gray according to increasing A_V. *Central panel* – Cumulative distribution $F(A_V)$ for the synthetic galactic stars in the dereddened stripe in the left panel. For each A_V, this function represents the fraction of candidate dereddened photometries with extinction $\leq A_V$. *Right panel* – Weighted probability density function $f(A_V)$ of the 40000 probabilistic A_V estimates for the representative case shown in the left panel, computed using a gaussian kernel. The weighted mean of the sample (dashed vertical line) is our best estimate of the true extinction affecting the star.

2.2. *Computation of the extinction map*

To compute the OMC-1 extinction map, we regularly grid the original survey area of R10 every 100 pixels (corresponding to 30″) both in the horizontal and vertical directions, and for each grid element we select the 20 closest contaminant stars within 1000 pixels (5′). The 1000 pixel limit is meant to prevent that, especially in regions with high extinction (A_V >15) where the density of contaminants can be extremely low, the algorithm collects sources over an area too widespread. By setting an upper limit for the distance of contaminant stars, we preserve the angular resolution of the extinction map, though smaller statistical samples will locally provide more uncertain results.

Each sample of 20 (or less) contaminant stars is averaged using an iterative sigma-clipping routine, deriving a robust estimate of both the mean local extinction $\bar{A_V}$ and standard deviation $\sigma_{\bar{A_V}}$. . This would represent the solution to our problem, except that we want to make sure that the surface density and brightness distribution of galactic sources is compatible with our estimated population of contaminants. To this purpose, we apply the $\bar{A_V}$ extinction to the synthetic galactic model and we correct the expected density for the completeness sensitivity computed by Ref. 5. This allows us to compare the predicted reddened population with the observed one.

This check allows us to consistently build a map which matches both the typical reddenening and source density of the galactic component, averaged over 5′, or less. The result is shown in Fig. 3, where we have interpolated the discrete gridding of the map using the IDL `GRIDDATA.pro` routine.

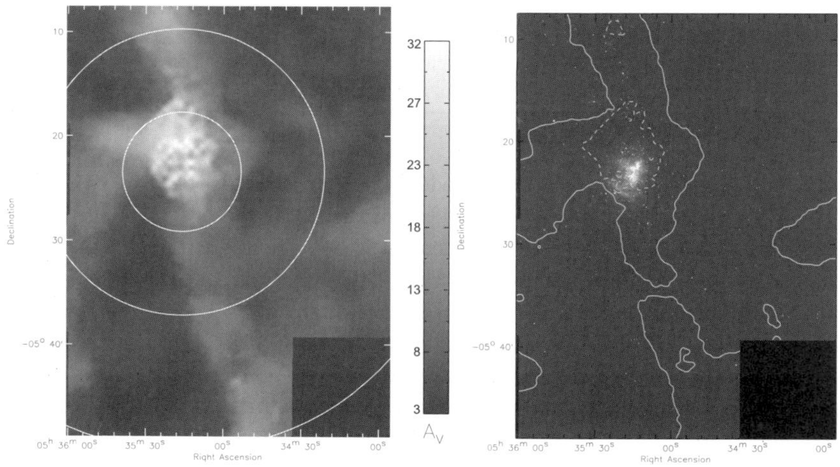

Fig. 3: *Left panel* - Extinction map of the OMC-1 derived as described in the text. The vertical bar shows the correspondence between the gray scale and visual extinctions. We also mark the boundaries of the three regions outlined by R10. *Right panel* - ISPI K_S mosaic of the surveyed ONC. The overplotted contours represent the A_V=10 (solid line) and the A_V=20 (dashed line) levels.

3. Conclusions

In Fig. 3 we show our derived extinction map and a contour plot of the same map overplotted on the K_S ISPI mosaic of the same region.

Along with the previous studies [6, 10–12], we find that the background extinction layer provided by the OMC-1 generally increases in the radial direction towards the Trapezium. In particular, we find that the Trapezium cluster is located in front of a high-extinction region ($A_V \gtrsim 30$) extending $\sim$10′ to the north of θ^1Ori-C. This region is delimited to the south-east side by a sharp edge, below which the optical thickness of the cloud decreases steeply by $\sim$20 magnitudes in a few arcminutes. This edge corresponds to the bright Orion Bar, clearly discernible in the ISPI images (right panel). The steep extinction drop can be confirmed even by a direct inspection of the images, as it is quite evident that the projected density of faint and red stars rapidly increases to the south of the Orion Bar. A similar extinction drop is found also to the north-eastern edge of the extinction peak, in correspondence of the dark structure, known as the Dark Bay or Fish Mouth, seen in absorption in the optical images of the ON. Elsewhere, the extinction smoothly decreases down to $A_V \gtrsim 4$ with increasing distance from the Trapezium cluster.

To compare the extinction map $A_V(SFD)$ of Ref. 10 to ours ($A_V(ISPI)$), we accordingly degrade our spatial resolution down to the $6'$ angular resolution. The pixel-by-pixel ratio $r_{A_V} = A_V(SFD)/A_V(ISPI)$ is plotted in Fig. 4 as a function of $A_V(ISPI)$. For low extinction values, we find that measurements of Ref. 10 generally overestimate the extinction by a factor of 1.5–2, compatibly to what is found by Ref. 11. Moreover, consistently to Ref. 12, our map suggests systematically smaller extinctions than Ref. 10, the ratio increasing in the $3 \lesssim r_{A_V} \lesssim 5$ range with the $A_V(ISPI)$ extinction. This is indicative of the fact that the map from Ref. 10 is not accurate either in high extinction regions or in regions with high extinction gradients.

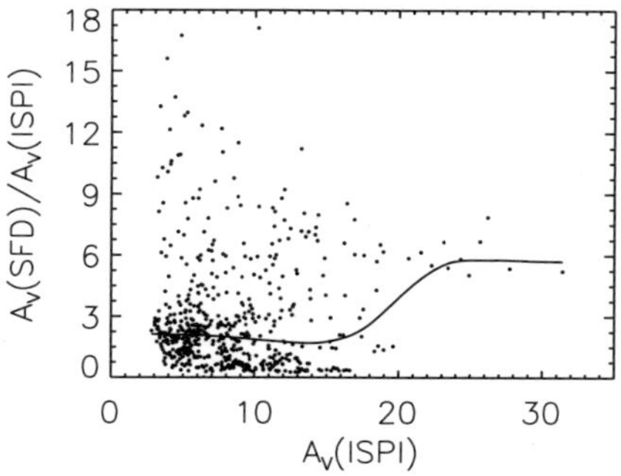

Fig. 4: Diagram of the ratio r_{A_V} as a function of $A_V(ISPI)$ computed over our extinction map. The solid line is a binned sample median, aimed at stressing the overall trend of the ratio. This diagram shows that for low extinctions our measurements and the extinction map of Ref. 10 are comparable, and that the latter becomes up to 3 times larger for higher extinctions, as found by Ref. 12.

References

[1] A. Muench, K. Getman, L. Hillenbrand, T. Preibisch, Star Formation in the Orion Nebula I: Stellar Content, *Handbook of Star Forming Regions, Volume I: The Northern Sky ASP Monograph Publications* **4**, eds. Bo Reipurth, (2008), p.483.

[2] C. R. O'Dell, A. Muench, N. Smith, L. Zapata, Star Formation in the Orion Nebula II: Gas, Dust, Proplyds and Outflows, *Handbook of Star Forming Regions, Volume I: The Northern Sky ASP Monograph Publications* **4**, eds. Bo Reipurth, (2008) p.544.

[3] A. A. Muench, E. A. Lada, C. J. Lada, J. Alves, *The Astrophysical Journal* **573**, 366 (2002).

[4] N. Da Rio, M. Robberto, D. R. Soderblom, N. Panagia, L. A. Hillenbrand, F. Palla, K. G. Stassun, *The Astrophysical Journal*, **722**, 1092 (2010).

[5] M. Robberto, D. R. Soderblom, G. Scandariato, K. Smith, N. Da Rio, I. Pagano, L. Spezzi, *The Astrophysical Journal* **139**, 950 (2010).

[6] L. A. Hillenbrand, J. M. Carpenter, *The Astrophysical Journal* **540**, 236 (2000).
[7] A. C. Robin, C. Reylé, S. Derrière, S. Picaud, *Astronomy and Astrophysics* **409**, 523 (2003).
[8] F. Allard, D. Homeier, B. Freytag, arXiv:1011.5405 (2010).
[9] B. W. Silverman, *Monographs on Statistics and Applied Probability* (Chapman and Hall, London, 1986).
[10] D. J. Schlegel, D. P. Finkbeiner, M. Davis, *The Astrophysical Journal* **500**, 525 (1998).
[11] H. G. Arce, A. A. Goodman, *The Astrophysical Journal Letters* **512**, L135 (1999).
[12] K. Dobashi, H. Uehara, R. Kandori,T. Sakurai, M. Kaiden, T. Umemoto, F. Sato, *Publications of the Astronomical Society of Japan* **57**, 1 (2005).

PART B

System: Structure and Behaviour

COMMON GROUNDS: THE ROLE OF PERCEPTION IN SCIENCE AND THE NATURE OF TRANSITIONS

G. BERNROIDER

University of Salzburg, Department of Organismic Biology
Neurosignaling Unit
5020 Salzburg, Austria
E-mail: gustav.bernroider@sbg.ac.at

The present work focuses on the relation of physics, phenomenology and brain organization. Classical dualities behind these concepts are mapped into the difference between the physical notion of states and the accompanying notion of transitions. It is argued that physics does not strictly deal with the concept of transitions and a more complete description of nature and reality is feasible if aspects of transition phenomenology are added to the physical concept of states. In living organisms transitions between a certain set of states has a particularly important meaning. These transitions are associated with the phenomenon of experience. In this view the issue of contextuality is seen as a local change in the similarity dimension within a partition of some fractal phenomenal dynamics. I argue that involving this view into physics can help to resolve the problem behind physical realism and the 'tuning problem' between life and cosmology.

1. Introduction

Science is made by Scientists and Scientists are consciously perceiving agents, experiencing, observing and analysing the world. From this simple and rather indisputable fact it is easy to see that we cannot discuss the emerging picture of our world without paying some attention to the way we perceive, experience and mentally construct the world. We build models of this world involving causality and complexity. As we abstract away from our primary experience by our mental abilities, we construct idealizations of the nature around us and about the role we play as perceiving and thinking agents. Our efforts however, are intimately interwoven with the pressing question to what degree the emerging picture of our nature truly reflects pre-existing properties of this world, independent of our own organization and the structure of our cognitive reasoning. Consensus is beginning to emerge that questions as this can no longer be exclusively assigned to the realm of philosophy, but must play a role in any radical formulation of natural science. To what extent do we construct the world? Is it all because we are here [1]?

However, in the context of these questions I do not intend to revitalize the dispute between an objectivistic versus constructivistic interpretation of science (for a more recent treatment see for example Ref. 2). Instead I will focus on two sets of pressing problems that have become central in physics as well as cognitive sciences. The first set is provided by the problem of contextuality in physics, and the second set is given by the problem behind computational theories of mind. In both domains we find enduring inconsistencies and in both domains the basic questions remain unanswered. In physics strict 'realism' is no longer a tenable position. It is for example generally violated by quantum mechanical predictions [3]. Consequently 'nature', i.e. the way we perceive the world, can be at variance with physics. Similar arguments apply to computational theories of mind, i.e. to theories that take the operation of computers as a root metaphor for cognition. There, we cannot find a clear causal relation between knowledge and 'reality' [2].

Interestingly but not surprisingly, the landscape of both sets of problems has something in common and this point will be the central issue of the present assay: It is the question about the '*meaning behind transitions*'. Any pure objectivistic interpretation of physics and cognition focuses on the conception of 'states', e.g. on the physical states of a system, on the representational states of our brain. Whereas the issue of 'states' has become a central concept of standard science, we are quite blind to the nature behind transitions between these states. Transitions cannot be captured and are resistant to consolidation and storage. They are even resistant to measurements, because measurements yield states, but not transitions. The conception of 'transitions' is not truly hosted by physical science at all. Looking at this situation from a different perspective, we do however find that, quite obviously, the dynamics between states cannot be described without the concept of transitions. Both concepts are intimately related, but only one of them, the conception of states is physically accessible.

Here I advocate the view that a more complete description of nature and reality is feasible if aspects of transition phenomenology (TP) are added to the physical state concept. Although the destinction between the two domains is usually sharpened by their dualistic philosophical background and the interpretation of the physical realm to be causally closed [4], I suggest a psychophysical model that provides a bridging principle between these two aspects. Similar to the organizational principles of states as attributes of systems in physics, also the phenomenology behind transitions has a rich and complex structure. I argue that the compositional principles of this phenomenology is largley and causally reflected by the compositional principles of states in living systems. The 'structure' of these composed worlds is found in living organisms, particularly in the brain of animals and humans and deserves close attention in order to proceed with our understanding about the relation of physics to nature, reality and cognition.

The present work is framed in terms of the relation of physics, phenomenology and brain organization. First, intuition suffices to expect that the key property behind the combination of transitions and states is likely to be scale-invariant, reflecting properties of self-similarity and discreteness. Any transition between two states can further be resolved into subsets of states and transitions (see Fig. 1). Thus embedding the structure of 'phenomenal symbolism' into a modellistic 'geometrical space' leads to an emerging geometry that is characterized by '*dimensional excess*', i.e. it constitutes a 'fractal set' in the sense of Mandelbrot [5].

I will shortly sketch the primitives of the underlaying state space and provide demonstrative examples about the physical realization of this space behind brain organization. There is a fascinating picture emerging from the present conception: The organization of the transition - state dynamics in the brain can itself provide us with indications about the way we perceive the world, about our abstractions and about the 'output of cognition', i.e. the mapping of experience onto our surroundings in terms of symbolized information, i.e. the signatures of culture and science.

2. Context

Can reality be independent of the measuring arrangement? Physical reality is usually interpreted as a set of pre-existing properties of objects and things, independent of the measurement-context. A long sequence of evidence, setting out with Bell's theorem [6], the Kochen-Specker theorem [7], Leggett's theorem [8] and subsequent experimental tests [3, 9, 10], provide clear evidence that this naïve interpretation of 'reality' is incompatible with quantum physics. But what is real? It is surprising to see that only strict objectivistic interpretations of physics impose reality demands on physics. From a constructivists point of view, the lack of 'reality' and the lack of locality seem to be much less surprising. Any measurement can not be independent of the existence of a cognitive subject. A view that is deeply grounded in Von Neumann's interpretation of quantum-classical transitions [11], as it allocates a physical role to the observing agent, making a psychologically intended action be part of physics. Henry Stapp's interpretation of von Neumann's process partitions has extended this view further [12].

In the present view a decisive step to deal with the problem of context and reality is the construction of a causal relation between the phenomenology of transitions and physical state space. I will set out with the definition of a contextual phenomenal partition that will subsequently be related to a physical state space.

2.1. *Fractal Contextuality*

We seek for an isomorphism between a physical state space and a minimal phenomenal partition that can capture the transition dynamics between the involved physical states. The conceptual minimum to obtain a physical realization is a two partite, two level composite system, i.e. where one mode (subsystem) represents the object to be measured and the second mode represents the measurement system.

The 'fractal' or 'similarity dimension' is just given by $\frac{\log N}{\log(\frac{1}{r})}$ if a unit length (or radius of a sphere as in the present model) is divided into $N = \gamma$ subsegments of length $r = \frac{1}{\gamma}$. The change in the 'similarity dimension' can be visaged as a change in the 'paving with segments' along some segment of a local partition and 'resolution' of the set $\perp$. For example the set $\perp$ demonstrated in Figure 1(b) gives

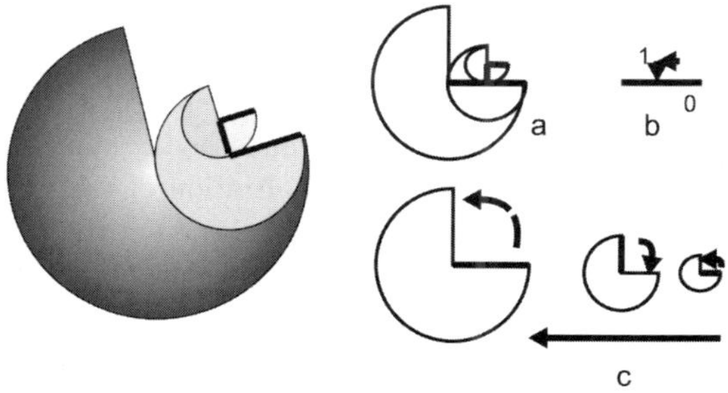

Fig. 1: A fractionally dimensioned subset of a transition-state space composed of spherical segments. The transition dynamics in this state space is defined as a set of operations that lead to openings or closing of spherical fractal segments (a-c). The characteristic fractal element for this dynamic is demonstrated by the U-shaped bold segment.

rise to an U shaped element with transitions into '⌋' and finally into '—' (bold segments) following the fractal spherical dynamics provided by circle segments. This will subsequently be used to symbolize a 'generating phenomenal partition' G for bivalent physical modes (i.e. a quantum bits). In Fig. 2 a stepwise demonstration of the local changes in the similarity dimension leading to a partition G that can be associated with a composed, two-level, two mode quantum system is demonstrated graphically.

2.2. *The fractal transition model applied to a two level composed quantum system*

We denote the interaction between a system and the measurement as H_{SA} and the two level system as $S : |\varphi_S\rangle = \alpha|\uparrow\rangle + \beta|\downarrow$ (e.g. control qubit) and a second two-level (target qubit or apparatus) system as $A : |\varphi_A\rangle = c_0|0\rangle + c_1|1\rangle$. We let S and A be pure with $|\alpha|^2 + |\beta|^2 = 1$ and $|c_0|^2 + |c_1|^2 = 1$ respectively. We know that the state space of the composite system is given by the tensor product of the single systems:

$$|\varphi_c\rangle = (\alpha|\uparrow\rangle + \beta|\downarrow\rangle) \otimes (c_0|0\rangle + c_1|1\rangle) \tag{1}$$

Even in this simple situation we can find a considerable amount of 'complexity' in the physical sense or 'dimensional excess' in the corresponding fractal geometrical view. This becomes apparent from the notion of a density operator $\rho^c = |\varphi_c\rangle\langle\varphi_c|$ associated with the composite state $|\varphi_c\rangle$. The operator in matrix notation gives the probability distribution for the outcomes of a measurement. The entries of this operator can be found from a 4×4 multi-linear Tensor map composed from the product of the two 2×2 density matrices associated with the states of S and A. The complete occupation of the matrix, in particular, the 'off-diagonal' entries, represent between system coherences, signalling a non-separable quantum correlation, i.e.

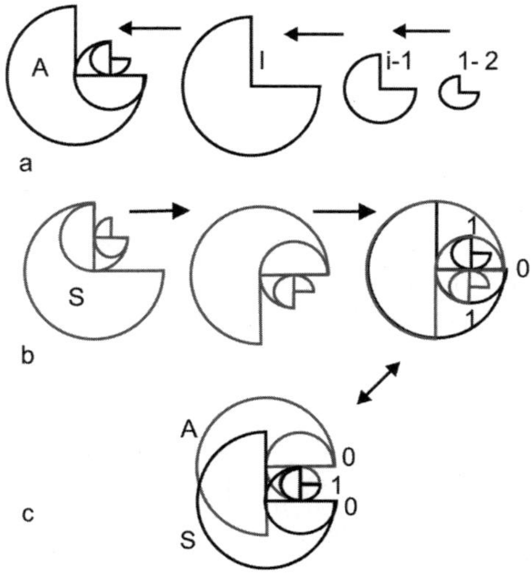

Fig. 2: The composition of a fractal dynamical set associated with interacting two-level (0,1), two partite (a,b) quantum systems (A,S) as demonstrated by equation (1) in the text.

entangled states (for example, for coefficients $\alpha = \beta = \frac{1}{\sqrt{2}}$ and $c_0 = c_1 = \frac{1}{\sqrt{2}}$, all entries would be equal to $\frac{1}{4}$).

If we let the integer i denote a given fractal 'resolution', the fractal set 'homes' the two modes and their two-level states within the fractal generations i, $i-1$ and $i-2$ of the local partition G ($\perp$) (going from right to left in Figure 2(a). If, for example, we let the systems state also encode logical states, i.e. $|\uparrow\rangle = |0\rangle$ and $|\downarrow\rangle = |1\rangle$, we can associate a 'transition algebra' to the state matrix underlaying the composed system in (2):

$$\begin{matrix} |0\rangle|0\rangle & |0\rangle|1\rangle \\ |1\rangle|0\rangle & |1\rangle|1\rangle \end{matrix} \longleftrightarrow \begin{vmatrix} - & \top \\ \perp & \| \end{vmatrix} \tag{2}$$

The right side of (2) symbolizes the transitions of the partition G that lead to the states shown on the right side of (2). The degrees of freedom associated with every single mode, the physical observables, are represented by $(N-1)$ fractal generations of the partition $G(\perp)$, where we have $N = \gamma$ subsegments of length $r = \frac{1}{\gamma}$. In the present demonstrative example we have $N = 3$ and $r = \frac{1}{2}$ indicating a similarity dimension of $D = \log N/\log(1/r) = 1.5849$. Thus we have $(N-1) = 2$ basic elements, '—' and '|' for the states of a single system. We can find their 'composites', placed along the off-diagonal entries of the state matrix, i.e. the elements '$\perp$', '$\top$' in the composed 2×2 matrix (2). Thus we have a total of $N + 1 = 4$ elements corresponding to $n = 4$ dimensions of the physical state space.

3. Context in the brain

A key step in making a concept as shown in (1) and (2) 'physical' is to resolve the '*basis ambiguity*' i.e. to select a 'preferred basis' for the expansion of the wave-function into its components. This can for example be achieved by introducing an 'environmental' delta-function as suggested by Zurek [13] expressing the orthogonal states of a third system (e.g. the environment of the composed system), $\langle e_i|e_j\rangle = \delta_{i,j}$. Because $\delta_{i,j} = 1$ for $i = j$ and 0 otherwise, only the states along the diagonal of the associated density matrix will survive and we obtain a table , which shows the resulting 'pre-decohered' joint system SA state, 'selected' by a third system (E) (*the von Neumann SAE chain*).

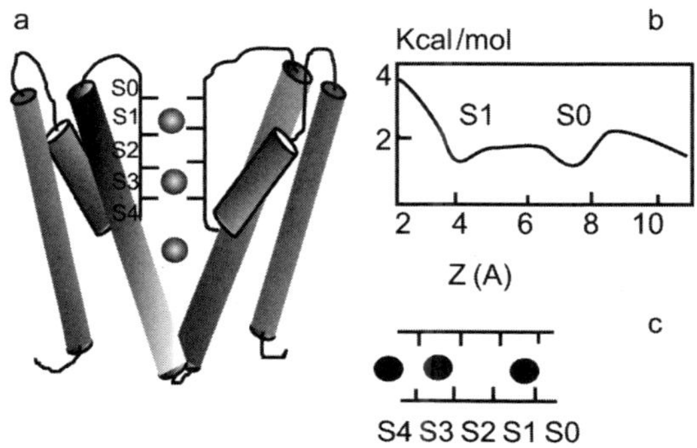

Fig. 3: Ion channel signature (A) with the ion hosting filter region, confining potentials (B) and ion-oxygen coordination states in a KcsA channel model (C). Ions and oxygen surrounding play the role of a system(S) and measurement (A) in the present context.

Henry Stapp was more explicit on the nature of this process, associating an intentional action of an observing 'agent' that is accompanied by conscious experience associated with Von Neumann's process 1 [11]. The physics of the brain would play the role of a '*template*' making this process possible. But how is this role instantiated in the brain?

4. Conclusion and Perspectives

The answer that I would suggest here is the following neurophysical outline: I indentify the primary process transitions, the fractal generations i, i-1, i-2 in Fig.2 and their associated physical states with ion-channel configurations that establish the key-signals along electrically excitable nerve cells, as shown previously [14, 15] (see Figure 3). The Von Neumann–environment mentioned above that selects a preferred basis for the physical state concept and gradually allows quantum state correlations to propagate into classical ion-channel state conformations is provided by the so-called '*segregation pattern*' behind perceptive neural organizations. This pattern represents the dominating functional maps in the neo-cortex [16] and is characterized by a highly specific geometry. This geometry seems to reflect properties of it's own coding equivalent, as for example recently shown in the arrangement of orientation columns in the visual cortex [17]. Others have argued that the phenomenal experience associated with perception is most probably due to some transitions characteristic at the scale of segregated activity patterns in the brain. Taking these aspects together, we might expect that some transition phenomenology hosting the fractal partition as shown in Figure 2, i.e. the fractal generation $i+1$ transition, that leads to some physical reality of the composed entangled system (1), can be a good candidate for hosting the phenomenology of experience that can be attributed to perceptive states.

The present conception of fractal transition models is similar to the 'Invariant set hypothesis' suggested by T.N. Palmer [18]. Whereas Palmer's 'Invariant Sets' are motivated and generated by gravitational systems, the present interpretation is founded within the phenomenology of life and experience (which I consider to be identical).

Acknowledgements

I greatfully acknowledge the assistance of my colleagues in discussing the matter and particularly to Roman Fuchs for his help to bring this paper to its final version.

References

[1] L. Susskind, *Cosmic Landscape: String theory and the illusion of intelligent design* (Little & Brown, 2005).
[2] J. Stewart, *Foundations of Science* **6**, 99 (2001), special issue on "The Impact of Radical Constructivism on Science" (ed. Riegler, A.).
[3] S. Groeblacher, T. Paterek, R. Kaltenbaek, C. Brukner, M. Zukowski, M. Aspelmeyer and A. Zeilinger, *Nature* **446**, 871 (2007).
[4] D. Chalmers, Consciousness and its place in nature, in *The Blackwell Guide to Philosophy of Mind*, eds. S. Stich and T. Warfield (Blackwell Publ., 2003) pp. 102–142.
[5] B. Mandelbrot, *Fractals, Form, Chance and Dimension* (W.H. Freeman & Company, 1977).
[6] J. Bell, *Physics* **1**, 195 (1964).
[7] S. Kochen and E. Specker, *J. Math. Mech.* **17**, 59 (1967).
[8] A. Leggett, *Found. Phys.* **22**, 1469 (2003).
[9] J. Clauser and M. Horne, *Phys. Rev. D* **10**, 526 (1974).
[10] D. Greenberger, M. Horne, A. Shimony and A. Zeilinger, *Am. J. Phys.* **58**, p. 1131 (1990).
[11] J. von Neumann, *Mathematical foundations of quantum mechanics* (Princeton Univ. Press, 1955).
[12] H. Stapp, *Mindfull Universe: Quantum Mechanics and the Participating Observer* (Springer, 2007). The Frontiers Collection.
[13] W. Zurek, *Los Alamos Science* **27**, 2 (2002).
[14] G. Bernroider and S. Roy, *Forma* **19**, 55 (2004).

[15] G. Bernroider and S. Roy, *Proc. SPIE* **5841**, p. 205 (2005).
[16] A. Shmuel and A. Grinvald, *Proc. Natl. Acad. Sci. USA* **97**, 5568 (2000).
[17] M. Kaschube, M. Schnabel, S. Löwel, D. Coppola, L. White and F. Wolf, *Science* **330**, 1113 (2010).
[18] T. Palmer, *Proc. R. Soc. A* **465**, 3165 (2009).

LOOKING THE WORLD FROM INSIDE: INTRINSIC GEOMETRY OF COMPLEX SYSTEMS

L. BOI

École des Hautes Études en Sciences Sociales, Centre de Mathématiques
190, avenue de France, 75013 Paris, France
Email: Luciano.Boi@ehess.fr

In this paper we shall address some meeting points between geometry and biology, in order to show that geometrical things and transformations take part intrinsically in the living systems. We focus on some features of macromolecular structures like DNA-proteins complexes. All things we speak about take place in the 3-dimensional space of a living cell and particularly in its nucleus, which of course interacts in many ways and at different levels with the whole cell, its cytoplasm and the organelles. Ideally, we think we should rather consider, instead of a 3-dimensional space, a configuration space characterized by all its phase spaces, since a living being is a very complex dynamical system, but this would be a too difficult, impossible task. This is of course a very partial view, an oversimplification, of what really happen in our organisms. Nevertheless, We believe that in biology we are today facing the following problem: how small or local changes in a living system do affect the global behaviour and response of the whole organisms? We search for an answer by arguing that mostly overall features of living systems are emergent properties of organization and regulation defined at the macroscopic level of their morphology and physiological behaviours, and also by showing that in complex living systems self-organization ensures robustness without loss of plasticity, in the sense that perturbations in the interactions properties of its single parts generally do not have damaging consequences on the living form as a whole.

1. Introductory Remarks on the Role of Topological Beings

Our physical and living macroscopic world abounds in knotted and linked structures. One of the most exciting developments of mathematical physics and cell biology in the last three decades has been the discovery of numerous intimate relationships between the topology and geometry of knot theory and the dynamics of many domains of classical" and "new" macroscopic and soft physics. Indeed, complex systems of knotted and entangled filamentary structures are ubiquitous in Nature and arise in such disparate contexts as electrodynamics, magneto-hydrodynamics, fluid mechanics (vortex structures), super-fluidity, dynamical systems, plasma physics, cosmic string theory, chaos of magnetic flows and non-linear phenomena, turbulence, polymer physics and molecular biology. In the recent years, mathematical tools have been developed to identify and analyse the geometrical and topological complex structures and behaviours of such systems and relate this structural and dynamical information to energy levels and stable states. The influence of geometry and topology on macroscopic physics and living matter has been especially fruitful in the study and comprehension of various important topics.

Let us here mention two examples. The applications of modern results from the topology and geometry of knots has led, in latest years, to new developments in the qualitative study of fluid dynamics. In particular, it has been shown recently that the helicity of a magnetic link (to be considered in a perfectly conducting, incompressible, viscous fluid), which is invariant under fluid flow, does reflect the topology and the geometry of the magnetic lines of forces within a magnetic link. In other words, the topology of the magnetic link $g_t L$, as expressed in knotting and linking, creates a barrier to the full dissipation of the magnetic link's energy, i.e., $E_M(g_t L)$ has a positive lower bound that results from the topology of $g_t L$. Thus, the magnetic link will reach a non-trivial stable and invariant energy state.

Another important example, related to the previous one, is the problem of finding the "best" shape, i.e. a shape of minimal "energy", for knots and links. More precisely, the objective is to find flows on isotopy classes of knots and links that terminate in a "best" shape for each given knot or link type, and then to use this "best" shape to develop new knot invariants; this method might be particularly useful to solve the problem of classification of trajectories in the n-body problem in Celeste mechanics or in a multivector-systemlike model of turbulent flows.

Now a property of stability similar to that we stated in the first example for incompressible fluids has been observed in the case of the DNA macromolecule, and more generally in the polymers, which can be expressed as follows: The number of linking numbers of the two strands of DNA has a positive lower bound, i.e. $= ALK(a)$ of the twists which create the intertwining of the two strands. The reach of stable structures is directly linked to the action of supercoiling in permitting an optimal compact form of DNA within the chromosome (for more details, see 14). Furthermore, the stable structures of such a DNA are those conformations that minimize a conformational energy subject to the constancy of the topological conditions.

2. Stable Structures and the Minimization of Energy; the Morphofunctional Space of Chromosomes

The stable shapes of polymer structures generally are those that minimize the kind of energy which is conserved under some invariant topological properties. This phenomenon gives rise to a range of variational problems. Experiments show that the stable structures of proteins minimize energy. Thus, in order to predict protein structures from sequences one must solve an optimization problem. This is actually very difficult to do, because there may be many thousands of degrees of freedom within a single molecule, so its configuration space is high dimensional.

Despite its immense length, the linear sequence map of human genome is an incomplete description of our genetic information. This is because information on genome function and gene regulation is also encoded in the way that DNA sequence is folded up with proteins in the chromosomes within the nucleus. This information cannot be portrayed in the DNA sequence alone. In the nucleus, individual chromosomes occupy discrete territories. So examining the spatial organization of human chromosomes and genes in the nucleus appear to be very important. It seems that this organization is changed notably during development but also in certain diseases. Consequently, the way the human chromosome is topologically organized might influence how abnormal chromosomes are formed. In other words, a great deal of experimental observations and many significant theoretical models allows us to assume that very likely there is a intimate relationship between the kind of topological organization to which living structures obey and their possible lost of functionality.

Still few words about the kind of morphofunctional space which characterize chromosomes. Using whole chromosome painting probes and fluorescence in situ hybridization (FISH), a territorial organization of interphase chromosomes has been demonstrated. Chromosome territories have irregular shapes and occupy discrete nuclear positions with little overlap. In general, gene-rich chromosome is located more in the nuclear interior while gene-poor chromosome territories are located at the nuclear periphery. In agreement with this, non-transcribed sequences were predominantly found at the nuclear periphery while active gene and gene-rich regions tended to localize on chromosome surfaces exposed to the nuclear interior or on loops extending from the territories (for a more accurate analyse of this important aspect, see our recent work 5–7).

3. The Structural Plasticity of Biomolecules and the Topology of Topoisomerases: Cutting and Reconnecting

It must be stressed that, essentially thanks to its topological properties, DNA is a very malleable, deformable molecule, being able to recombine through a series of phases. It is maybe this complex property of malleability or deformability that characterize the DNA molecule, and which might distinguish the living organisms from the inanimate matter. Moreover, one of the most striking structural properties of DNA molecule is its *flexibility* or *plasticity*, which influence in a fundamental way its biological functions. Thus, the molecule can freely move about, although under certain chemical and geometrical constraints, in the space of the cell's nucleus and convert itself into several shapes without losing a certain energetically (biochemical) favourable condition. This movement is twofold: the three-dimensional two-strands helical structure of DNA molecule can extend and compact.

The extended (unfolded) conformation of DNA, which put it under tension as if the molecule was subjected to shear one dynamic force, seems to be especially required for DNA replication. This process uses each of the two strands of DNA as a template for the formation of a complementary DNA strand. The original strands therefore may remain intact through many cell generations.

DNA compaction inside cells occurs by successive orders of coiling. One can show that a DNA double helix is compacted in about four successive steps corresponding to a family of very complex topological deformations (see 7, 8). Only the first step of nucleosome formation is quite well understood. In this step, DNA coils twice in a left-handed helical fashion around a set of proteins called histones. The nucleosomes are then coiled successively to give the final forms, called a chromosome. In the phase of this process, namely the recombination, the knot type of DNA molecule is actually changed. The whole process, from the original splicing to the recombination, is the result of the action of a single enzyme/catalyst called a topoisomerase. To be more precise, all these nuclear processes that occur during an entire cell cycle need to be properly and continuously orchestrates by a family of topoisomerases each of one has a specific task, although they are at the same time severely interconnected.

The process of mutation due to topoisomerases can be described in simple terms as follows: First a strand of the DNA is cut at one place, then a segment of the double-helix passes through this cut, and finally the DNA reconnects itself. So surgery cutting and self-recognition are always two consecutive closed related operations. There exist two examples of the action of a topoisomerase on a DNA molecule:

1. The simple strand has a single cut due to a topoisomerase and the DNA passes through it and recombines; this is called a *type I topoisomerase.* However, topoisomerases I is also capable to tied complex knots; in fact, the enzyme produces every knot theoretically possible. Thus, the requirement for excess enzyme to form complex knots suggests a role for topoisomerase I in contorting the DNA in addition to promoting strand passage. For example, it has been shown that *Escherechia coli* topoisomrase I can catenate circles in nicked, duplex DNA. Furthemore, it is also able to tie a knot by inverting either a (+)- or a (−)-node during strand passage. In other words, the DNA must fold to provide an inversion node, at which topoisomerase I passes one DNA segment through a transiently introduced break in the other DNA strand and thereby inverts the topological sign of the node. The inversion node divides the ring into two domains, and there must be at least two nodes between these two domains. Intradomainal nodes alone never lead to knots, although they can contribute. Even a complicated knot can be tied by just one strand passage.

2. A cut in a double-strand DNA, due again to a topoisomerase, allows a double-strand DNA to pass through it and recombine, this is as expected called a *type II topoisomerases.* In other words, type II topoisomerases are essential enzyme that pass one DNA through another and thereby remove DNA entanglements. They make a transient double-stranded break in a gate segment (G segment) that allows passage by another segment (T segment) of the same or another DNA molecule. Thus, these enzymes have the potential to convert real DNA molecules into phantom chains that freely pass through themselves to generate an equilibrium distribution of knots, catenanes, and supercoils. In fact, they reduce the fraction of knotted and catenated circular DNA below thermodynamic equilibrium values. To do that, enzymes use the energy of ATP hydrolysis. Active topology simplification

by topoisomerases II has an important biological consequence. It helps explain how topoisomerases can remove all DNA entanglements under the crowded cellular conditions that favour the opposite outcome. Recently, a new model was designed to explain this surprising finding, in which eukaryotic and prokaryotic topoisomerases bend DNA sharply upon binding. In this model, bending is a local manipulation, which is responsible for the change of the global, topological configuration of circular DNA. The challenge, however, is to ask oneself whether there may be some kind of global process which could be responsible for the topological compaction of chromatin within the chromosome (for some hints on this problem, see 8).

4. Observations on the Geometry of Complexity

Let us start this section by some general definitions of the concept of complexity. We take a complex system to be one composed by a large number of parts, and whose properties are not fully explained by an understanding of its components parts. Studies of complex systems recognized the importance of "wholeness", defined as problems of organization, phenomena non resolvable into local events, dynamics interactions in the difference of behaviour of parts when isolated or in higher configuration, etc., in short, systems of various orders not understandable by investigation of their respective parts in isolation. In a complex system it is essential to distinguish between 'global' and 'local' properties. Theoretical physicists in the last two decades have discovered that the collective behaviour of a macro-system, i.e. a system composed of many objects, does not change qualitatively when the behaviour of single components are modified slightly.

There are many universal classes which describe the collective behaviour of the system, and each class has its own characteristics; the universal classes do not change when we perturb the system. The most interesting and rewarding work consists in finding these universal classes and in spelling out their properties. This conception has been followed in studies done in the last twenty years on second order phase transitions. The objective, which has been mostly achieved, was to classify all possible types of phase transitions in different universality classes and to compute the parameters that control the behaviour of the system near the transition point as a function of the universality class.

This point of view is not very different from the one expressed by Thom in the introduction of *Structural Stability and Morphogenesis* (1972). It differs from Thom's program because there is no *a priori* idea of the mathematical framework which should be used. Indeed Thom considers only a restricted class of models (ordinary differential equations in low dimensional spaces) while we do not have any prejudice regarding which models should be accepted.

One of the most interesting and surprising results obtained by studying complex systems is the possibility of classifying the configurations of the system taxonomically. It is well-known that a well founded taxonomy is possible only if the objects we want to classify have some unique properties, i.e. species may be introduced in an objective way only if it is impossible to go continuously from one specie to another; in a more mathematical language, we say that objects must have the property of ultrametricity. More precisely, it was discovered that there are conditions under which a class of complex systems may only exist in configurations that have the ultrametricity property and consequently they can be classified in a hierarchical way. Indeed, it has been found that only this ultrametricity property is shared by the near-optimal solutions of many optimization problems of complex functions, i.e. corrugated landscapes in Kauffman's language. These results are derived from the study of spin glass model, but they have wider implications. It is possible that the kind of structures that arise in these cases is present in many other apparently unrelated problems.

Before to go on with our considerations, we have to pick in mind two main complementary ideas about complexity. (i) According to the prevalent and usual point of view, the essence of complex systems lies in the emergence of complex structures from the non-linear interaction of many simple elements that obey simple rules. Typically, these rules consist of 0–1 alternatives selected in response to the input received, as in many prototypes like cellular automata, Boolean networks, spin systems, etc. Quite intricate patterns and structures can occur in such systems. However, what can be also said is that these are toy systems, and the systems occurring in reality rather consist of elements that individually are quite complex themselves. (ii) So, this bring a new aspect that seems essential and indispensable to the emergence and functioning of complex systems, namely the coordination of individual agents or elements that themselves are complex at their own scale of operation. This coordination dramatically reduces the degree of freedom of those participating agents. Even the constituents of molecules, i.e. the atoms, are rather complicated conglomerations of subatomic particles, perhaps ultimately excitations of patterns of superstrings. Genes, the elementary biochemical coding units, are very complex macromolecular strings, as are the metabolic units, the proteins. Neurons, the basic elements of cognitive networks, themselves are cells.

In those mentioned and in other complex systems, it is an important feature that the potential complexity of the behaviour of the individual agents gets dramatically simplified through the global interactions within the system. The individual degrees of freedom are drastically reduced, or, in a more formal terminology, the factual space of the system is much smaller than the product of the state space of the individual elements. That is one key aspect. The other one is that on this basis, that is utilizing the coordination between the activities of its members, the system then becomes able to develop and express a coherent structure at a higher level, that is, an emergent behaviour that transcends what each element is individually capable of.

5. Complex Systems

There are many different definitions of a complex system. It may range from the classical algorithmic complexity (Kolmogorov, Chaitin) to more recent and sophisticated definitions, such as: chemical definitions, statistical-physics definitions, topological-dynamic definitions, biological definition, ect. We already have given the most common and general definition in the literature of complex systems. It should be clear, however, that any given definition (especially a mathematical one) couldn't capture all the complex meaning we associate with the word complexity.

One interesting definition rest on the basic idea that the more complex the system, the more can be said about. I am excluding the factual description of the system, which may be very long. I refer only to the global characteristics. A few examples will help clarify this point. If I have a sequence of randomly tossed coins, 50% probability head, I have described the system. The only improvement would be the knowledge of the sequence itself. If on the contrary the sequence of bits represents a book, there is much more information, such as style, choice of words, the plot and so on. If the book is really deep, complex, there are a very large number of things that can be said about it. Sometimes the complexity is related to the existence of different levels of description: one can describe an *Escherichia coli* at the molecular level, at the biochemical level, and at the functional level.

If we move towards a mathematical definition, we must realize that the concept of complexity, like entropy, is of probabilistic nature and it can be defined more precisely if we try to define the complexity of ensembles of objects of the same category. This is related to the notion of classification. The meaning of a complex classification is quite clear intuitively: a classification is very complex if there are many levels (i.e. orders, families, genera) and there are many elements in each level. Consequently a reasonable mathematical definition of the complexity of a classification should be possible.

6. External and Internal Complexity

Let's now introduce the notions of *external* and *internal* complexity of complex adaptive systems. These concepts are especially useful to analyze relations between an adaptive system and its environment. All open systems, let they be either thermodynamics, biological or cognitive, are chiefly concerned with this relation.

A *complex adaptive system* is situated in an environment. That environment is always more complex than the system itself, and therefore, it can never be completely predictable for the system, but the system depends on regularities of the environment for maintaining its energy supply needed to support its internal structure.

One important hypothesis one can suggest is that complex adaptive systems try to increase their external complexity and to reduce internal complexity. Each of two processes will operate on its own scale, but they are also intricately linked and mutually dependent upon each other.

The increase of internal complexity can for example occur through the creation of redundancy, e.g. duplication of some internal units or structures. The property of redundancy is very important in biological systems at the genetic level as well as at other more complex levels; for example, a same gene may realize different functions and, on the other hand, many genes may accomplish the same function. Upon this redundancy, a process of differentiation or specialization can operate, through controlled random mechanisms or internal selection, so that the system will become able to handle more diverse input and thereby increase its external complexity. Once this happened, the system can then again try to represent this newly acquired input more efficiently and thus decrease its internal complexity. Conversely, for the decrease of internal complexity, the system can also find some of its input as irrelevant and meaningless for its purposes and thus decrease the external complexity.

As first definition, one can say that *external complexity* measures the amount of input, information, energy obtained from the environment that the system is capable of handling, processing. It is important that this can be measured as an entropy – and therefore, terms like "energy" need some qualification when employed in this context. In this sense, external complexity is data complexity.

Internal complexity can be defined as what that measures the complexity of the representation of this input by the system. In this sense, internal complexity is model complexity. The system will try to increase (or maximize) its external complexity, and to reduce (or minimize) its internal complexity.

We now proceed to give formal definitions of our complexity notions based on the concept of entropy from statistical mechanics and information theory. Given a model θ, the system can model data as $X(\theta)$, with $X = X_1, \ldots, X_k$, and we assume that $X(\theta)$ introduces an internal probability distribution $P(X(\theta))$ so that an entropy can be computed in (1) bellow. Our hypothesis then is that the system will try to maximize the external complexity,

$$-\sum_{i=1}^{k} P(X_i(\theta)) log_2 P(X_i(\theta)) \tag{1}$$

The purpose of the probability distribution $P(X(\theta))$ is simply to qualify the information value of the data $X(\theta)$. In principle, this quantification is also possible through other means, for example, through the length of the representation of the data in the internal code of the system. If we assume optimal coding, however, which is a consequence of the minimization of internal complexity, then the length of the representation of a datum $X_i(\theta)$ behaves like $log_2 P(X_i(\theta))$ (a code is god if frequent inputs are represented by short code words).

The system can try to increase the amount of information $X(\theta)$ that is meaningful within the given model θ on a short time scale, or it can adapt the model θ on a larger time scale so as to be able to process more inputs as meaningful. When the input is given, however, for example when the system has gathered input on a time scale when the distribution of input patterns ϵ^* becomes stationary, then the model should be improved to handle that input as efficiently as possible, i.e. to decrease the internal complexity which we now define as follows

$$-\sum_{i=1}^{k} P(\epsilon_i|(\theta))log_2 P(\epsilon_i|(\theta)) - log_2 P(\theta) \tag{2}$$

The variation gives

$$min_\theta(-log_2 P(\epsilon|(\theta)) - log_2 P(\theta)) \tag{3}$$

The expression to be minimized now consists of two terms, the first measuring how efficiently the data are encoded by the model, and the second one how complicated the model is. Of course the probability $P(\theta)$ assigned to a model depends on the internal structure of the system, and in principle, that internal structure then also became subject to optimization, in the sense that frequently used or otherwise important models get higher probabilities than obscure ones.

7. Pattern Recognition in a Neural Network

We only mention the principle according to which a neural network can recognize patterns on the basis of a selective evaluation of inputs features via an internal feedback loop. (No detailed description will be presented here.)

We assume that the network or system has stored or identified a collection of patterns labeled by $i = 1, \ldots, n$. These patterns might correspond to faces, visual shapes or other geometric objects; for thinking about this example, it is probably useful to think about patterns to be recognized in visual scenes. Also, on its input, the system can evaluate certain features $\alpha = 1, \ldots, m$, like edges, corners, or better, features of a somewhat higher level, like specific distribution of input pixels on some small sub-regions of the retina, or relative distances between certain conspicuous points of the scene. It is important for understanding the purpose of the network (the system) that we assume to in a situation where the network is not capable of evaluating all the possible features simultaneously in its inputs, simply because there are typically far too many possibilities.

Rather the idea is that the network will selectively perform observations, that is, evaluate those features that have the highest potential for discriminating between these patterns that are probable candidates on the basis of the observations already performed. Thus, the basic design principle is a feedback loop between observations that affect the probability distribution in the space of patterns and the selection of further observations on the basis of that probability distribution.

*We use a different letter now to denote the inputs because we are now considering patterns on a different time scale.

We first need to implement the relationship between patterns and features. This can be done on the basis of supervised learning as is standard in neural networks. So, the observed values x^α of the features induces activations y^i of the patterns:

$$y^i : f(\sum_\alpha w_{i\alpha} x^\alpha) \tag{4}$$

where f might be a sigmoid function $f(s) = 1/1 + e^{-ks}$, where for our purpose a rather large value of the parameter k might be best so as to get a sharp threshold later on. Namely, we call a pattern i activated if $y^i > \theta$ is some threshold that we can turn to our convenience, perhaps again by supervised learning. The $w_{i\alpha}$ are weights that can likewise be learned through supervised Hebbian learning. The essential point is that they should be positive, and perhaps large, if feature α occurs in pattern i, and 0 or negative if not.

8. Examples of Complex Systems

Let us now give briefly some examples of complex systems took from different disciplines: chemistry, biology and physics. In all this examples, understanding how parts of a living system – genes or molecules – interacts is just as important as understanding the parts themselves.

In Chemistry the word *complexity* present some ambiguity and it is highly dependent on context. In one characterization a complex system is: (i) one whose evolution is very sensitive to initial conditions or to small perturbations; (ii) one in which the number of independent interacting components is large; (iii) or, one in which there are multiple pathways by which the system can evolve. Analytical descriptions of such systems typically require nonlinear differential equations. In chemistry, almost every thing of interest is complex by one or both definitions.

We are here concerned with "tractable complexity": a subset of complex problems (for example, oscillating reactions) provides classical examples of complex systems in the sense that they can be described analytically by relative simple sets of nonlinear differential equations. But, there are other complex problems of general importance for which there are no simple general solutions.

In the sequence of complexity – from static equilibrium, to dynamic steady state, to dynamic complexity, to chaos – there are chosen sets of chemical reactions whose properties make them appropriate as case studies in complexity. Oscillating reactions of the type represented by the Belousov-Zhabotinsky reaction are perhaps the best-known example. This class of chemical reactions has the characteristic that the simultaneous operation of two processes, reaction and diffusion, results in a system in which the concentration of reactants and products oscillate temporally and spatially and in which this oscillation can result in ordered patterns. In other words, coupled chemical reactions cause changes in concentration of the reagents that, in turn, cause local changes in the oxidation potential of the solution. These potentials can be visualized as oscillating travelling waves in such a reaction.

These reactions can be described mathematically by a system of nonlinear equations of greater or lesser complexity, but equations bellow represent a minimum set of two reaction-diffusion equations

$$\partial u/\partial t = F(u,v) + D_u \nabla^2 u \tag{5}$$

$$\partial v/\partial t = \epsilon G(u,v) + D_v \nabla^2 v \tag{6}$$

Here, u is the concentration of a species that catalyze reaction; v is the concentration of a species that inhibits reactions; $\partial u/\partial t$ and $\partial v/\partial t$ describe changes in concentration of u and v, respectively withy time; $F(u,v)$ and $\epsilon G(u,v)$ characterize reactions between u and v, respectively; and D_u and D_v are the diffusion coefficients of u and v, respectively.

An important motivation in chemistry of studying complexity has been to learn about processes in living systems. One of the most striking characteristics of cells is the sheer complexity of metabolism. The human genome probably has on the order of 105 expressed gene products; many of these proteins are enzymes, receptors, and members of signalling sequences, that is, functional parts of metabolism. Understanding a system with this many interacting components is clearly out of the question. A more tractable problem is to examine discrete, relatively self-contained sections of metabolism. One metabolic cycle that has been studied in substantial detail is glycolysis, that is, the conversion of glucose to pyruvate with the production of adenosine 5′-triphosphate and the reduced form of nicotinamide adenine dinucleotide (NAD)[†]. This sequence of reactions involves 10 enzymes, with various levels of modulation of the catalytic activities of some of these enzymes by the products of others.

The second example of complexity I would like to mention concerns *biological signalling systems.* Biological signalling pathways interact with one another to form complex networks. Complexity arises from the large number of components, many with isomorphs that have partially overlapping functions; from the connections among components; and forms the spatial relationship between components.

Signalling in biological systems occurs at multiple levels. Already compartmentalization introduces several levels of complexity. First, many signalling components and their substrates are anchored in the plasma membrane. The plasma membrane provides a milieu for biochemical reactions that is quite distinct from the cytoplasm in its properties. The lipid environment enables a new class of reactions involving

---

[†]Nicotinamide adenine dinucleotide (NAD) and its relative nicotinamide adenine dinucleotide phosphate (NADP) are two of the most important coenzymes in the cell. NADP is simply NAD with a third phosphate group attached. NAD partecipates in many redox reactions in cells, including those in glycolysis, and most of those in the citric acid cycle of cellular respiration. NADP is the reducing agent produced by the light reactions of photosynthesis, consumed in the Calvin cycle of photosynthesis, and used in many other anabolic reactions in both plants and animals.

hydrophobic interactions. Organelle formation leads to a further expansion of the possible cellular microenvironments, each with different biochemical properties and signalling capabilities. Second, the separation of reactions in space allows the same molecules in the same cell to carry entirely different signals. In other words, we already have signalling "wires" distinguished by the identity of the molecules in the pathways. Compartmentalization duplicates these existing wires and separates them in space. This multiplies the number of signals they can carry about.

In addition to sub-cellular compartmentalization recent research has highlighted the role of molecular scaffolds that provide regional organization by assembling signalling components into functional complexes. The cytoskeleton is a dynamic framework on which the cell builds this regional organization. The most dramatic example of its dynamism is cell division. In the quiescent cell, it is both the substrate and the scaffold for signalling processes.

A prime example of its dual role is its synapse. Here the cytoskeleton, in particular the pre- and postsynaptic structures, are the anchors for a wide array of synaptic signalling molecules. Consequently, modifications of the synaptic cytoskeleton are a likely candidate for causing long-term changes in synaptic efficacy.

To conclude this section, we point out some theoretical remarks about characteristic properties of complex living systems. Let's start with some observations.

9. Network Behaviours and Emergent Properties

1.Today, it is clear that the specificity of a complex biological activity does not arise from the specificity of the individual molecules that are involved, as these components frequently work in many different processes. For instance, genes that affect memory formation in the fruit fly encode proteins in the cyclic adenosine monophosphate (cAMP)[‡] signalling pathway that are not specific to memory. Biological specificity results from the way in which these components assemble and work together. Interactions between the parts, as well as influences from the environment, give rise to new features, such as network behaviour.

2.Consequently, "emergence" has appeared as a new concept that complements "reduction" when reduction fails. Emergent properties resist any attempt at being predicted or deduced by explicit calculation or any other means. In this regard, emergent properties differ from resultant properties, which usually can be predicted from lower-level information. For example the resultant mass of a multi-component protein assembly is simply equal to the sum of the masses of each individual component. However, the way in which we test the saltiness of sodium chloride is not

[‡]Cyclic adenosine monophosphate (cAMP, cyclic AMP or $3'-5'$-cyclic adenosine monophosphate) is a second messenger important in many biological processes. cAMP is derived from adenosine triphosphate (ATP) and used for intracellular signal transduction in many different organisms, conveying the cAMP-dependent pathway. One important intracellular signal transduction is the transferring of the effects of hormones like glucagon and adrenaline, which cannot pass through the cell membrane. It is involved in the activation of protein kinases and regulates the effects of adrenaline and glucagon. It also regulates the passage of Ca 2+ through ion channels.

reducible to the properties of sodium and chlorine gas. An important aspect of emergent properties is that they have their own causal powers, which are not reducible to the powers of their constituents. For instance, the experience of pain can alter human behavior, but the lower-level chemical reactions in the neurons that are involved in the perception of pain are not the cause of the altered behavior as the pain itself has causal efficacy. It should be added that the concept of emergence implies "down-ward causation" by which higher-level systems influence lower-level configurations.

3.The constituents of complex systems interact in many ways, including negative feedback and feed-forward control, which lead to dynamic features (i.e., evolving in time and changing with time) that cannot be predicted satisfactorily by linear mathematical models that disregard cooperativity and non-additive effects.

4.Robustness is another essential property of biological systems. Understanding the mechanisms and principles underlying biological robustness is necessary for an in-depth understanding of biology at the system level. The phenomenological properties exhibited by robust systems can be classified into three areas: (i) *adaptation*, which denotes the ability to cope with environmental changes; (ii) *parameter insensitivity*, which indicate a system's relative insensitivity to specific kinetic parameters; (iii) *graceful degradation*, which reflects the characteristic slow degradation of a system's functions after damage, rather than catastrophic failure. In other systems, such as fluid-mechanics systems, and also engineering systems, robustness is attained by using the following properties: (a) the capability to form a system control such as negative feedback and feed-forward control; (b) *redundancy*, whereby multiple components with equivalent functions are introduced for backup; (c) structural stability, where intrinsic mechanisms are built to promote stability; and (d) modularity, where sub-systems are physically or functionally insulated so that failure in one module does not spread to other parts an lead to system-wide catastrophe.

10. Remarks about the Property of Structural Stability and on Self-organization

It remains an open question whether the property of structural stability used in the biological context present some similar characteristics with respect to the concept of structural stability as it has been defined in differential topology in the 1960s by R. Thom and S. Smale.

Intuitively, a *phase portrait* (i.e. all the qualitatively different trajectories of the system) is structural stable if its topology cannot be changed by an arbitrarily small perturbation to the vector field. For instance, the phase portrait of a saddle point is structurally stable, but that of a center is not: an arbitrarily small amount of damping converts the center to a spiral. Related to the concept of structural stability, we have the notions of *attractor* and strange *attractor*. The term attractor is difficult to define in a rigorous way. Loosely speaking, an attractor is a set to which all neighbouring trajectories converge. More precisely, we define an attractor to be a closed set A with the following properties: (i) A is an invariant set: any trajectory $x(t)$ that starts in A stays in A for all times; (ii) A attracts an open set of initial conditions: there is an open set U containing A such that if $x(0) \in U$, then the distance from $x(t)$ to A tends to zero as $t \to \infty$. This means that A attracts all trajectories that start sufficiently close to it. The largest such U is called the *basin of attraction of A*; (iii) A is minimal: there is no proper subset of A that satisfies conditions (i) and (ii).

Finally, we define a strange attractor to be an attractor that exhibits sensitive dependence on initial conditions. Examples of strange attractors are fractal sets and also chaotic attractors. Just to conclude this section, let us remark that the four properties listed above are also found in biological systems. Bacterial chemotaxis[§] is an example of negative feedback that attains all three aspects of robustness. Redundancy is seen at the gene level, where it functions in control of the cell cycle and circadian rhythms, and at the circuit level, where it operates in alternative metabolic pathways in *Escherichia coli*. Structural stability provides insensitivity to parameter changes in the network responsible for segment formation in Drosophila. And modularity is exploited at various scales, from the cell itself to compartmentalized yet interacting signal-transduction cascades.

Lastly, the concept of self-organization in cellular architecture is linked to the complexity of biological systems. A central question in modern cell biology is how large, macroscopic cellular structures are formed and maintained. It is unknown what determines the different shapes and sizes of cellular organelles, why specific structures form in particular places, and how cellular architecture is affected by function and vice-versa.

Two fundamentally different mechanisms exist to generate macromolecular structures: *self-assembly* and *self-organization*. Whereas self-assembly involves the physical association of molecules into an equilibrium structure, self-organization involves the physical interaction of molecules in a steady-state structure. For example, virus and phage proteins self-assemble to true equilibrium and form stable, static structures. In contrast, most cellular structures (i.e., the cytoskeleton, nuclear compartments, or endocytic compartments) are open for exchange of energy and matter and are governed by steady-state dynamics.

The concept of self-organization is based on observations of chemical reactions far from equilibrium, and it is well established in chemistry, physics and ecology.

[§]Chemotaxis is the phenomenon in which somatic cells, bacteria, and other single-cell or multicellular organisms direct their movements according to certain chemicals in their environment. This is important for bacteria to find food (for example, glucose) by swimming towards the highest concentration of food molecules, or to flee from poisons (for example, phenol). In multicellular organisms, chemotaxis is critical to early development (e.g. movement of sperm towards the egg during fertilization) and subsequent phases of development (e.g. migration of neurons or lymphocytes) as well as in normal functions. In addition, it has been recognized that mechanisms that allows chemotaxis in animals can be subverted during cancer metastasis.

Self-organization in the context of cell biology can be defined as the capacity of a macromolecular complex or organelle to determine is own structure based on the functional interactions of its components. In a self-organized system, the interactions of its molecular parts (and not the molecular parts them-selves) determine its architectural and functional features. The processes that occur within a self-organized structure are not underpinned by a rigid architectural framework; rather, they determine its organization.

For self-organization to act on macroscopic cellular structures, three requirements must be fulfilled: (i) a cellular structure must be dynamic; (ii) matter and energy must be continuously exchanged; (iii) overall stable configuration must be generated from dynamic components.

Recent studies indicate that many cellular structures fulfil the requirements for self-organization. Particularly, self-organization seems to play an important role in the construction and dynamics of three major macroscopic cellular structures: namely the cytoskeleton, the cell nucleus, and the Golgi complex.

Why self-organization? And why self-organization is related to complexity? To answer these questions, consider the followings facts. Macroscopic cellular structures are characterized by two apparently contradictory properties. On one hand, they must be architecturally stable; on the other hand, they must be flexible and prepared fro change. Self-organization ensures structural stability without loss of plasticity. Fluctuations in the interactions properties of its components do not have deleterious effects on the structure as a whole. However, global and persistent changes rapidly result in morphological transformations. The basis for the responsiveness of self-organized structures is the transient nature of the interactions among their components.

The dynamic interplay of components generates frequent windows of opportunity during which proteins can change their interaction partners or be modified. The effective availability of components is controlled by posttranslational modifications via signal transduction pathways.

Self-organization is an elegant, efficient way to organize complex structures. The properties that determine the organization are the intrinsic properties of the structure's components. In protein polymers, the protein-protein interaction properties determine the architecture; in membrane structures the flow of membranes determines the architecture. Thus, self-organization is a simple but effective way to optimally organize cellular structures.

The study of complex dynamic behaviour of cellular structures requires new tools. The behaviour of dynamic cellular structures cannot be described accurately by conventional equilibrium dynamics or by static models. To understand the behaviour of complex (dynamic) systems, the kinetic and topological characteristics of their components must be known. In contrast to the study of molecular mechanisms, isn't must sufficient to understand in detail the behaviour of single molecules; rather, the rules that govern the global and collective behaviour of systems must be uncovered.

We would like to conclude by stressing the fact that a new theory of living organisms not only requires new mathematics, new physics, and new epistemology, but also that some striking new ideas and methods borrowed from each of these disciplines be worked out together and merged into a meaningful interdisciplinary and global explanation of biological systems. Such an approach might contribute to restore to our fragmented biological sciences the kind of integration and unity they strongly need. The plethora of recent biological observations and theoretical models may pave the way to discover new mathematical objects and concepts and to open new frontier-problems, and conversely, biology may benefit from these mathematical structures in organizing experimental data and clarifying their descriptions. This gathering of deep mathematical concepts, physical principles and epistemological analysis is necessary for constructing a kind of relational biology, which will enable us to explore interactions among the systems, properties and behaviors of living organisms. An attempt should be made to construct a theory in which the merely informational language (code, program, computation) so characteristic of molecular biology be completed by (and somehow translated into) the language of dynamical systems (phase space, bifurcations, attractors) and the language of topology (deformations, plasticity, forms). These considerations lead ultimately to the suggestion that our traditional modes of system representation, involving fixed sets of sequential states, together with imposed mechanical laws, strictly pertain to an extremely limited class of systems that can be called simple (static) systems or mechanisms. Biological systems are not in this class, and they must be called complex or dynamic. Complex systems can only be in some sense approximated, locally and temporally, by simple ones, and so require a new set of mathematical ideas to be described. Such a fundamental change of viewpoint leads to a number of theoretical and experimental consequences, some of which were briefly addressed in this discussion.

References

[1] P. W. Anderson,*Science*, **177** , 393–396 (1972).
[2] F. M. Atay, J. Jost, *Complexity*, **10** (1), 17–22 (2004).
[3] A. Bates, A. Maxwell, *DNA Topology*, (Oxford University Press, 1993).
[4] M. Ballerini et al., Interaction Ruling Animal Collective Behaviour Depends on Topological rather than Metric Distance: Evidence from a Field Study, *Proc. Nat. Acad. Sci. USA*, **105** (2008), 237-248.
[5] L. Boi, *International Journal of Bifurcation and Chaos*, **21** (3), 815–867 (2011).
[6] L. Boi, Plasticity and Complexity in Biology: Topological Organization, Regulatory Protein Network, and Mechanisms of Genetic Expression, in *Information and Biological Systems. Philosophical and Scientific Perspectives*, G. Terzis and R. Arp (eds.), The MIT Press, Cambridge, MA, 234-279 (2011).
[7] L. Boi, When Topology Meets Biology 'For Life'. The Interaction Between Topological Forms and Biological Functions, in *New Trends in Geometry. Their Role in the Natural and Life Sciences*, C. Bartocci, L. Boi, C. Sinigaglia (eds.), Imperial College Press, London, 241–302 (2011).
[8] L. Boi, *Biophysical Reviews and Letters*, **2** (3/4), 287–299 (2007).

[9] L. Boi, *JP Journal of Geometry and Topology*, **8** (2), 151–184 (2008).
[10] L. Boi, *Rivista di Biologia/Biology Forum*, **101**, 405–442 (2008).
[11] L. Boi, "Topological knot theory and macroscopic physics", in *Encyclopedia of Mathematical Physics*, J.-P. Françoise, G. Naber and T.S. Sun (Eds.), (Elsevier, 2006), 271–277.
[12] L. Boi, "Géométrie, dynamique et auto-organisation dans la nature et le vivant", in *Symétries, Brisures de Symétries et Complexité en Mathématiques, Physique et Biologie*, L. Boi (Ed.), 9–19, (Peter Lang, 2006) .
[13] L. Boi, "Sur quelques propriétés géométriques globales des systèmes vivants", *Bulletin d'Histoire et d'Epistémologie des Sciences de la Vie*, **14** (1), 71–113 (2007).
[14] L. Boi, *Geometries of Nature, Living Systems and Human Cognition: New Interactions of Mathematics with Natural Sciences and Humanities*, (World Scientific, 2005).
[15] A. Cornish-Bowden, M.L. Cardenas, *Biochemical Society Transactions*, **33**, 516–519 (2005).
[16] P. De Gene, *Scaling Concepts in Polymer Physics*, (Cornel University Press, 1979).
[17] G. Del Re, Organization, Information, Autopoiesis: From Molecules to Life, in *The Emergence of Complexity in Mathematics, Physics, Chemistry, and Biology*, B. Pullman (Ed.), Pontificia Academia Scientiarum, Vatican City/Princeton University Press, 277–305, (Princeton, 1996).
[18] E. Eigen, *Naturwissenschaften*, **58**, 465–523 (1971).
[19] S.C.R. Elgin and J.L. Workman, *Chromatin Structure and Gene Expression*, (Oxford University Press, 2000).
[20] Gell-Mann, *Complexity*, **1**, 16–19 (1995).
[21] J. Jost, *Theory in Biosciences*, **117**, 161–171 (1998).
[22] E. Karsenti, *Nature Reviews*, **9**, 255–262 (2008).
[23] L.H. Kauffman, *Knots and Physics*, (World Scientific, 2001).
[24] S. Kauffman, *The Origins of Order: Self-Organization and Selection in Evolution*, (Oxford University Press, 1993).
[25] S. Kauffman, *At Home in the Universe: The Search for Laws of Self-Organization and Complexity*, (Oxford University Press, 1995).
[26] J.-M. Lehn, *Reports on Progress in Physics*, **67**, 249–265 (2004).
[27] R. Li and B. Bowerman, *Cold Spring Harbor Perspective in* Biology, **2**, 1–5 (2010).
[28] F. Nedelec, T. Surrey and E. Karsenti, *Curr. Opin. Cell Biol.*, **15** (1), 118–124 (2003).
[29] T. Mastelli, *The Journal of Cell Biology*, **155** (2), 181–185 (2001).
[30] R. M. May, *Nature*, **261**, 459–464 (1976).
[31] H. Meinhardt, Biological Pattern Formation as a Complex Dynamic Phenomena, in *Pattern Formation in Biology, Vision and Dynamics*, A. Carbone, M. Gromov, P. Prusinkiewicz (Eds.), World Scientific, Singapore, 97–132 (2000).
[32] G. Nicolis and I. Prigogine, *Exmploring Complexity: An Introduction*, (Freeman, New York, 1989).
[33] G. Parisi, *Statistical Field Theory*, (Addison-Wesley, 1988).
[34] I. Prigogine and G. Nicolis, *Self-Organization in Non-Equilibrium Systems*, (Wiley, 1977).
[35] D. Rolfsen,*Knots and Links*, Publish or Perish, Math. Lecture Series, Houston, (1990).
[36] S. Smale, *The Mathematics of Time. Essays on Dynamical Systems, Economics Processes, and Related Topics*, (Springer-Verlag, 1980).
[37] S.H. Strogatz, *Nonlinear dynamics and chaos. With applications to physics, biology, chemistry, and engineering*, (Perseus Publishing, 1994).
[38] R. Thom, *Stabilité structurelle et morphogenèse*, (W.A. Benjamin, 1972).
[39] D.A.W. Thompson, *On Growth and Form*, (Cambridge University Press, 1942).

[40] G. Tononi, O. Sporns, G. Edelman, A measure for brain complexity: relating functional segregation and integration in the nervous system, *Proc. Nat. Acad. Sci. USA*, **91** (1944), 5033-5037.
[41] L. Wang et al., *Journal of Bacteriology*, **187** (6), 2066–2076 (2005).

THE BUTTERFLY AND THE PHOTON: NEW PERSPECTIVES ON UNPREDICTABILITY, AND THE NOTION OF CASUAL REALITY, IN QUANTUM PHYSICS

T.N. PALMER

Department of Physics, University of Oxford
Clarendon Laboratory, Parks Road, OX1 3PU, Oxford, UK
E-mail: t.n.palmer@atm.ox.ac.uk

This essay discusses a proposal that draws together the three great revolutionary theories of 20th Century physics: quantum theory, relativity theory and chaos theory. Motivated by the Bohmian notion of implicate order, and what in chaos theory would be described as strange attractor, the proposal attributes special ontological significance to certain non-computable, dynamically invariant state-space geometries for the universe as a whole. Studying the phenomenon of quantum interference, it is proposed to understand quantum wave-particle duality, and indeed classical electromagnetism, in terms of particles in space time and waves on this state space geometry. Studying the EPR experiment, the acausal constraints that this invariant geometry provides on spatially distant degrees of freedom, provides a way for the underlying dynamics to be consistent with the Bell theorem, yet be relativistically covariant ("nonlocality without nonlocality"). It is suggested that the physical basis for such non-computable geometries lies in properties of gravity with the information irreversibility implied by black hole no-hair theorems being crucial. In conclusion it is proposed that quantum theory may be emergent from an extended theory of gravity which is geometric not only in space time, but also in state space. Such a notion would undermine most current attempts to "quantise gravity".

1. Introduction

20th Century theoretical physics saw three great revolutions: quantum theory, relativity theory and chaos theory. To a great extent, these revolutions are still very much disjoint. For example, our ongong inability to make quantum theory and general relativity theory compatible is famous. Moreover, it is generally believed that the unpredictability of quantum measurement on the one hand, and of chaotic evolution on the other, have little in common: for one thing chaotic systems are profoundly nonlinear whilst the Schrödinger equation is linear, and for another, any underpinning deterministic dynamic would have to be constrained by the Bell Theorem and hence could not be (special) relativistic covariant. Finally, even chaos theory and relativity theory are not readily compatible with one another. For example, the "butterfly effect" which characterises a chaotic system is generally associated with the existence of one or more positive Lyapunov exponents. However, the notion of state-space trajectories diverging exponentially in time is not a relativistic one —

merely move to a frame where time has been transformed logarithmically and the same trajectories are now diverging linearly in time.

This essay[a] discusses a new proposal [1] which may draw these three great theories together and at the same time move theoretical physics forward. It is essentially a geometric proposal, but where the geometry in question lies in state space, rather than space time. It is motivated by the Bohmian notion of undivided wholeness and implicate order [2], and of the notion of the strange attractor in chaotic dynamics. Whilst strange attractors are pretty much "common or garden" in dynamical system's theory, they have not received much attention in fundamental physics circles. One reason for this is that such geometries are intrinsically linked to irreversibility. Insofar as fundamental physics is generally considered Hamiltonian, such geometries are irrelevant to fundamental physics. However, we argue below that black-hole no-hair theorems provide the basis for associating such geometries with an extended causal theory of gravity. The principal conclusion of this paper is that, the Bell Theorem notwithstanding, quantum theory could be emergent from such an extended causal theory of gravity, with the Schrödinger and Dirac equation playing the role of the Liouville equation for conservation of probability (necessarily linear) on this geometry.

Section 2 describes the proposal in more detail. Section 3 describes the phenomenon of quantum interference, and quantum nonlocality from this perspective. Some conclusions are given in Section 4.

2. The Invariant Set Postulate

Let us start by recalling some of David Bohm's thoughts on physics and the nature of space time at the deepest level. Bohm had a strong belief that space and time are not the dominant factors determining the relationships of dependence or independence between different elements of reality. Rather, he argued that an entirely different basic connection of elements is possible, from which our ordinary notions of space and time, along with those of separately existent material particles, are abstracted as forms derived from a deeper order. These ordinary notions in fact appear in what is called the "explicate" or "unfolded" order, which is a special and distinguished form contained within the general totality of all the implicate orders [2].

I want to use this idea to illustrate the basis for the geometric proposal which I call "The Invariant Set Postulate". I will apply this postulate to delineate states of physical reality of our universe, from those of physical unreality. However, before going this far, let us start by considering, from a Bohmian perspective, the relationship between the chaotic fractionally dimensioned zero-volume Lorenz attractor on the one hand, and the Lorenz equations on the other (Lorenz, 1963). Since the Lorenz equations map points within the attractor, I will henceforth use the somewhat more general phrase "invariant set" or "invariant geometry" to describe this attractor.

[a]Based on an invited talk at the SCIENCE:IMAGeINAcTION workshop, Erice, 2011

Using Bohmian language, I would like to describe the invariant geometry as the implicate-order representation of an "undivided wholeness" and the dynamical differential equations as the form which arises from the "unfoldment" of the implicate to the explicate. To understand this, recall that physics in general, whether classical or quantum, treats dynamical evolution equations as primitive (in this respect the Lorenz equations are just truncated versions of Newton's Second Law and the Laws of Thermodynamics, applied to a spatially continuous thermally convecting fluid).

The relationship between the dynamical equations on the one hand, and the invariant geometry on the other, is an extremely profound one. Isaac Newton would have understood the Lorenz equations as coupled nolinear differential equations. However, he would have had no conception of the type of invariant geometry that these equations generate. This is not suprising, since these (fractionally dimensioned) geometries are formally non-computable: that is to say, there exists no finite algorithm that can determine whether a point in the Euclidean state space of the differential equations actually lies on the geometry [3]. More generally, the structure of the invariant geometry in some neighbourhood of state space, in particular the lacunarity of the geometry (see below) cannot be deduced by integrating the dyanmical equations over short intervals of time from points around that neighbourhood. Indeed, these invariant sets can be thought of as providing a very tangible link between the calculus of Newton (and hence theoretical physics) and the most profound theorems of 20th Century mathematics, describing the undecidability and incompleteness of formal logic. As an example, Dube [4] has shown that some of the standard undecidable propositions of computing theory can be cast in terms of invariant geometry. For example, the Post Correspondence Problem can be recast by asking whether there exists a finite algorithm which can determine whether a given line intersects the invariant set of an iterated function system. This fact alone, that there is a link here between the bread and butter of physics, the differential equation, and perhaps the most profound of ideas emerging from 20th Century mathematics, alone suggests that such fractionally-dimensioned invariant geometries are objects deserving of more attention in fundamental physics.

We live in a universe with many more degrees of freedom than the three of the Lorenz equations. However, size is not especially relevant in discussing these foundational issues. So, motivated by the Lorenz invariant set and these more general Bohmian notions of wholeness and implicate order, let us postulate the "unbroken wholeness of the Universe" as some non-computable geometry, which for our own convenience we can imagine existing in some larger dimensional Euclidean Space (the Whitney Embedding Theorem always allows this). The dynamical equations which describe evolution on individual trajectories of this primitive geometry describe the Bohmian unfoldment of this geometry to provide the physically explicate.

A number of possible objections may be considered to this notion. First of all, the Lorenz invariant set has zero volume in the Euclidean state space of the three components of the Lorenz system. The fact that arbitrary state-space volumes shrink to zero arises from the fact the Lorenz equations are dissipative. We do not believe such energy dissipation to be a fundamental process in physics. Hence, why should such zero-volume geometries play such a profound role in fundamental physics?

To answer this, note that there is a place where irreversibility may play a fundamental role in physics — in black hole dynamics, as revealed by the no-hair theorem of classical general relativity. There is some debate as to whether quantum field theoretic effects associated with Hawking radiation restores reversibility, but here I strongly support the view of Roger Penrose that they do not [5]. Consider the Hawking Box [5–7] as a metaphor for the Universe as a whole. Assuming convergence of trajectories associated with black hole information loss[b], then the Liouville theorem for conservation of probability imply that general state space volumes must shrink asymptotically to zero. However, "zero volume" needn't mean "shrink to a point", and if we use dynamical systems theory as a guide, one could imagine that the asymptotic end state of the state space of the Hawking Box will be a dynamically complicated non-computable invariant geometry of the type discussed above.

As mentioned, The Hawking Box can be taken as a metaphor for the Universe as a whole. Of course in this respect one would not take the Big Bang as representing some beginning which cannot extended mathematically further back in time, nor the Big Crunch (if such exists) as some end which cannot be extended mathematically into the future. Rather the current evolution of the universe would merely describe some trajectory portion of this putative non-computable geometry. Penrose [7] has recently proposed a new perspective on this through the notion of conformal cyclic cosmology (though here I would not interpret the word "cyclic" as meaning "periodic".

The key point here is that gravity provides all the basic ingredients to account for the putative existence of the non-computable invariant geometry of the Universe: gravity is nonlinear, has ubiquitous instabilities and, through the no-hair theorem, supports of a form of irreversibility of information. We therefore propose that states of physical reality correspond precisely to (the points of) this geometry — that is the Invariant Set Postulate.

3. Towards a Realistic and Causal Perspective on Quantum Theory

Feynman always maintained that the real mystery of quantum theory was embodied in the notion of quantum interference. In the first part of this section we discuss this phenomenon from the invariant set perspective outlined above. This will naturally lead to a discussion of a new perspective on nonlocality and causality in the second section.

---

[b]Penrose considers the Hawking box gedanken experiment to support his view that the Liouville Theorem requires a physical process which generates divergence of phase space trajectories to compensate for the convergence of trajectories associated with black hole information loss. Personally, I do not believe the Liouville Theorem in its general form implies the need for such compensating divergence.

3.1. *Wave-Particle Duality. Particles in Space, Waves on the Invariant Geometry*

Consider a single photon in a Mach-Zehnder interferometer. It is well known that it is impossible to describe the path of the photon through the interferometer in terms of classical particle physics. For example, if we try to localise the photon in the upper arm of the interferometer at some instant in time, then how on earth does the photon know what the configuration is of the lower arm of the interferometer. If the lower arm is blocked, the photon can travel to either of the two final photon detectors with equal probability. If the lower arm is not blocked and has the same length as the upper arm, one can predict with probability one, in which of the two final photon detectors the photon will be detected. The two arms can in principle be light years apart, suggesting unsurmountable causality problems for any classical "particle" theory of light. Quantum theory describes the state of the photon in terms of superposed states in complex Hilbert Space, and these states are physically interpreted by saying that the photon has both wave and particle properties in physical space. Quantum theorists learn rapidly not to worry about such physical interpretations, and rather to have complete faith in the Hilbert Space description.

The invariant set postulate provides a profoundly new perspective on this age-old conundrum of physical interpretation and wave-particle duality. To understand this perspective, let us assume as above that states of physical reality lie precisely on some potentially large (but fractional) dimension zero volume, non-computable geometry, and consider the intersection of this geometry with various two-dimensional Euclidean subsets of state space. One such two-dimensional space could be associated the position and momentum of some particle A (a conventional phase space), though this is not a subset we shall focus on here. Another two-dimensional space could be associated with the position of particle A and the position of Particle B, where A and B are potentially far apart in physical space. More generally imagine a two dimensional associated with two "degrees of freedom" of the universe which need not be spatially local in any sense.

The Invariant Set Postulate implies that the values of these two "degrees of freedom" which correspond to physical reality are precisely those on the invariant set. If the invariant set intersected this two dimensional space in something approximating a straight line, then the values of these two degrees of freedom would in reality be linearly correlated. More generally, the constraint implied by the invariant geometry imposes some form of functional relationship between the two variables.

A key property of a fractionally-dimensioned set (fractal) is lacunarity or gappiness [8]. For the two dimensional cross sections I want to consider here, I will assume some sort of relationship between measures of lacunarity in one direction and lacunarity in the other: $lacunarity_1 \times lacunarity_2 > k$, where k is some constant which ultimately I would link to Planck's constant. This relationship is reminiscent of the Heisenberg Uncertainty Principle, and implies that if lacunarity is small in one direction, it must be large in the other.

Let us return to the interferometer, consider again the issue which any realist has to face: how does a photon in the upper arm "know" whether the lower arm is blocked or not? Of course using the word "know" is a bit anthropomorphic, so let us rephrase the question differently. Imagine that in the real world the lower arm is not blocked and the upper and lower arm have the same length. In this case it is predictable which detector will detect the photon. Now imagine a hypothetical world, a counterfactual world, in which the properties of that photon in the upper arm were unchanged, but the lower arm was blocked. If that counterfactual world was an allowable state of physical reality, and if for the sake of argument, the lower arm was light years away from the upper arm, then simple causality would suggest that it will remain predictable which detector will detect the photon.

However, we know that in reality when the lower arm is blocked, it is not predictable which detector will detect the photon. How do we interpret this?

The Invariant Set Postulate provides a new perspective on this problem. If the invariant set is fundamentally lacunar in some state space direction X, where, we will suppose, $X < 0$ if the lower arm is blocked and $X > 0$ if the lower arm is not blocked. In the real world, by construction, $X > 0$. We want to consider a counterfactual perturbation which maps $X > 0$ to $X < 0$ keeping other components of the state of the universe (and in particular the components Y which describe the state of the photon considered as a particle in the upper arm) fixed. Given the supposed lacunarity of the invariant set in the X direction, such a counterfactual perturbation will move the state of the universe off the invariant set to a state of physical unreality.

We could look at the problem slightly differently and ask whether there exists a state of the universe which *is* a state of physical reality, and where $X < 0$. We may be able to find such a state on the invariant set, but, by the above discussion, such a state will not have the same Y value as when $X > 0$, ie in such a counterfactual universe, we will not be dealing with the same photon, with the same photon properties Y as before, but with a different photon, with different Y properties, properties that will ensure the photon is not predictably detected by just one detector.

We have in this picture a way to describe how the photon does "know" about the state of the lower arm, without in any way violating the notion of physical causality and hence without violating the bedrock of relativistic covariance. We will pursue this further in the next subsection. For now, let us consider the question: where does the "wave" of "wave-particle duality" fit in to such a picture?

My view is that in the concept of "wave particle duality" the wave is a wave on the invariant geometry. To be a bit more specific, consider a one dimensional fractal and colour points on the fractal (say red and blue) according to which of two possible basins of attraction the points evolve towards. These basins of attraction represent specific measurement outcomes, or in the current situation, photon detectors. The use of such basins of attraction can be thought of a way to represent in a dynamical systems context the notion of gravitationally-induced "objective reduction" discussed by Penrose [5]. At one end of the fractal we can imagine all points coloured red and at the other end all coloured blue. As we move from one end to the other, the ratio of red to blue points will vary sinusoidally. In general in any neighbourhood on the fractal, there exist both red and blue points, finely interleaved with one another. This would represent a fairly generic situation in nonlinear dynamics, where basins of attraction are riddled or intertwined with respect to one another [9, 10]. The proposal is that it is this sinuoidal wave of colour variation on the invariant geometry that is the "wave" of "wave–particle duality" in quantum theory. (Without going into details, it is also possible to generate quaternion structures on the invariant set using permutations on multiple bit strings, and interpreting the bits values in terms of symbolic dynamic strings; [11]).

Although the mathematics of this has to be developed more fully, some basic consequences of this idea can be discussed already. Although the discussion above was about single photons, in principle there is no difference if the photon source for the interferometer were emitting multiple photons, i.e. that we are considering in effect a classical electromagnetic wave. If the wave of a single photon is a wave on the invariant geometry, then so is the classical electromagnetic wave. Perhaps this solves a difficult conundrum which predates quantum theory — the null result of the Michaelson-Morley experiment with respect to the "medium" on which electromagnetic waves were previously supposed to propagate. That there is no "*aether*" in physical space is of course, entirely consistent with special relativity. But this doesn't quite answer the question which bothered nineteenth century physicists — on what, or with respect to what, is the electromagnetic wave propagating? Interestingly, this question doesn't arise when one considers the gravitational waves of general relativity — they are waves on the geometry of space time. The Invariant Set Postulate provides a possible solution to this nineteenth century conundrum — like single photon waves, classical electromagnetic waves are actually representations of waves on the geometry of the invariant set!

The language of fractal invariant sets and riddled basins is the language of nonlinear dynamics. However, the Schrödinger equation is linear, and the Schrödinger equation describes quantum interference perfectly. So is there some inconsistency here? No. If we study the evolution of a probability distribution on the invariant set of a nonlinear dynamical system, it will be described by a Liouville equation, which is precisely linear in the probability density (irrespective of now nonlinear the underlying dynamics are). Dirac many times commented on the formal similarity of the Schrödinger equation and the Liouville equation for a classical Hamiltonian. There are differences of course such as the use of abstract operator algebra in the case of quantum systems and this will be discussed in the next part of this section.

3.2. *Nonlocality without Nonlocality*

This "Wheelerism" of the title was used because I suspect people use the word "nonlocality" to mean different things. At one level there is unquestionably something "nonlocal" about quantum physics, and indeed there is something unquestionably "nonlocal" about the geometric constraints implied by the Invariant Set Postulate above. However, the word "nonlocal" was originally coined as shorthand for "not locally causal" and used to imply that any hidden-variable model postulated as a deterministic underpinning of quantum theory, cannot be locally causal, and hence cannot be consistent with relativity theory. That is, by saying that any putative deterministic reality is nonlocal, means that such reality is not relativistic. In this form, nonlocality remains a profound conundrum. Indeed Roger Penrose has written [6]:

> "My own view is that to understand quantum nonlocality, we shall require a radical new theory. This new theory will not just be a slight modification of quantum mechanics but something as different from standard quantum mechanics as General Relativity is different from Newtonian Gravity."

I agree with this and in this respect want to make a claim that will seem outlandish. *Bell's Theorem does not imply that hidden-variable theories must be nonlocal.* More precisely, I believe the Invariant Set Postulate opens up a massive loophole (if that is the right word) in the Bell Theorem through which deterministic dynamical systems with perfectly locally–causal dynamics can pass. Bell's Theorem does not imply that objective reality is incompatible with the notion of relativistic covariance.

To understand this point, let us assume that the assumed dynamical evolution equations (the explicate order, in our Bohmian language of Section 1) which describe evolution on trajectory segments of the invariant geometry of the universe are perfectly relatvistically invariant. We now consider a standard EPR experiment performed within such an assumed causal universe, and ask whether the Bell Inequalities are necessarily satisfied.

A crucial assumption in deriving the Bell inequalities is the assumption that the orientation of Bob's polariser does not depend on the hidden variable of Alice's photon. Consider this assumption in the light of the Invariant Set Postulate. The argument is actually very similar to the one we used in analysing the Mach Zehnder interferometer. So again consider a two dimensional cut through the invariant geometry, and let X denote the "degree of freedom" associated with the orientation of Bob's polariser, and let Y denote the "degree of freedom" associated with the hidden-variable of Alice's photon which can be assumed to be arbitrarily distant from Bob's polariser. Again, assume large lacunarity in the invariant set in the X direction, so that a counterfactual perturbation in the X direction, which keeps Y unchanged, moves a point, on the invariant set, off it, to a state of the universe of physical unreality.

Conversely, if we consider a perturbation to the X component of the state vector which keeps the state of the universe on the invariant set, then the Y component of the state vector can no longer be unchanged. That is to say, we cannot change the state of Bob's polariser without altering the state of Alice's photon, and the Bell Theorem is nullified! There is clearly nonlocality of a sort here, but crucially, the underpinning dynamics are causal and hence relativistic. The Invariant Set

Postulate allows us to return to the type of objective causal realism that Einstein argued so strongly for, Bell's Theorem notwithstanding.

In order to show that one can actually generate the right quantum statistics from the Invariant Set Postulate requires us to consider again the wave–like structures associated with the riddled–basin structure on the invariant set. The key point is that the invariant geometry implies potential "acausal" constraints between spatially distant parts of the universe. The word "acausal" here does not mean "not causal" but rather "outside causality", or putting it more blatantly, it means "in a sense in which causality is an irrelevance".

One fundamental objection to the argument above could be the following. Surely Bob is free to set his polariser as he likes? It is true that the Invariant Set Postulate is not compatible with the notion that we can treat the orientation of Bob's polariser as a free variable.

Does that mean Bob does not have free will? It may be worth spending a few words exploring this concept further. When I get up in the morning, I have to decide what shirt to wear. I often wear polo shirts, unless I have to go to some important meeting when I will wear a shirt (and if the meeting is especially important a tie too). If I wear a polo shirt, the chances are it will be the shirt on the top of the pile of polo shirts. However, occasionally I will choose a different one to the top one, perhaps because I take a fancy to its colour on that particular day, it's hard to tell. Maybe I took a fancy to that colour because a brightly coloured bird flew past my bedroom window just before I pulled the shirt out that I would wear. Now you could interpret my behaviour as mostly following a set of rules, but occasionally I display my individuality and "do my own thing". Now many would say "doing my own thing" means free will, in the sense that I could have done otherwise. Could I have worn the blue polo shirt, even though I wore the red?

Now a key aspect of the Invariant Set Postulate is that such counterfactual situations (keeping everything fixed except for the polo shirt I wear) are likely to lie off the invariant set and therefore not be states of physical reality. So for me, "doing my own thing" has to mean something different to "I could have done otherwise". My pragmatic interpretation of "free will" is this: I defy anyone to define a finite algorithm that can always predict which shirt I will wear on any given day. An algorithm can be written which might be right most of the time, but not all of the time. If there exists no algorithm to describe my behavious, I would say I have free will for all practical purposes. In this sense Bob can be said to have free will, even though he probably could not have done other than to set the polariser the way he did.

Now we can see the potential relevance of the non-computability of the invariant set. If it was computable, then one could construct a finite algorithm for predicting which polariser setting Bob will use for any given EPR experiment. This really would present a paradox, because according to my definition, if you can predict Bob's behaviour, he doesn't have free will. However, if you tell Bob how he will set his polariser, you can be sure he will do otherwise! The non-computability of the invariant set provides the means to avoid this paradox, whilst at the same time allowing Bob and me to exercise free will in any practical shape or form.

On the other hand, in quantum theory itself, Bob is assumed to be free to choose any orientation for his polariser in the more traditional "could have done otherwise" sense. In my view, this literal interpretation of the notion of free will is the origin of the need for abstract operator theory in quantum mechanics. On the invariant set we can define probabilities in terms of simple frequentism; off the invariant set it seems we can define mathematical objects which have the same algebraic structures as probabilities, but they have no underpinning frequentist interpretations. That suggests (to me) that the structure of quantum theory implies that non-computability of the invariant set is somehow maximal, but I don't wish to expand on this idea at present.

4. Conclusions

What is the relevance of all this for physics? The Invariant Set Postulate provides a possible framework for finding a deterministic causal underpinning for quantum theory including quantum field theory. We have argued that this underpinning describes a causal theory of gravity which extends general relativity by being geometric not only in space time, but also in state space. In this picture, the Schrödinger and Dirac equations would emerge as Liouville equations (necessarily linear) for probability densities on the invariant geometry. The irreversibility of the black-hole no-hair theorem appears to play a key role in generating zero-volume non-computable geometries which are invariant under dynamical evolution operators but not under generic dynamically unconstrained counterfactual perturbations.

If these ideas are right, they fundamentally undermine current approaches to quantum gravity, such as string theory, which are based on applying standard quantisation rules to some model for gravity (such as general relativity or some modification thereof). Rather, I expect these quantization rules to be emergent from this extended theory of gravity, providing a computational albeit probabilistic theory of quantum physics, from an essentially deterministic but non-computational one.

In this essay I have tried to show that there is a place for chaotic dynamics as an underpinning to quantum theory, thereby providing a new approach to reconciling quantum theory and relativity theory. To close the circle, we return briefly to the issue of the non-relativistic invariance of the notion of Lyapunov exponents. Reconciling chaos and relativity is also straightforward if we think (as Cornish did [12]): define chaos in terms of geometric properties of the invariant set (eg its fractional dimension) and the definition is now perfectly relativistic.

References

[1] T. Palmer, *Proc. R. Soc. A* **465**, p. 3165 (2009).
[2] D. Bohm, *Wholeness and the Implicate Order.* (Routledge, 1980).
[3] L. Blum, F. Cucker, M. Shub and S. Smale, *Complexity and Real Computation* 1997.

[4] S. Dube, *Complex Systems* **7**, 423 (1993).
[5] R. Penrose, *The Road to Reality: A Complete Guide to the Laws of the Universe.* (Jonathan Cape, 2004).
[6] R. Penrose, *The Large, the Small and the Human Mind.* (Cambridge University Press, 1997).
[7] R. Penrose, *Cycles of Time: An Extraordinary New View of the Universe.* (Random House, 2010).
[8] B. Mandelbrot, *The Fractal Geometry of Nature.* (W.H. Freeman, 1977).
[9] E. Ott, J. Sommerer, J. Alexander, I. Kan and Y. J.A., *Phys. Rev. Lett.* **71**, 4134 (1993).
[10] T. Palmer, *Proc. R. Soc. A* **451**, 585 (1995).
[11] T. Palmer, *Proc. R. Soc. A* **460**, 1039 (2004).
[12] N. Cornish (1997), gr-qc/9709036.

SELF-REPLICATED WAVE PATTERNS IN NEURAL NETWORKS WITH COMPLEX THRESHOLD EXCITATION

V.I. NEKORKIN

Institute of Applied Physics of RAS,
46 Uljanov Street, 603950 Nizhny Novgorod, Russia
E-mail: vnekorkin@neuron.appl.sci-nnov.ru
www.ipfran.ru

This paper reviews some recent investigations of localized structures in one- and two-dimensional neural-like networks with complex threshold excitability.

Keywords: Pattern formation, Bifurcation, Nonlinear dynamics, Localized patterns

1. Introduction

Formation of stable spatially localized structures (patterns) in nonequilibrium media is one of the striking phenomena of nonlinear science. Such structures consist of one or more regions in one state surrounded by other part in a qualitatively different state. Localized structures have been found experimentally in a wide range of systems from most branches of science. For example, such structures are observed in planar gas-discharge systems with high ohmic barrier [1–3], granular materials [4] and optical systems [5, 6], in modified Belousov-Zhabotinsky reactions [7–11], in catalytic reduction of NO with hydrogen on a Rh(110) surface [13] and oxidation of CO on Pt(110) [12]. Localized structures are believed to play key role in information processing in the brain and, for example, are correlated with working memory tasks [14], orientation or feature selectivity in visual system [15], activity in the head-directional [16], somatosensory [17] and movement coordination [18] systems. Some neurological disorders, such as epilepsy and migraine, are also characterized by space localized activity [19].

In this review we present some recent [20, 21] investigations resulting from the modeling of localized structures in neural networks.

We consider the following network composed of identical neural-like units or cells coupled locally in a nearest-neighbor fashion:

$$\begin{cases} \dot{u}_{j,k} = f(u_{j,k}) - v_{j,k} + \dots \\ \qquad \dots + d(u_{j-1,k} + u_{j+1,k} - 4u_{j,k} + u_{j,k-1} + u_{j,k+1}), \\ \dot{v}_{j,k} = \varepsilon(\, g(u_{j,k}) - v_{j,k} - b), \end{cases} \tag{1}$$
$$j = 1, 2, \dots, N, \quad k = 1, 2, \dots, M,$$

where

$$f(u) = u - u^3/3, \quad g(u) = \begin{cases} \alpha u, u < 0 \\ \beta u, u \geq 0 \end{cases} \tag{2}$$

The variable $u_{j,k}$ describes the evolution of the membrane potential of the neuron located at spatial site (j, k) and the variable $v_{j,k}$ mimics the dynamics of outward ionic transmembrane currents which are responsible for restoring the membrane resting potential [22]. The parameter $\varepsilon\,(\varepsilon > 0)$ controls the rate of changing ionic transmembrane currents and the parameters α and β $(\alpha, \beta > 0)$ define the nonlinear properties of these currents. The parameter b controls depolarization level of membrane of the neuron and, hence, defines its dynamical mode. The parameter d characterizes the strength of inter-neuron coupling. We assume that units are coupled electrically. The action of such coupling is based on the existence of so-called gap junction between neighboring neurons [23]. Such junction leads to direct connection of the cytoplasm of the neurons and allows various molecules and ions to pass freely between them. Hereupon the current passing through the gap junction obeys Ohm's law and depends directly on gap conduction (or inversely on gap resistance) and directly on difference in electric potential between neurons. Thus the electrical coupling is well-approximated by linear resistive coupling. Therefore in (1) the coupling terms which, in fact, account for ingoing currents from neighboring neurons are expressed in the form of standard two-dimensional discrete Laplacian.

We apply the periodic boundary conditions to lower and upper boundaries of the system $(u_{j,k+M}(t) \equiv u_{j,k}(t))$. For left and right boundaries we use either periodic

$$u_{j+N,k}(t) \equiv u_{j,k}(t) \tag{3}$$

or zero-flux (von Neumann) boundary conditions

$$u_{0,k}(t) \equiv u_{1,k}(t), u_{N+1,k}(t) \equiv u_{N,k}(t)\ . \tag{4}$$

The dynamics of the local unit is described by the FitzHugh-Nagumo system with complex threshold excitation. This model provides a wealth of dynamical behaviors and can describe different modes of excitability including multi-spike excitation [24, 25].

Different dynamical modes of the local unit are illustrated in its phase plane (Fig. 1) for corresponding intervals of the control parameter, $\varepsilon > 0$.

The main feature of the regimes are simultaneous existence of several excitation thresholds.

For example, the regimes depicted on Fig. 1a and b have two thresholds. The first threshold is defined by the unstable limit cycle l^u which, in fact, determines the basin of attraction of O_1. The overcoming of this threshold leads to fast transition of the neuron to another stationary state O_3. The second threshold is defined by the unstable limit cycle L^u in the case of Fig. 1b or by the stable separatrices w_1^s and w_2^s of the saddle point O_2 in the case of Fig. 1a. This threshold separates the basins of attractions of stable point O_3 and the stable limit cycle L^s. Thus the overcoming

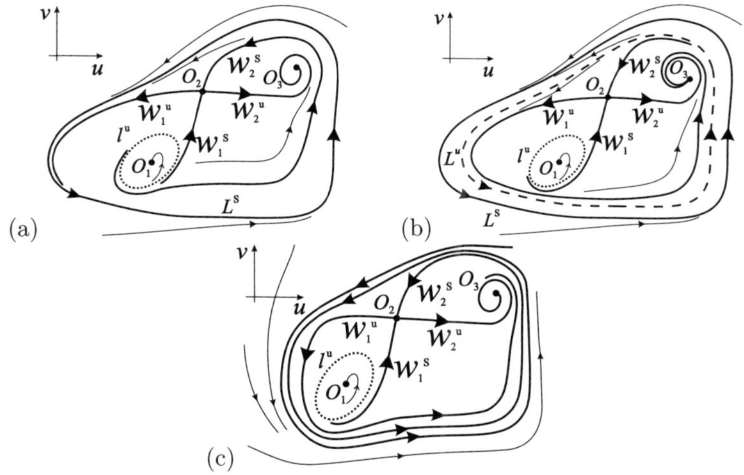

Fig. 1: Phase planes of different regimes of local unit of the system (1) possessing bistability (multistability) properties and oscillatory excitation thresholds. Parameter values: $\alpha = 0.9$, $\beta = 0.8$, $b = -0.025$ and (a) $\varepsilon_1 < \varepsilon < \varepsilon_2$, (b) $\varepsilon_2 < \varepsilon < \varepsilon_3$, (c) $\varepsilon_3 < \varepsilon < \varepsilon_4$, where $\varepsilon_1 = 0.5837$, $\varepsilon_2 = 0.6318$, $\varepsilon_3 = 0.6332$, $\varepsilon_4 = 0.6606$.

of this threshold leads to the regime which is characterized by periodic activity of the neuron.

Consider the regime shown on Fig. 1c. It has more complicated threshold properties. Like the previous regimes, there is one threshold defined by the cycle l^u but, due to oscillatory behaviour of incoming separatrix W_2^s (see Fig. 1c), there are additional thresholds which number grows with decreasing of the parameter ε and approaches to infinity for $\varepsilon = \varepsilon_3$. Indeed, the separatrix makes a number of rotations around the three fixed points before going to infinity. In this case the different spirals of the separatrix correspond to the different thresholds. The overcoming of any of such thresholds leads to a generation by the neuron of definite number of response pulses before going to the O_3.

Note that bistability is typical property of many real neurons such as thalamocortical cat neurons [26].

2. Self-replicated Solitary Waves, Bound States and Complex Wave Patterns in a One-dimensional Neural Network

Let us consider the system (1) with $M = 1$ and the boundary conditions (4). In paper [20] we develop an analytical approach stating that complex wave dynamics can be associated with the appearance of co-dimension 2 heteroclinic contours where a number of solitary waves (fronts, pulses, pulse trains) may co-exist simultaneously. A particular type of these waves can be realized from definite initial perturbation or as the result of wave collisions. Such contours can be treated as indicators of the emergence of complex wave patterns in reaction-diffusion systems.

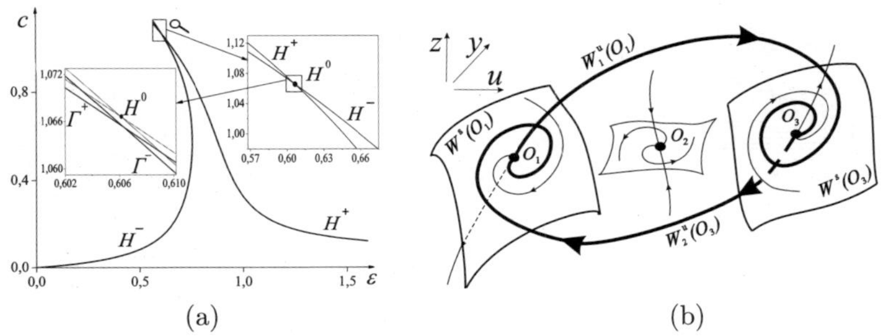

Fig. 2: (a) Wave velocity c as a function of the parameters - bifurcation sets H^+, H^- and Γ^+, Γ^- in neighborhood of the point $H^0(\varepsilon^0 = 0,60593, c^0 = 1,0664)$; (b) heteroclinic contour, corresponding to H^0 ($\alpha = 0,9$; $\beta = 0,8$; $I = 0,024$).

It was shown that there exists the bifurcation set H^+ corresponding to a heteroclinic orbit formed by separatrix $W_1^u(O_1)$ and manifold $W^s(O_3)$ in the phase space of system for traveling waves (Fig. 2). Similarly, there exists the bifurcation set H^- corresponding to a heteroclinic orbit formed by separatrix $W_2^u(O_3)$ and manifold $W^s(O_1)$.

In the plane (ε, c) the curves H^+ and H^- have the intersection at some point $H^0(\varepsilon^0, c^0)$. This point, H^0, ensures the existence of heteroclinic contour C composed of two heteroclinic orbits formed by separatrices $W_1^u(O_1)$ and $W_2^u(O_3)$, $W^s(O_3)$ and $W^s(O_1)$ (Fig. 2 (b)). It has been also found that in the neighborhood of H^0 there exist another bifurcation curves. Let us show that for some parameters in phase space of system for traveling waves the appearance of a homoclinic orbits (separatrix loops) formed by manifolds of fixed points O_1 and O_3 is possible.

Due to the existence of the heteroclinic contour for values of parameters taken in the neighborhood of H^0, the system for traveling waves exhibits extremely complex dynamics. Furthermore one can expect complex wave dynamics of original distributed system for traveling waves. The set of homo- and heteroclinic orbits provides a wealth of traveling wave profiles that can propagate with approximately the same velocities. In numerical simulations we examine the stability of such waves relative to step-like and pulse-like initial perturbations. The wave fronts of simple shapes (kink and anti-kink solutions) have shown to be stable. They are realized from step-like perturbations for the same parameter values. Note, that such a heteroclinic contour provides the possibility of a "composite" pulse-like solutions (bound states) composed of two fronts traveling together. In terms of system for traveling waves they corresponds to homoclinic orbits that make a number of rotations near the opposite saddle point (O_1 or O_3) before returning to the original

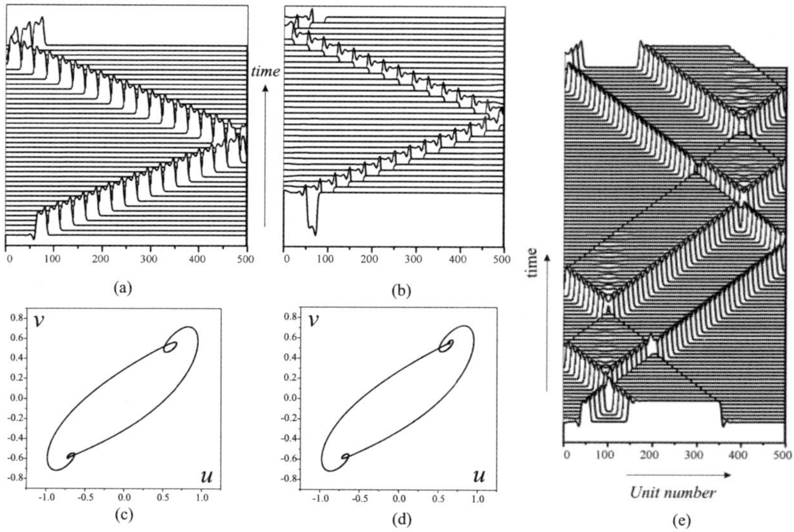

Fig. 3: (a), (b) Space-time plot (u-variable) of two-hump bound states (composite solitary pulses) propagating along the homogeneous states O_1 and O_2, respectively, in system (1); (c), (d) plots of homoclinic orbits in projection to the (u, v)-plane corresponding to the two-hump excitations shown in (a) and (b) pictures, respectively; (e) wave pattern formed by two-hump pulse, kink and anti-kink self-replicated in collisions. Parameter values: $\varepsilon = 0.585, \alpha = 0.9$; $\beta = 0.8$; $I = 0.024$; $d = 1$.

state. Furthermore, the profiles of such wave have a number of humps with decreasing amplitude at the peak values. Figure 3(a), (b) illustrates the evolution of two-hump solutions formed relative to the stable homogeneous states O_1 and O_3, respectively. The noticable feature of such pulses is that they can recover its shape after collisions with boundaries (Fig. 3(a),(b)) and in head-on collisions with another solitary waves (Fig. 3(e)). The shapes of corresponding homoclinic orbits in projection of phase space of system for traveling waves to the $(u(\xi), v(\xi))$-plane are shown in Fig. 3(c), (d). Indeed, they are very close to the heteroclinic contour. Figure 3(e) shows the example of co-existence in the system of three waves comprising the two-hump pulse, kink and anti-kink that can cross each other self-replicated in collisions. The number of such co-existed solitary waves is defined by the initial perturbations and by system "volume" (the number of cells and the value of the diffusion coefficient).

3. Polymorphic and Regular Localized Activity Patterns in a Two-dimensional Neural Network

It was shown in [21] that, under certain conditions, the system of Eqs. (1), (2) supports formation of two sorts of stable moving localized structures even without any extra stabilizations. These are regular asymmetric and polymorphic structures. The regular structures preserve their shape and velocity under propagation while the shape and velocity of polymorphic ones show complex (even chaotic) behaviour. The polymorphic structures are novel and have not yet been reported in the literature. Also the regular structures have not been observed before in typical homogeneous two-component reaction diffusion systems having no any specific stabilizing actions. Indeed, the regular structures have been observed only in multi-component systems and two-component systems with complex enough organization including either different types of feedback loops or grid-like inhomogeneity or strong anisotropy of diffusion. Note that shape of regular structures found in our system is very similar to shape of stabilized regular structures obtained in aforementioned two-component reaction-diffusion systems. However, mechanisms of their formation are completely different. On the other hand, regular structures, which are observed in multi-component systems, are not asymmetric but have radial or almost radial symmetry.

3.1. *Regular localized patterns*

In this case for numerical investigations we apply the boundary conditions (3) and initial conditions in the form of fragment of plane wave (plane wave segment). To form wave segment, for definiteness, we use a plane wave which propagates in j-direction. The width of the segment can be chosen arbitrary by appropriate cutoff of the wave. Let P_W be the segment with initial width W. Note that, in addition to excitatory region, such segment includes also refractory region. Due to the presence of refractory region the excitation propagates only in one predefined direction which in our case coincides with j-direction. Notice that the localized structures considered in this paper can be also formed from another initial conditions. For example, one may use the initial conditions in the form of localized excitations which have rectangular section.

The study of evolution of segments P_W with different initial widths and for different values of parameters ε and d has shown that the segments can demonstrate three possible types of spatially-temporal behaviours (Fig. 4). For any initial width of the segment and fixed d there are two critical values of parameter ε. Below the first critical value, wave segment grows until it reaches the boundary of the system (Fig. 4a). Once pinned at the boundaries, the wave takes on the features of a plane wave, which then propagates indefinitely in the system (1). Above the second critical value of ε the segment contracts and eventually disappears (Fig. 4b) setting the spatially-homogeneous steady state in the system (1). More interesting behaviour takes place for values of ε lying between the two critical ones. The evolu-

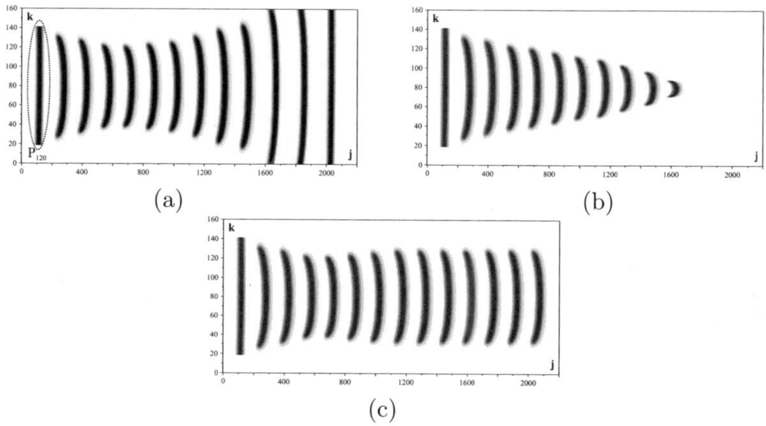

Fig. 4: Spatially-temporal dynamics of wave segment P_{120} for different values of ε and $d = 1$: (a) - evolution to plane wave ($\varepsilon = 0.64851$); (b) - decaying to spatially-homogeneous state S_1 ($\varepsilon = 0.64852$); (c) - formation of regular localized structure ($\varepsilon = 0.648515$). The simulations is carried out in a lattice of $2200 * 160$ units. Each image here represents an overlay of snapshots taken every 155 time units.

tion of wave segment in this case is shown in Fig. 4c. After short transient process the form of the segment and its velocity become stable and do not vary anymore.

We have found that, by changing the width of the initial wave segment and appropriate choice of parameters ε and d, one can obtain stable moving segments of any predefined width. Next we concentrate our attention on such stable moving (or wave) segments and refer to ones as regular localized moving structures. We denote such structures by C_W where W stand for the width of the structures (see below on how W is precisely defined). The typical snapshot of localized regular structure is shown on Fig. 5a. One can see that such structure has asymmetric crescent-like shape. That is, it has curvature which strongly depends on spatial coordinates. Note that regular structures obtained in two-component reaction-diffusion systems with different stabilizing actions [10, 11, 27–29] has similar asymmetric crescent-like shape. On the other hand, regular structures obtained in multi-component reaction-diffusion systems [30–32] are not asymmetric. They have radial or almost radial form. That is, their curvature is almost constant.

Figure 5b depicts amplitude profile $A(k)$ of the structure from Fig. 5a computed along its crest line. One can see that the amplitude reaches maximum (A_{max}) in the center of the structure and decays sharply enough on its edges. So we can simply introduce the width W of the structures to characterize their spatial properties. The W is calculated on 10% level from the maximum amplitude value A_{max}.

We have studied the regions of existence of regular structures on (ε, d)-parameter plane. For example, Fig. 6 shows the regions of existence for the structures C_{22}, C_{86}

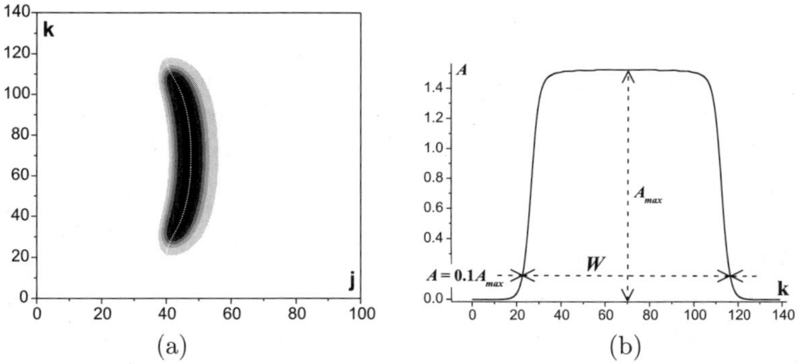

Fig. 5: (a) Snapshot of single two-dimensional spatially-localized regular structure; (b) amplitude profile $A(k)$ of the structure computed along its crest line. The lattice consist of $100 * 140$ units. Parameter values: $\varepsilon = 0.648515, d = 1$.

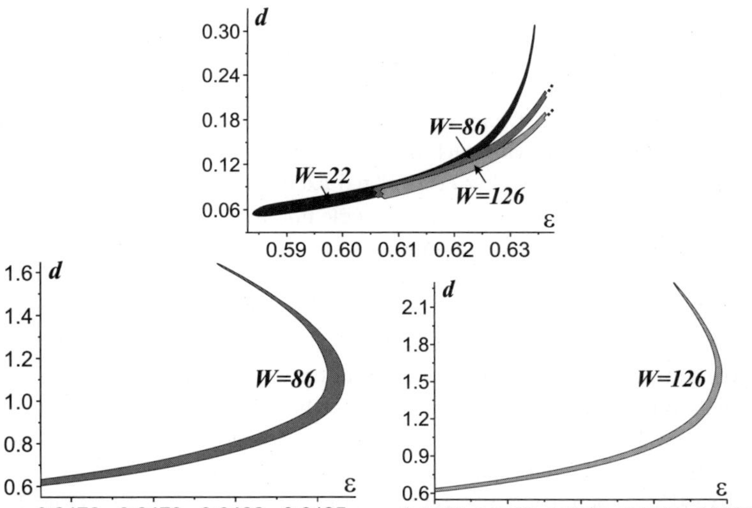

Fig. 6: Regions on (ε, d) - plane corresponding to the existence of regular localized structures C_{22}, C_{86} and C_{126}. Bottom figures are in continuation of top figure along ε axis.

and C_{126}.

We have found that for each set of parameters α, β and b, there is a critical regular structure which can exist in the system (1). All the structures with width less then the width of critical one are unstable. For the parameters chosen in this paper the critical structure is C_{22}. Its region of existence on parameter ε is located on the left of the regions of existence of other regular structures (see Fig. 6). Then the regions of existence follows each other when increasing the width of the structures. In the limit (for large width structures) they border the boundary of existence of plane waves. We computed the velocity of the structures and discovered that it increases with the increase of the width of the structures and, for large structures, it also approaches to the velocity of plane wave.

3.2. *Polymorphic localized patterns*

We have found that, in addition to regular structures, the system (1) supports formation of another type of localized structures. The shape and velocity as well as other integral characteristics of such structures show more complex (even chaotic) behaviour. Therefore we refer to them as polymorphic structures. Note that such type of localized structures have not been observed before neither in two-component nor in multi-component reaction-diffusion systems.

Here, we concentrate our attention on single polymorphic structures. Such structures consist of one growing propagating localized excitation which has teardrop-shaped asymmetric form (see, for example, Fig. 7a). As opposed to regular structures, existence and properties of polymorphic structures depend on type of the boundaries used along propagation direction of localized excitations. The single polymorphic structures exist only in the case of von Neumann boundaries. We will show later that interaction with such boundaries provides periodic neutralization of the effects caused by the growing of localized excitation.

Here, for definiteness, we consider the single polymorphic structures which propagates in j-direction. Numerical simulations are carried out on a lattice of $35 * 140$ units with the boundary conditions given in eq. (4).

Figure 7a and b show snapshots of a typical single polymorphic structure taken at its different propagation phases. In particular, Fig. 7a corresponds to propagation of the structure to the right after its interaction with left boundary while Fig. 7b corresponds to propagation of the structure to the left after its interaction with the right boundary. To characterize the spatial properties of the structure, we introduce characteristic width W and height H_k (see Fig. 7c and d). Here $W = \max W_j, j = \overline{1..N}$, where W_j for definition are computed on the level $u_{j,k} = -0.1$ (Fig. 7c). The height H_k is the greatest value of the variable $u_{j,k}$ by j computed relatively to level corresponding to spatially-homogeneous steady state (Fig. 7d).

The study has indicated that there exist three different types of single polymorphic structures. These are periodic, quasiperiodic and chaotic ones (see Fig. 8, Fig. 9, Fig. 10 respectively).

4. Conclusion

The spatio-temporal dynamics a network system modeling collective behavior of electrically coupled neurons with complex threshold excitability have been investigated.

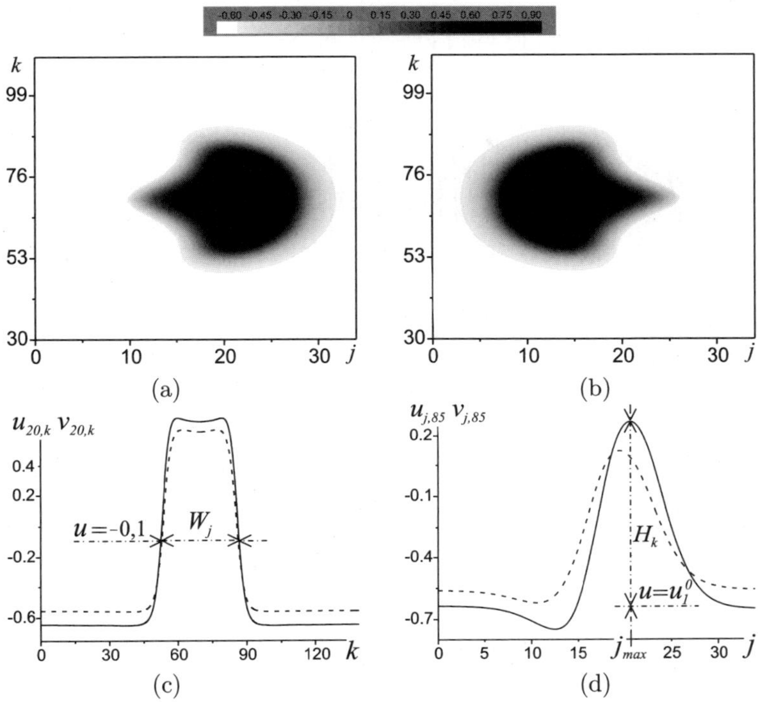

Fig. 7: (a),(b) Two snapshots of typical single polymorphic structure corresponding to its different propagation phases; (c),(d) typical distribution of variables $u_{j,k}$ (solid line) and $v_{j,k}$ (dashed line) corresponding accordingly to vertical ($j = 20$) and horizontal ($k = 85$) "slices" of the structure.

In one-dimensional case it has been found that the activity of the network essentially depends on the value of parameter ε controlling the rate of changing of ionic currents passing through the neuron membrane (v-variable). For sufficiently slow or rapid changing of v the system behaves similarly to reaction-diffusion models with autowave dynamics. In this case wave fronts (kinks and anti-kinks) are the key wave processes. However, we have found that for some conditions the wave dynamics become highly non-trivial. In particular, wave fronts and solitary bound states display particle-like behavior in collisions what may lead to the appearance of complex self-reproducale spatio-temporal patterns. It has been found that the complexity can be associated with the existence of heteroclinic contour in the phase space of three-dimensional nonlinear system describing the profiles of traveling waves. This heteroclinic contour is formed by manifolds of two saddle-foci with unstable leading direction.

In the case of two-dimensional network it has been found that formation of two types of stable two-dimensional spatially localized moving (or wave) structures are

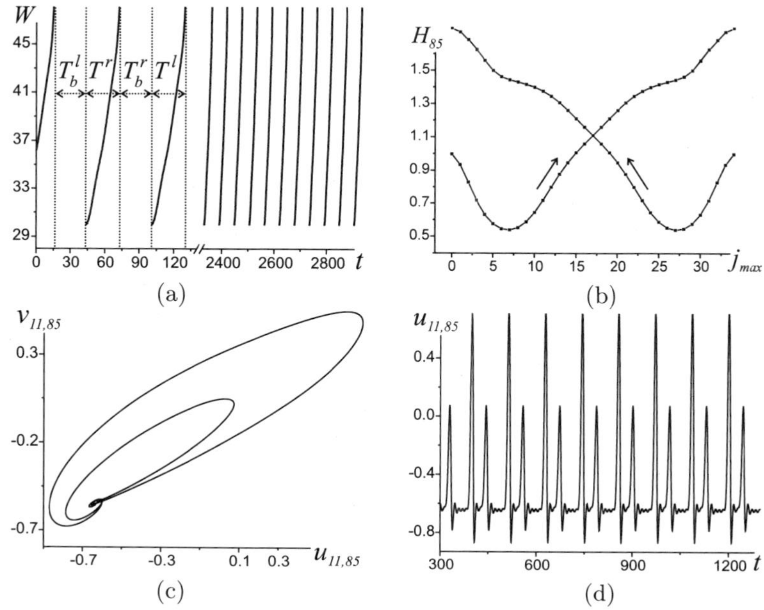

Fig. 8: Spatially-temporal characteristics of periodic polymorphic structure: (a) - time evolution of width W; (b) - spatial evolution of height H_k in section $k = 85$; (c) - projection of phase trajectory of the system (1) corresponding to the structure on local phase plane of the unit located in the spatial node ($j = 11, k = 85$) and (d) - time evolution of its $u_{j,k}$ - variable. Parameter values: $\varepsilon = 0.6262, d = 1$.

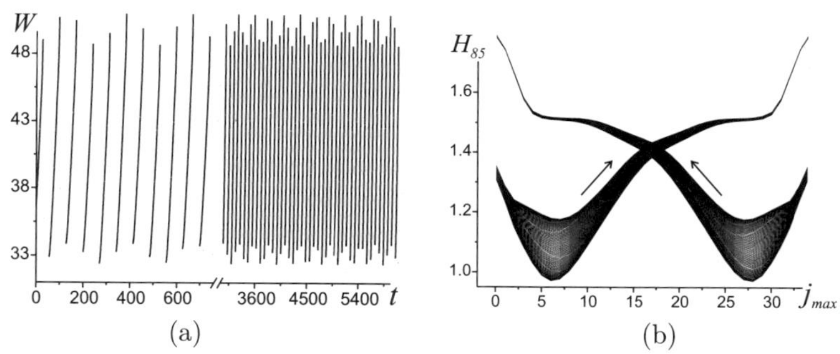

Fig. 9: Spatially-temporal characteristics of quasiperiodic polymorphic structure: (a) - time evolution of width W; (b) - spatial evolution of height H_k in section $k = 85$. Parameter values: $\varepsilon = 0.627519, d = 0.5397$.

possible without any external stabilizing actions. These are regular and polymorphic structures. The regular structures preserve their shape and velocity under propaga-

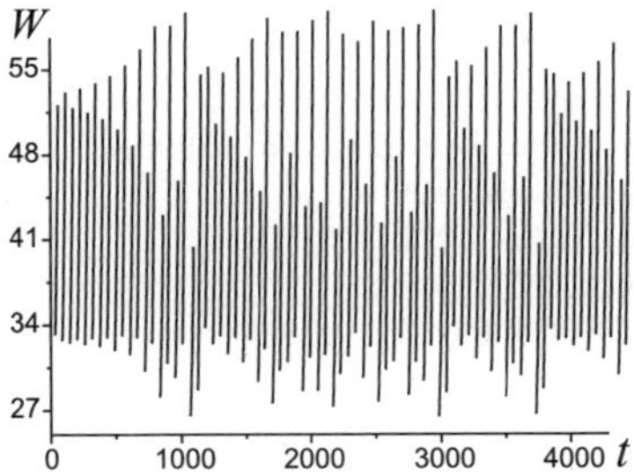

(a)

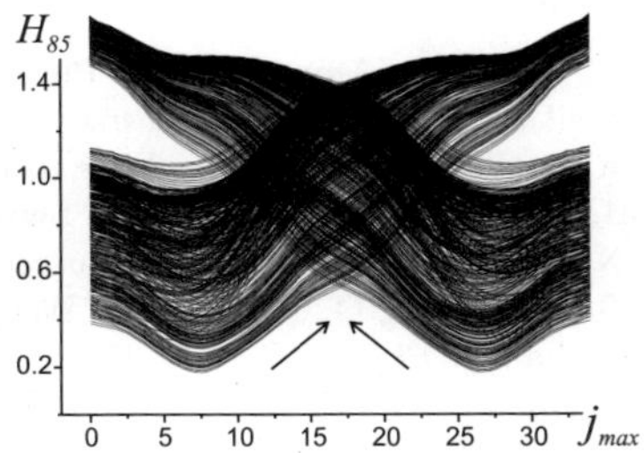

(b)

Fig. 10: Spatially-temporal characteristics of chaotic polymorphic structure: (a) - time evolution of width W; (b) - spatial evolution of height H_k in section $k = 85$. Parameter values: $\varepsilon = 0.6241, d = 1$.

tion. In contrast with regular structures, integral characteristics of polymorphic ones such as width, height, profile and velocity change in time periodically, quasiperiodically and even chaotically. Note here that polymorphic structures are novel and have not yet been reported in the literature. Also, regular structures have not been observed before in typical homogeneous two-component reaction-diffusion systems which have not any specific stabilizing actions. In addition, the study of regions of existence of regular structures have indicated that such structures possess high multistability which is not typical for regular localized structures obtained in other (even multi-component) reaction-diffusion systems. Due to such multistability, the structures along with typically observed interaction scenarios (annihilation, elastic particle-like behaviour) show inelastic particle-like behaviours. The mechanisms of formation of regular and polymorphic structures is discussed. In particular, it has been shown that polymorphic structures are formed due to the balance between the processes of their "dissipation" under interactions and the processes of their growing because of inherent instability. The correspondence between the structures and trajectories in multidimensional phase space associated with the system (1) is given. Bifurcation mechanisms leading to loss of stability of regular structures as well as to transition from one type of polymorphic structures to another are indicated. In particular, we have shown that regular structures correspond to a stable limit cycle. They lose stability when the limit cycle undergoes tangential bifurcation. We established that periodic polymorphic structures correspond also to a stable limit cycle. On the other hand, quasi-periodic structures emerge in the result of a Neimark-Sacker bifurcation and correspond to two-dimensional torus. The chaotic structures appear in the result of period doubling bifurcations cascade and correspond to Feigenbaum attractor.

This work has been supported by the Russian Foundation for Basic Research (Grants 09–02–91061, 09–02–00719, 10–02–00643) and the Ministry of Education and Science (Governmental contracts P942, 02.740.11.5188 and 14.740.11.0348).

References

[1] Y.A. Astrov, E. Ammelt, H. G. Purwins, *Phys. Rev. Lett.* **78**, 3129 (1997).
[2] I. Muller, E. Ammelt, H. G. Purwins, *Phys. Rev. Lett.* **82**, 3428 (1999).
[3] Y.A. Astrov, H.G. Purwins, *Phys. Lett. A* **283**, 349 (2001).
[4] P.B. Umbanhowar, F. Melo, H.L. Swinney, *Nature* **382**, 793 (1996).
[5] N.N. Rozanov, *Optics and Spectroscopy* **102**, 255 (2007).
[6] V.B. Taranenko, G. Slekys, C.O. Weiss, *CHAOS* **13**, 777 (2003).
[7] V.K. Vanag, *Physics-Uspekhi* **47**, 923 (2004).
[8] A. Kaminaga, V.K. Vanag, I.R. Epstein, *J. Chem. Phys.* **122**, 174706 (2005).
[9] V.K. Vanag, I.R. Epstein, *Phys. Rev. Lett.* **92**, 128301 (2004).
[10] Aaron J. Steele, M. Tinsley, K. Showalter, *CHAOS* **18**, 026108.1 (2008).
[11] E. Mihaliuk, T. Sakurai, F. Chirila, K. Showalter, *Faraday Discuss* **120**, 283 (2001).
[12] H.H. Rotermund, S. Jakubith, A. von Oertzen, G. Ertl, *Phys. Rev. Lett.* **66**, 3083 (1991).
[13] F. Mertens, N. Gottschalk, M.Bar, M. Eiswirth, A. Mikhailov, R. Imbihl, *Phys. Rev. E* **51**, R5193 (1995).
[14] X.-J. Wang, *Trends Neurosci.* **24**, 455 (2001).
[15] R. Ben-Yishai, R. L. Bar-Or, H. Sompolinsky, *Proc. Natl. Acad. Sci. USA* **92**, 3844 (1995).
[16] Jeffrey S. Taube, Joshua P. Bassett, *Cerebral Cortex* **13**, 1162 (2003).
[17] J-Y Wu, Li Guan, Yang Tsau, *J. Neurosci.* **19**, 5005 (1999).
[18] E. Leznik, V. Makarenko, R. Llinas, *J. Neurosci.* **20**, 2804 (2002).
[19] P. Jung, J. Milton, *Epilepsy as a dynamical disease* (Springer, New York, 2003).
[20] V.I. Nekorkin, D.S. Shapin, A.S. Dmitrichev, V.B. Kazantsev, S. Binczak and J.M. Bilbault, *Phisyca D* **237**, 2463 (2008).
[21] V.I. Nekorkin, A.S. Dmitrichev, S. Binczak and J.M. Bilbault, *Physica D* **239**, 972 (2010).
[22] A. Scott, *Neuroscience: A mathematical premier* (Springer-Verlag, Berlin, 2002).
[23] E.R. Kandel, J.H. Schwartz, T.M. Jessell, *Principles of neural science* (Prentice-Hall Intern. Inc., New Jersey, 1991).
[24] V.B. Kazantsev, *Phys. Rev. E* **64**, 056210 (2001).
[25] V.I. Nekorkin, A.S. Dmitrichev, D.S. Shapin and V.B. Kazantsev, *Mathematical Modeling* **17**, 75 (2005) (in Russian).
[26] S.W. Hughes et al., *J. Physiol.* **517**, 805 (1999).
[27] V.S. Zykov, K. Showalter, *Phys. Rev. Lett.* **94**, 068302 (2005).
[28] A. Kothe, V.S. Zykov, H. Engel, *Phys. Rev. Lett.* **103**, 154102 (2009).
[29] Irene Sendina-Nadal et al., *Physical Review Letters* **86**, 1646 (2001).
[30] M. Bode, A.W. Liehr, C.P. Schenk, H.G. Purwins, *Physica D* **161**, 45 (2002).
[31] Y. Nishiura, *CHAOS* **15**, 047509 (2005).
[32] A.N. Zaikin, *Fizicheskaya Mysl' Rossii* **1**, 54 (1995) (in russian).

A LOCAL EXPLICATION OF CAUSATION

GIOVANNI BONIOLO*

IFOM, Firc Institute of Molecular Oncology, via Adamello 16 - 20139 Milano (Italy), Dipartimento di Medicina, Chirurgia ed Odontoiatra, Università di Milano, via di Rudini, 20122 Milano (Italy)

* *E-mail: giovanni.boniolo@ifom-ieo-campus.it*

ROSSELLA FARALDO†

Department of Physics, University of Padova, via Marzolo 8 - 35100 Padova (Italy), INFN Padova Section

† *E-mail: faraldo@pd.infn.it*

ANTONIO SAGGION‡

Department of Physics, University of Padova, via Marzolo 8 - 35100 Padova (Italy), INFN Padova Section

‡ *E-mail: antonio.saggion@pd.infn.it*

In this paper, we propose a Generalized Explication of Causality (GEC) with an approach based both on an integral and on the local-causes point of view.

The language of Newton based on dynamical point of view of fluents and fluxions is the basic formalism (borrowed by analogy from Newton) for visualisation and quantitative description of the local-causes account.

We use the formalism of continuous media both from the Lagrangian and the Eulerian (local) point of view. The role of Observer in the definition of casual connection will be emphasized. The definition and the role of Extensive quantities in the GEC will be discussed.

1. Introduction

Causation is concept very widely discussed by philosophers, epistemologists, metaphysicians and by scientists in various different fields. In a recent paper by Boniolo, Faraldo and Saggion [1] (hereinafter BFS) we entered this field of investigation by trying to improve our understanding of what we mean when we speak of causation and casual interactions. In particular, we will offer an explication *à la* Carnap[a]. Note that Carnap [3] discussed two kinds of explication. Both of them are obtained by tuning a somewhat imprecise and ambiguous term (the *explicandum*) into a precise and unambiguous term (the *explicatum*). In the first case, the explication is

[a]We are well aware of the objections of Bontly [2] to the explication process, but this is not right place to discuss them as deeply as they would merit.

reached by inserting the *explicatum* into a well-constructed system of logical mathematical concepts, i.e. by adopting a logical-mathematical language. Whereas in the second case, the explication is obtained by inserting the *explicatum* into a system of empirical concepts, i.e. by adopting an empirical language pertaining to empirical sciences such as physics, biology and other fields. It is this second kind of Carnapian explication that we will apply to the notion of 'causation'.

The starting point in BFS was the proposal by Dowe [4, 5] of a "conserved quantity theory of causation". Dowe put together the best part of the "transference theory of causation" by Aronson [6] and Fair [7] and the "mark transmission theory of causation" by Salmon [8, 9]. In BFS we provide a formal structure and a generalization of Dowe's proposal introducting a *generalized explication of 'causation'* (GEC), as we wish to call it. In our GEC the key role for understanding causation is played by extensive quantities both conserved and non conserved.

In this paper we want to go deeper into the main features of the strictly scientific side of that proposal and to stand out some implications connected with the description of reality in terms of dynamics of continuous media. In particular the local point of view of the Observer is emphasized. In order to do this we shall adopt conceptual framework, and the consequent language, borrowed from Physics but this does not mean that the use of a physical language will limit the resulting empirical explication only to the field of Physics. Second, and more importantly, the empirical explication of a notion has no direct implications for the reality referred to in that notion.

Note that, in what follows, we shall refer to *causation as a relation* between different Systems. In this way we include the very large variety of equilibrium and non-equiliberium stationary situations (and hence in an apparent absence of events) in which a casual connection has to be recognized.

The acknowledgment of a casual connection between two Systems is equivalent to say that two Systems are in interaction. For this reason it seems reasonable to look at Physics as the best starting point to derive a more general account.

Before beginning, it is worth adding two further remarks, which are essential in order to better understand what we are going to propose. The first concerns the kind of empirical language we will adopt. It is the typical language used to describe the continuous variations of quantities, for instance as that adopted in fluid-dynamics but which can be found in all those physical domains where we have to speak in terms of continuity equations. Terms like "flux", "density", "exchange" commonly used in such specific fields of Physics will be used metaphorically when dealing abstract extensive quantities.

In all physical domains like e.g. fluid-dynamics, thermodynamics, theory of elasticity in which the formalism of continuous media is adopted, the apparent antinomy between the perspective of continua and that of discrete structure derived from the atomic theory of matter is usually overcome by the assumption that in performing the limit for δV becoming "sufficiently" small (δV being a small volume around some point P under consideration), the number of elementary constituents within

δV still remains very large.

This would exclude the domain of quantum mechanics in which this assumption is conceptually contradictory and the result is a total separation between the atomic domain and the description of continuous media the latter being confined only to the so-called macroscopic systems.

We want to stress that the interpretation in terms of field dynamics (or, in other words, dynamics of continua) constitutes a perspective which is applicable at the most general level no-matter if we occasionally go, for reasons of opportunity, to a particle description[b].

The second remark regards the conceptual framework in which we will move. Our conceptual background is *not* based on the idea that we have a casual interaction whenever we have an intersection of the world-lines of two bodies with an exchange of conserved quantities as in Salmon-Dowe's account. The acknowledgment of a causal connection between system does not require any a-priori spatial-temporal relation as shown in the discussion and by the examples below.

2. Interactions in Mechanical Systems

The term mechanical interactions is frequently used to indicate a particular type of interaction between systems in which each system exerts (or applies) a force on the other system. We have invented many different denominations for many different kind of forces: contact forces, forces mediated by specific material media, actions at a distance or (when we like to stay within a relativistic context) forces mediated by fields and so on.

The *name* "force" was introduced by Newton as a mathematical concept to represent mathematically the fluxion of the fluent named "quantity of movement" but it is the latter quantity (and it is an extensive quantity) which plays the central

---

[b]One may guess an analogy between a description based on flux-dynamics and the Newtonian formalism developed in "De methodis serierum et fluxionum" (see 10).
Newton defines an infinitesimal quantity named "moment": "*an infinitesimal increment acquired by a finite quantità which varies in time*".
According to this concept, the objects to which Newton's algorithm is applied are quantities which 'flow' in time. For instance the motion of a point generates a line, the motion of line generates a surface. The quantities generated by 'flow' are 'fluents'. Their instantaneous speeds are called 'fluxions'. The 'moments' of fluent are 'the infinitely small additions by which those quantities increase during each infinitely small interval of time' ("Addimenta infinite parva quibus illae quantitates per singular temporis infinite parva intervalla augentur" Mathematicak Papers3.80-1. Traslation by Witheside). Therefore, consider a point which flows with variable speed along a straight line Nweton further observed that moments of the fluent quantities 'are as their speeds of flow' (i.e. as the fluxions. "sunt ut fluendi celeritates". Mathematical Papers,3:78-9). His reasoning is based on the idea that during an 'infinitely small period of time' the fluxion remains constant, therefore the moment is proportional to the fluxion.
In this passage we see (clearly we think) how deeply Newton's process of construction of the mathematical algorithm was anchored to a geometrical thinking. Newton's idea of the flowing of points or of lines for the generation of multidimensional extensions recalls the process of substance generation in a geometrical description. From here we may speak of Substance Like Quantities (SLQ) to mean an Extensive Quantity in our dynamical picture [11].

role in his Principia [12]. When we do not need to differentiate the specific kinds of interaction taking place between the systems, we correctly agree to call all these forces as "Newtonian forces or mechanical interactions".

All kinds of forces have one feature in common: they designate the change of momentum per unit time. Hence we may say that in the Newtonian theory the application of a force by system P^T on system P'^T is a way to say that P^T is transferring momentum at a certain rate to P'^T. Exactly the same view has been maintained by Einstein in the formulation of relativistic mechanics. Newton's third law is equivalent to say that momentum transferred by P^T to P'^T is opposite to that transferred from P'^T to P^T and this is precisely what is expressed in our integral (and local) explication of causation. The adoption of this prespective for mechanical forces is the basis of our GEC which can be generalized to all phenomena in the way we are proposing. The possibility of regarding to mechanical interactions in terms of momentum currents between systems is very appealing and deserves some deeper consideration.

This alternative point of view was suggested by Planck [13], Weil [14] and in a set of papers by Herrmann & Schmid [15–17], Heiduck, Herrmann & Schmid [18] and Garbois & Herrmann [11] in which several specific examples are examined.

Since momentum is a vector quantity the transfer of it from one system to another will be described by a tensor (the transfer of each component is described by a vector field).

In order to have a clear representation of the momentum currents it is convenient to examine the flux lines for each component of the momentum because these lines will be represented by the usual three-dimensional lines of flux.

Herrmann & Schimd [17] discussed the case of electromagnetic interaction. The Maxwell's stress tensor can be written as [19]:

$$T_{\alpha\beta} = E_\alpha E_\beta + B_\alpha B_\beta - \frac{1}{2}(\mathbf{E}\cdot\mathbf{E} + \mathbf{B}\cdot\mathbf{B})\delta_{\alpha\beta} \tag{1}$$

where:

$$dP_\alpha = T_{\alpha\beta} n_\beta dk \tag{2}$$

is the flow of the component of momentum through the surface element with area dk and normal unit vector $\mathbf{n}$.

If we consider a volume V surrounded by a closed surface K, the total flow of momentum entering the volume V is:

$$C_{P_\alpha} = \int_K dP = \int_K T_{\alpha\beta} n_\beta dk \tag{3}$$

This is equal to the time derivative of the *total* momentum contained within the volume limited by the closed surface K.

By total momentum we mean the sum of the component associated with matter (and which is usually called *mechanical* momentum) and the component *associated to the field* whose density can be written as:

$$\mathbf{g} = \frac{1}{c^2}(\mathbf{E} \times \mathbf{H}) \tag{4}$$

In the examples discussed by Herrmann & Schmid [17] some momentum currents originate from one body and terminate on the other: this means that some amount of the selected component of the momentum leaves the first body at that point and is transferred to another point of the other. Some originate and terminate in the same body. The former describe the interaction between the two systems while the latter describe the interaction among points of the same body. They describe the stress state of the body from which the deformation can be inferred if we know their elastic properties.

The case of gravitational field has been investigated by Heiduck, Herrmann & Schmid [18]. they examined the momentum current distribution for each component in the case of the Earth-Moon gravitational interaction treated in the weak field, static approximation. The procedure is similar to the case of e.m. fields and the resulting pictures of the currents of the different components of momentum are shown. The presence of momentum currents connecting either different points of the same body or points of the two bodies is confirmed and is a general feature to be expected whenever a continuous density of the field sources is assumed to describe the object at stake.

It is important to recall that in this perspective there is no room for point charges or point masses but a continuous distribution of field sources are assumed. In this picture, mechanical interactions are represented by a complex current distribution all over the whole Universe; the resulting current distribution is, at each point, determined by the entire distribution of the sources of the field (e.g. gravitational or e.m.). It is only when we consider interactions between limited portions of the whole extension that we have the necessity of defining closed and open system and we determine our level of complexity (Theoretical context T). This description of mechanical interactions (by means of momentum currents) offers a deeper representation and allows the observer to move from an "action at a distance" analysis in which only integral quantities are grossly involved to a local interaction view (which contains the former). The difference is not only in the detail of the description but principally in the conceptual point of view.

3. General Explication of Causation

3.1. *Def*$_1$*: System and state of the system*

Let us assume a theoretical context T and a system P. We can know, inside a given theoretical context T, which system we are speaking of by indicating the set of quantities $\{P_1, \dots, P_r\}$ (e.g. velocity, mass, charge, etc.) defining it in T. Let $P^T = \{P_1, \dots, P_r\}$ be the system at stake. Since P^T is a empirical system, knowing, via measurements, the instantiations of $\{P_1, \dots, P_r\}$ becomes necessary. This means knowing the state of the system in a given space-time point x^i (with i=0,1,2,3, where x^0 is the time coordinate and x^1,x^2,x^3 are the spatial coordinates). Therefore $P^T_{x^i}$={p_1,...,p_n} is the state of the system P^T in x^i, where {p_1,...,p_n} are the instantiations of the quantities {P_1, ... , P_r}[c].

The possibility of knowing the above instantiations requires an abstraction by the Observer: instantiations are the end-point of complex chain of operations and they have to be realized in a "condition of simultaneity" (cfr A. Einsten [20]). The required abstraction concerns the necessity of defining the instantiations at one instant at all parts of the volume of the system. In a eulerian point of view this problem is overcome because it reduces to a local measurement in an (infinitely) small volume. In a lagrangian point of view the "abstraction of simultaneity of operations" has to be carried out throughout the whole space generated by the system *and requires the previous definition of simultaneity.* In other words it requires the pre-construction of the space-time.

3.2. *Remark 1*

Clearly the number r indexing the {P_1, ... , P_r} can be rather large if we want to specify all the quantities we are interested in, but in general, there will be relations among some of them. For instance, in a black body the peak frequency of the emitted radiation, the energy density, and the temperature are mutually dependent. Let us agree to select one set of *independent* quantities that are necessary and sufficient to define the system in T. In general there are various different possible sets and the specific choice will depend on the nature of the problem the observer is dealing with.

As mentioned, *'system' and 'state of the system' are theoretical context dependent notions.* By this we mean simply that different theories deal with these notions differently, providing, therefore different representations. The particular choice of the theoretical context may be naturally suggested by what we want to observe, or by the constraints on the system. In other situations, it may be determined by more fundamental conditions, as in the case of physics in which we have to shift from a classical to a relativistic, or to a quantum-mechanical, description. A paradigmatic example of context dependency of the notion of 'state of the system' concerns neutrons and protons. They have to be considered as two different system in a theoretical context emphasizing the electromagnetic interactions, but they have to be regarded as two quantum states of the same system (the nucleon) if we move to a theoretical context emphasizing the strong interaction[d].

---

[c]Note that $n \leq r$, since it is not said that all quantities can be instantiated.

[d]The notions of 'system' and 'state of the system', and the fact they are theoretically context dependent are extremely useful for our aims. In particular the usefulness concerns the problems regarding the so-called "timewise gerrymanders" and "spacewise gerrymanders". The former would be "putative objects defined over a time interval where the definition changes over time", while the latter would be "putative objects consisting of many independent objects"(Dowe 1992, 2000). We could avoid the difficulties concerning them simply by ruling them out of our analysis. But it

The quantities $\{P_1, ..., P_r\}$ are of two kinds: they may be extensive quantities or intensive quantities.

3.3. *Def₂: Extensive quantities*

Our approach to the notion of causation is based on the notion of extensive quantity. There are various definitions of the latter. Some authors define it as a quantity which scales with mass, others as a quantity which scales with the quantity of matter; some other refer to additivity. Let us adopt the most general definition, since is the only one suitable for generalizations to arbitrary space-time metrics or to arbitrary spaces: Given a system P^T and given the quantities defining it, by *extensive quantity* we mean any quantity E whose value is given by the volume integral of function $\epsilon(x^i)$, called *density*, of E in the point x^i. Note that in the *definiens* above, volume plays the fundamental role: it is, by definition, *the extensive reference quantity*; it is extensive, homogeneous and isotropical thanks to the homogeneity and isotropy of space-time.

Having this definition in mind, the "additivity property" often used to define extensive quantities (which would be a better choice in a quantum domain) acquires a rigorous and general meaning: if we considered the system P^T as if it were subdivided into n sub-systems, *in the sense of volume* (n sub-volumes), the value of E would be given by the "sum" of the n values of E within the n sub-systems (by the additivity of integrals).

---

is not that easy.
According to the definition above, a particle beam is a spacewise gerrymander, and of course, we cannot eliminate it from our discussions. Similarly we cannot eliminate another spacewise gerrymander, that is, a population of living beings such as a flock of birds or a school of fish. Nevertheless we must eliminate a spacewise gerrymander such as 'my pen + mu jacket + the light bulb in front of me + the hostess in a restaurant 2000 km away from me'. But we cannot eliminate 'my pen in my pocket + my jacket + the light bulb in my hand + the hostess in the hall of the restaurant I am just enetring' if I must physically discuss the impact between me and the hostess. I must take into account all the masses, all the velocities, etc. How can I demarcate the "good" spacewise gerrymanders from the "not-good" ones? The same goes for the timewise gerrymanders. Usually, the timewise gerrymanders should be ruled out. But why? Simply because we do not like them? Should all of them be banned? Let us think about the whole phylogenetic tree starting more or less $3{,}5\times10^9$ years ago and arriving at our age. Are not the definitions of the different living species valid for a certain time period (of course here at issue is not the notion of species *qua* category but *qua* taxon)? Do they not change over time? How can we demarcate a "good" timewise gerrymander from a "not-good" one? The solution is pragmatically contained in the proposed notion of 'system' and 'state of system'. Both are theoretically contextualized, and note that the system is something which is structured in a certain way, that is, it is characterized by its elements and by its relations among the elements. Yet which are the relations that can be considered "good" relations so that the system can be considered a "good" system, and the timewise or spacewise gerrymander a "good" timewise or spacewise gerrymander? The answer is: the theoretical context. That is, if we have a "good" theory about that system (be it or not a timewise or spacewise gerrymander), then that system is a "good" system. We are perfectly aware that the way out we are proposing could suffer of the same objections that could be raised to Goodman's solution to the green-blue theory *versus* grue-bleen theories on emeralds. but this is not the right place to embark ourselves on this question.

The mass of a system is a paradigmatic example of extensive quantity, since it can be thought of as the volume integral of the mass density. Only in this sense we can affirm that the mass of the whole is the "sum" of the parts; this would no longer be true, in general, if we "put together" the parts (for instance if one affirms that a deuton is composed by one neutron and one proton he should *not* expect that the mass of the deuton is the sum of the two masses).

Other extensive quantities are the total energy, the momentum and the angular momentum, the electric charge, the total number of moles, entropy, etc.

3.4. *Def₃: Intensive quantities*

Given a system P^T and given the quantities defining it, by intensive quantity we mean any quantity I whose value is defined in every point of the system, but not for the system as a whole.

3.5. *Remark 2*

An instance of intensive quantity is given by the velocity, which is defined at a point. For, strictly speaking, we cannot talk of the velocity of an extended body unless it is rigid and in translational motion. Likewise for the temperature: it is defined in every point. Only in the case of the isothermal body do we use the convention of speaking of "the temperature of the body". The pressure, the electric potential, the field intensities, the chemical potential, and, of course, all the densities of the various extensive quantities are examples of intensive quantities.

Each intensive quantity is defined as the partial derivative of an extensive quantity with respect to another extensive quantity (of course keeping some others constant; the latter depends on the choice of the variables we consider as independent). That is, given an intensive quantity I, two extensive quantities E and E' exist so that

$$I = \frac{\partial E}{\partial E'} \tag{5}$$

For example, temperature τ is defined as

$$\tau = \frac{\partial U}{\partial S} \tag{6}$$

where U is the energy and S is the entropy (U and S are two extensive quantities and the derivative must be carried out the constant volume).

This is a crucial point for our proposal and we must dwell on it. It appears that the E-quantities play a fundamental role in the definition of a system while the I-quantities can be considered as sort of "second rank" parameters. Actually the latter tell us how rapidly an extensive parameters varies when another is varied in a given way. Nevertheless, an extremely relevant consequence of the above regards the definition of state of the system. Now it can be fixed by resorting only to extensive quantities. For a quantity can be either extensive or intensive and in the latter case it can be defined as a function of the former. That is $P^T_{xi} = \{E_1, ..., E_n\}$ (where the E_1, ... E_n are mutually independent).

Needless to say, what we are affirming and what we will affirm in terms of extensive and intensive quantities can be applied also to non-physical domains, such ad, for example, chemistry and biology. It is sufficient to individuate, for that particular chemical or biological system, the suitable sets of extensive and intensive quantities.

3.6. *Th_1: Extensive quantities and the balance equation*

Given a theoretical context T and a system P^T, any variation of an extensive quantity E of P^T can be described by a balance (or continuity) equation, whose integral form is:

$$\frac{dE}{dt} + C_E = \Sigma_E \qquad (7)$$

where C_E is outgoing flux of the extensive quantity E, and Σ_E is its production rate within the volume occupied by the system.

3.7. *Remark 3*

The proof concerning the above balance equation comes as an application of a general theorem called Transport Theorem or Reynolds' Theorem, which is well known in the mathematical theory of continuous media (see 21). The flux term of the balance equation, C_E, is an integral over the surface K surrounding the physical system P^T. The surface K is oriented outward the system. Instead the production rate term, Σ_E is an integral over the volume V bounded by the surface K. Both terms contribute to representing the observed change in the extensive quantity E but in substantially different way. The production rate term, Σ_E, describes the contribution due to processes occurring *in* the system. It depends only on the state of the system and does not depend on the state of the external environment (other system). The C_E term describe a "flux" of the quantity E across the surface and it describes the "exchange" of the quantity E with the external environment in which P^T is embedded.

Note that, actually, "flux", "exchange" and "transmission" must be interpreted metaphorically: there is no quantity which flows, nor which is exchanged or transmitted! Hereinafter this extremely important question will be better clarified.

3.8. *Def_4: Conserved quantity*

Given a theoretical context T and a physical system P^T, an extensive quantity E of P^T is a conserved quantity if and only if

$$\Sigma_E = 0 \qquad (8)$$

From which:

$$\frac{dE}{dt} + C_E = 0 \tag{9}$$

3.9. *Remark 4*

One of the most paradigmatic examples of a conserved quantity is given by energy. In our account, the principle of energy conservation states that the total energy of any system P^T can vary only by "exchanges" with the environment in which P^T is embedded. Hence any change of energy in P^T must be exactly and numerically balanced by the opposite change in its environment. The same applies to momentum, angular momentum, electric charge, etc.

Among the non-conserved quantities we may recall the total number of moles n and the numbers of moles of various components n_i. In this case, if the measured variation of n_i is not numerically balanced with the C_n term (in particular, this happens when the system cannot exchange matter), then something must be occurring within the system and accounting for the difference.

Entropy (S) is a particularly relevant non-conserved extensive quantity. In this case the Σ_S term of the balance equation contains the core of the Second Principle of Thermodynamics. For:

(1) $\Sigma_s = 0$ for changes in a system in internal equilibrium. This can be conceived only in an ideal, asymptotic, case. In real processes equilibrium and change are incompatible concepts at least on a macroscopic scale;
(2) $\Sigma_s > 0$ for any real case;
(3) Σ_s can be written as the sum of various terms, that is,

$$\Sigma_s = \sum_{i=1}^{n} \frac{dE_i}{dt} \chi_i \tag{10}$$

Though with different notations this is a fundamental relation in non-equilibrium thermodynamics (see, for instance [22, 23]). Note that each term on the right-hand side is written as the product of the time derivative of an extensive quantity E_i times a suitable intensive quantity χ_i, called the entropy-conjugate quantity of the extensive quantity E_i.This equation is extremely general, since it connects entropy and each possible extensive quantity E (to be more precise, their time variations). For example, if the extensive quantity is the energy U, one of the terms can be given by the scalar product $\mathbf{j}\cdot \text{grad}(1/\tau)$, where $\mathbf{j}$ is the flux of energy (supposed, in this case, as a heat flux), and τ is the absolute temperature.

The three points above are of fundamental importance. Point a) characterizes equilibrium states, that is, those ideal states in which every small change is, by definition, reversible. Actually reversible changes do not exist in the real processes, but the former can approximate the latter with the highest accuracy in a very large number of cases. Point b) separates allowed directions from forbidden directions in all real processes. It states a sort of thermodynamical arrow of time. Point c), as just observed, explains the profound meaning of the distinction between extensive and intensive quantities.

3.10. *Remark 5*

The "exchange" of an extensive quantity is completely symmetrical between the two system, owing to the balance condition. The direction of the "exchange" is merely conventional since any situation can be depicted as an "exchange" of the opposite quantity in the opposite direction. To illustrate this symmetric scenario, one could note the analogy with the content of the third law of Newtonian physics in the context of mechanics, in which the interactions consist in the "exchange of quantity of movement" (momentum). As we will show, our explication of the 'causal interaction' will be given in a more general sense, and the explication will not imply a time succession, as can be also intuitively inferred from what has just been said on the symmetry property of the "exchanges".

Let us recall some basic definitions which will be used in the following and in particular the definition of closed, open and isolated system.

(1) Closed systems: given a system P^T within a theoretical context T we say that P^T is closed with respect to an extensive quantity E if we know that the boundary of the system does not allow the quantity E to be exchanged with the external world. For instance if the Observer considers the boundary as rigid, he knows that P^T cannot exchange volume. If he knows that the boundary does not allow a certain chemical species γ to pass through, he will say that P^T is closed with respect to n_γ where n_γ the mole number of present within the volume under consideration.

(2) Open systems: given a system P^T within a theoretical context T if the observer has no reason to say P^T is closed with respect to the extensive quantity E, P^T will be said to be open with respect to E.

(3) Isolated system: given a system P^T will be said to be an isolated system if it is closed with respect to all the extensive quantities defining its state.
The above definitions are necessary in the lagrangian point of view in which we follow the evolution of the system thought of as an object provided with some boundary; they will be not essential in a eulerian point of view (followed in this paper) when we consider the evolution of the extensive properties in a certain volume V.

(4) Stationary states: given a system P^T within a theoretical context T, P^T will be said to be in a stationary state in the time interval Δx^0 if we observe that $dE/dt = 0$ in Δx^0 for every extensive quantity defining its state. Note that, as a consequence dI/dt will be zero also for every intensive quantity I i.e. every state variable will be constant in time within Δx^0.

4. Explicating the Notion of "Causation"

4.1. *The integral explicatum*

Let us fix a theoretical context T and let us have two system P^T and P'^T in the states $P^T_{x^i}$ and $P'^T_{x'^i}$, respectively, determined by the respective instantiations of the extensive quantities E_1, ... , E_n. We are allowed to speak of *integral causal interaction* between P^T and P'^T, in the time interval $\Delta x^0 = |x^0 - x'^0|$, *if and only if*:

(1) We observe, relatively to an extensive quantity E_i, that $C_{E_i} = -C'_{E_i}$ and $C_{E_i} \neq 0$ in Δx^0;
(2) P^T and P'^T form a system G that is closed with respect to the quantity E_i.

We call this 'integral *explicatum*', since we want to stress that it is written in integral form and by this we mean that the "exchange" of the extensive quantity E_i between the two systems is given by the total amount of E "flowing" through the whole surface separating the two systems.

Let us dwell on the relations between the two requirements above. Suppose that, as a result of our observations, we measure the two fluxes C_{E_i} and C'_{E_i} and we see that 1) is fulfilled. Is this sufficient to claim that P^T and P'^T are in casual interaction? The answer is negative: 2) also has to be satisfied, for we have to be sure that the "exchange" of E takes place *between* P^T and P'^T only.

Suppose now that 2) is fulfilled. We expect that 1) is satisfied as a natural consequence if we have that $C_{E_i} \neq 0$ in the case of conserved quantities. For instance if two bodies P^T and P'^T form an isolated system G with respect to energy, the non-zero variation of the energy in one of the two is compensated by the opposite variation in the other, and 1) is automatically satisfied.

In the case of non-conserved quantities a deeper knowledge of the occurring processes will be necessary, i.e. we should know the values of Σ_{E_i} and Σ'_{E_i} which may now be non-zero. Hence 1) may fulfilled provided that the adopted theoretical context is absolutely *safe*, which means: first, the closure of the total system G with respect to E_i is insured and, second, our knowledge of the two source terms is also secure. If 1) is not verified then the entire description is to be re-examined and nothing can be said about causal interactions.

One further comment must be devoted to the "closure condition" explicitly required in point 2) of the above integral *explicatum*. We have seen that this condition is necessary in order to prove that some extensive quantity has been exchanged.

To exemplify even more what just said, let us consider two colliding billiard balls within the theoretical context of classical mechanics. *Prima facie*, the system can be considered isolated with respect the momentum, but we could wonder how our account works if we release the closure condition. This is exactly what happens in real situations if we consider that something else is interfering with the two balls' interactions; for instance, the friction with the billiard table, the presence of the air, etc. In this case, the balance equation between the two balls is no longer satisfied and we have to consider a more complex situation in which the "exchange" of the extensive quantity is among three, or more systems.

Some effect on the new partners can be estimated by observing their modifications and, hopefully, we can argue that some momentum has been "exchanged" between the two balls; in any case we must come to a quantitative estimation no matter how large the error will be.

Concerning the supposed restrictive nature of the closure condition, let us anticipate that this condition of complete dependence from it will be completely abandoned, when we shall formulate our account in the local-causes point of view. This latter statement will represent the complete generalization of the integral account we are discussing here in the sense that we shall seek for "fluxes" of extensive quantities point by point instead of global "fluxes".

So we may have local interactions even if we have zero global interactions, and we do not need any closure requirement. However let us remind that concepts like 'closure' or 'isolation' are still included in the definition of *conserved and non-conserved* quantities. In the case of non-conserved quantities E, the determination of the internal production rate Σ_E also has to be considered.

Suppose we are given a container which is exchanging molecular oxygen with another container in such a way that the two containers form a system which is closed with respect to O_2. Let us suppose that the extensive quantity we are observing is the amount of molecular oxygen O_2 inside it.

Actually, for our convenience, we will deal with the mole number n_{O_2} and measure its variation, say Δn_{O_2}, in a definite time interval. *Further* we know that the following chemical reaction is at work inside the system:

$$2C + 2C_a + 3O_2 \longrightarrow 2C_aCO_3 \qquad (11)$$

It is important, for what follows, that we know that a definite process (in this case "case" chemical reaction) is at work within the system. For instance, in the above example, we could have a monitoring of the temporal growth (positive or negative) of both calcium and calcite. Then we are able to measure the production rate of oxygen *within the system* by measuring the amount of calcium Δn_{Ca}, destroyed in the same time interval (negative production). Now we can evaluate the flux term, i.e., the $C_{n_{O_2}}$, and have a complete understanding of the balance equation *with respect to the variations in molecular oxygen.* At this point, according to our proposal, we are able to speak in terms of causal interaction since the two requirements indicated above regarding the concept-*explicatum* 'causation' are satisfied. This means we are allowed to claim that there is a casual interaction *even if there are no conserved quantities at stake.*

4.2. *The local explicatum*

In the above example of the containers exchanging molecular oxygen we may add a third container at stake. In order to acknowledge a causal interaction between two of the three containers with respect to the extensive variable n_{O_2} the Observer has to know enough about the variations of the state variables and the source terms in each system; moreover it is necessary to consider each overall flux $C_{n_{O_2}}$ as composed by the sum of the two contributions and verify by means of the balance equations that condition 1) is realized. This trivial example introduces a substantial change in the level of complexity at which our causality account will be formulated.

In the balance equation introduced before, the flux term C_E is, by definition, an integral quantity or, in other words, it takes into account the amount of extensive quantity E "transferred", per time unit, as a whole to the system under consideration. In the above example C_E has a structure: it is the sum of two contributions, each describing the flux of the quantity E through two different regions of the boundary.

From a more general point of view we may see that the structure of the global flux C_E may have two different origins: one with respect to the different regions of the boundary surrounding the system and one concerning the possible superposition of different fluxes at the same point.

As far as the first structure is concerned we may write the total flux in the form:

$$C_E = \int_K \mathbf{G}_E \cdot \mathbf{n} dk \tag{12}$$

where $\mathbf{G}_E$ is the flux density (a vector) at each point of K; K denotes the surface surrounding the volume occupied by the system and dk the area of the surface element $\mathbf{n}$ being the unit vector normal to the element dk and directed outward towards the volume.

Hence if we know the flux density $\mathbf{G}_E$ at every point of the surface we can describe the "exchange" of a given extensive quantity point by point through the entire surface, limiting the volume. It follows, and by taking into account all that has been said until now, that we are able to speak in terms of "local causation", and thus to study causal interactions between small (as small as we wish) subsystems.

The second kind of structure is concerned with the possibility of considering the local flux density $\mathbf{G}_E$ as the result of the superposition of different flux densities of the extensive quantity E flowing through the same area and *with the possibility of separating the various components.*

In this case, how can we distinguish between two sub-fluxes flowing through the same surface element?

We can do it *only if we know* (and this depends on out T) that the different sub-fluxes are "exchanging" the quantity E between *different parts* of our system P^T and *different parts* of the other system P'^T, the candidate to be in causal interaction with P^T. *We could say, more simply, that the sub-fluxes are "connecting"* P^T *with different subsystem of* P'^T.

Let us now propose the local *explicatum.*

Let us fix a theoretical context T and let us have two system P^T and P'^T in the states $\mathrm{P}^T_{x^i}$ and $\mathrm{P}'^T_{x^i}$ respectively, determined by the respective instantiations of the extensive quantities E_1, ... , E_n. Let dk be the area of an infinitesimal element of the surface separating P^T and P'^T; let $\mathbf{n}$ and $\mathbf{n}'$ be the two corresponding normal unit vectors both perpendicular to the surface element at that point and directed outward towards P^T and P'^T respectively; and let $\mathbf{G}_E$ be the flux density in that point x^i. By defining the (infinitesimal) flux entering P^T and P'^T we have that $\mathbf{G}_E \cdot \mathbf{n} = - \mathbf{G}_E \cdot \mathbf{n}'$. Thus, we are able to speak of *local causal interaction* between P^T and P'^T, *if and only if*:

$$\mathbf{G}_E \cdot \mathbf{n} \neq 0 \tag{13}$$

at least at one point of one surface selected to separate the two systems.

This implies $dC'_E = -dC_E$, where dC_E and dC'_E are the infinitesimal fluxes leaving P^T and P'^T respectively, through the element dk of the surface K at point x^i. Needless to say, now we no longer need to introduce the condition according to which P^T and P'^T form an isolated system G together.

Evidently this discussion implies the ability, by the Observer, to separate the total flux C_E into different fluxes and each flux into different sub-fluxes and *this is entirely dependent on the adopted theoretical context T.*

This last comment implies, for instance, that the existence of causal connection between two systems may appear or not according to the selected theoretical context T (as will be clear in the example of equilibrium systems).

The consequence of this is that the existence of an interaction between two systems is always *Observer dependent* in the sense that we are aware only of the interactions that we are able to acknowledge.

Another comment which should be born in mind refers to the choice of the extensive quantity E.

The possibility of speaking in terms of causal interaction between two systems depends on the choice of the (conserved or non-conserved) extensive quantity.

Differently from Dowe [4, 5], who argued for a "conserved quantity theory of causation" our account introduces also a generalization to non-conserved quantities.

This is conceptually necessary and not only practically useful.

Indeed the property of being conserved for an extensive quantity E, is dependent on the theoretical context T selected by the Observer and may be assumed or abandoned accordingly.

Another point regarding the necessity to move to a theory of causation which deals with extensive quantities as a whole (conserved and non-conserved) concerns the impossibility of reducing non conserved quantities to conserved ones; as an example we refer to many situations which are very common in chemistry where the number of molecules is non conserved by effect of chemical reactions while the number of atoms may be thought to be reasonably conserved at least in many terrestrial conditions.

The problem is that a molecule is an object which is quite different from the mere sum of the atoms by which we say it is composed. In particular we have to refer to the *functional property* which makes one kind of molecules different from another and this difference is due not only to the chemical formula but also to the space structure of the compound (and for complex biological systems to different shapes). This introduces us to the interesting perspective of considering the "information" transferred by different material support (different molecular species) in a truly convective mechanism.

5. Causation in Equilibrium and Non-equilibrium Stationary States

Let us briefly discuss the problem of stationary states from the point of view of their possible causal connections with other systems. Is it correct to claim for the presence of interactions if the state of the system changes? Vice versa if the state does not change in time (stationary state) is it correct to claim that the system is not in interaction with other systems?

In both cases the answer is negative the two cases should be examined with some care. Both in mechanics and in thermodynamics we need to state, by principle, what a "situation of non interaction" is. In the former case we need to define "inertial motion" i.e. those dynamical configurations which need interactions to be modified (in Newtonian language we could say that their fluents have null fluxions); in the latter we define equilibrium states as those configurations which are realized in isolated systems i.e. in which interactions with external systems are prevented. Conversely if an inertial motion is observed can we affirm that the system is not in interaction with something else? The answer is obviously negative but it is worth pointing out that this conclusion is entirely dependent on the theoretical context T (both in a substantial or local point of view) adopted by the Observer.

Let us consider, as an example, of a glass leaning on a horizontal table. Its momentum is constant (zero) and therefore the *total* flux of momentum entering the glass from the external world is zero. According to the integral account, our glass is not causally connected with it.

Within a more complex theoretical context we separate the external world into parts: in our case we may start with two: the table and the Earth and we know (because our Theoretical context is complex enough) that there are currents of momentum between the glass and the Earth. From this we may affirm that a convenient momentum current is flowing between the glass and the table and then they are causally connected.

This simple example shows that causal connections between systems can or cannot be acknowledged by the Observed according to the complexity of T in use or, in other words, we may say that the Observer sees only those interactions that he is allowed to see according to the whole knowledge he has created until then (and according to the constraints he has adopted).

Another example, taken from thermodynamics, deserves some attention because in this case it is common to have stationary non-equilibrium states. For instance a resistor (electric heater) traversed by a constant electric current may stay in a stationary state in which all state parameters (both extensive and intensive ones) maintain constant values. In this example entropy is produced at some rate within the system (non conserved quantity $\Sigma_s \neq 0$) and is transferred to the external world at the same rate: the integral account is enough to recognize that the system is causally connected with the room through entropy.

The same integral account fails in finding causal connections with respect to energy. Since it is a conserved quantity and its value is constant in time we conclude that the flux of energy towards the external world is zero. This zero flux is the sum of two opposite fluxes of energy: one between the system and the room (positive) and another between the system and some electric plant somewhere hence we may prove that our system is interacting *with the room* also with respect to energy.

From this very simple example we learn two things; one is that new causal connections are "discovered" according to our theoretical context (as in the previous example) but, differently from before, we have a criterion to separate equilibrium from non equilibrium stationary states. This criterion is given by entropy.

6. Conclusions

In the above, we have given a rigorous and unambiguous empirical explication of the notion of 'causation'. We have arrived at this result by offering a strongly consistent analysis and by resorting to an approach based on extensive quantities and on the balance equation.

In this way we have proposed what we have called a *generatized explication of 'causation'* (GEC), both in the integral, less complex form and in the local, general form in which both conserved and non-conserved quantities have the same role. In particular the local account frees the Observer from the constraints required in the integral point of view.

Note that the possibility of dealing with non-conserved quantities is of fundamental importance. We can easily consider several examples in which we naturally have to deal with non-conserved quantities, as occurs in chemical processes or in population dynamics, but this generalization is necessary for conceptual reasons.

For the GEC, which is a *theoretical context sensitive approach*, enables us to assert that two systems, discussed inside a given theoretical context, are causally connected if and only if a flux of extensive quantities is quantitatively observed.

Hence we have shown that a causal connection between systems can be recognized by the Observer according to his level of complexity.

The role of the Observer is fundamental even (we could say, in some sense, previously) in the definition of boundaries of Systems as was discussed by Boniolo, Faraldo and Saggion [24]. Definition of boundaries and the acknowledgments of causal connections between Systems constitutes the core of the abstraction performed by the Observer in his interaction with external world.

Note also that in our GEC time has no role: causation is a temporal (Riemann 1867). The role of time in causal connection is widely discussed in scientific and philosophical domains but depends on the definition of space-time.

To conclude, we believe that the GEC provides a well-grounded platform which allows us to understand precisely and unambiguously what we are affirming whenever we speak of causation in many different contexts, provided that there is the possibility of characterizing the system under discussion in terms of extensive quantities. In these cases, our GEC can be applied (and it can be applied without any reductionist or reconstructionist purpose).

The formal algorithm and the visualization by means of flux changes can be brought back to Newton's geometrical vision of fluents and fluxions.

References

[1] G. Boniolo, R. Faraldo and A. Saggion, *Explicating the notion of causation: the role of extensive quantities*, in *Causality in the sciences*, (in P. McKay Illary, F. Russo, J. Williamson (eds.), Oxford University Press, 2011), pp. 502–525.
[2] T. Bontly, *Synthese* **151**, 177 (2006).
[3] R. Carnap, *Logical Foundations of Probability* (University of Chicago Press, 1956).
[4] P. Dowe, *Philosophy of Science* **59**, 195 (1992).
[5] P. Dowe, *Physical Causation* (Cambridge University Press, 2000).
[6] J. Aronson, *Synthese* **22**, 414 (1971).
[7] D. Fair, *Erkenntnis* **14**, 219 (1979).
[8] W. Salmon, *Scientific Explanation and Causal Structure of the World* (Princeton University Press, Princeton, 1984).
[9] W. Salmon, *Philosophy of Science* **64**, 461 (1997).
[10] N. Guicciardini, *Reading the Principia: The Debate on Newton's Mathematical Methods for Natural Philosophy from 1687 to 1736* (Cambridge University Press, 2003).
[11] F. Herrmann, *Eur. J. Phys.* **21**, 49 (2000).
[12] I. Newton, *Philosophiae Naturalis Principia Mathematica* (Cantabrigiae, 1712).
[13] M. Planck, *Phys. Z.* **9**, p. 828 (1908).
[14] H. Weyl, *Die Naturwissenschaften* **12**, 197?(1924).
[15] F. Herrmann and G. Schmid, *American Journal of Physics* **52** (1984).
[16] F. Herrmann and G. Schmid, *European Journal of Physics* **6**, p. 16 (1985).
[17] F. Herrmann and G. Schmid, *American Journal of Physics* **53**, p. 415 (1985).
[18] G. Heiduck, F. Herrmann and G. Schmid, *European Journal of Physics* **8**, 41 (1987).
[19] J. D. Jackson, *Elettrodinamica classica* (Ed. Zanichelli, 2001).
[20] A. Einstein, *Annalen der Physik* **17**, 891 (1905).
[21] M. Gurtin, *An Introduction to Continuum Mechanics* (Academic Press Inc., 1981).
[22] H. Callen, *Thermodynamics and an Introduction to Thermostatistics* (John Wiley & Sons, 1985).
[23] I. Prigogine, *Thermodynamics of Irreversible Processes* (John Wiley & Sons, 1955).
[24] G. Boniolo, R. Faraldo and A. Saggion, *Foundation of Science* **14**, 181 (2009).

EVOLVING COMPLEXITY, COGNITION, AND CONSCIOUSNESS

H. LILJENSTRÖM[a,b]

[a] *Biometry and Systems Analysis, Swedish University of Agricultural Sciences*
Uppsala, SE-750 07, Sweden
Email: hans.liljenstrom@slu.se

[b] *Agora for Biosystems*
Manfred Björkqvists allé 2-4, Box 57, SE-193 22, Sigtuna, Sweden

All through the history of the universe there is an apparent tendency for increasing complexity, with the organization of matter in evermore elaborate and interactive systems. The living world in general, and the human brain in particular, provides the highest complexity known. It seems obvious that all of this complexity must be the result of physical, chemical and biological evolution, but it was only with Darwin that we began to get a scientific understanding of biological evolution. Darwinian principles are guiding in our understanding of such complex systems as the nervous system, but also for the evolution of human society and technology. Living organisms have to survive in a complex and changing environment. This implies response and adaption to environmental events and changes at several time scales. The interaction with the environment depends on the present state of the organism, as well as on previous experiences stored in its molecular and cellular structures. At a longer time scale, organisms can adapt to slow environmental changes, by storing information in the genetic material carried over from generation to generation. This *phylogenetic learning* is complemented by *ontogenetic learning*, which is adaptation at a shorter time scale, occuring in non-genetic structures. The evolution of a nervous system is a major transition in biological evolution and allows for an increasing capacity for information storage and processing, increasing chances of survival. Such neural knowledge processing, cognition, shows the same principal features as non-neural adaptive processes. Similarly, consciousness might appear, to different degrees, at different stages in evolution. Both cognition and consciousness depends critically on the organization and complexity of the organism. In this presentation, I will briefly discuss general principles for evolution of complexity, focussing on the evolution of the nervous system, which provides organisms with ever increasing capacity for complex behaviour, cognition and consciousness. I will also discuss some computational approaches, as tools for understanding relations between structure, dynamics and function of the nervous system.

1. Evolving Complexity

Since the origin of the universe, it appears as if structures and interactions have increased in complexity. This increasing complexity involves the formation of matter, of galaxies and solar systems, of planets and of life. As far as we know, living systems in general, and the human brain in particular, may be the most complex systems in the entire universe. The complexity of our brains is the result of billions of years of biological evolution, an evolution which was preceded by a geological and chemical evolution on planet Earth. It seems appropriate to link these evolutions to the cosmological evolution that provided the accumulation of heavy minerals and organic molecules as substrates for life. Even though evolution in itself does not necessarily imply increasing complexity, in general it seems to be the case when we regard the evolution of the universe and of life. Indeed, even Darwin recognized an increasing complexity as a sign of evolution. In his book *On the Origin of Species* [11] Darwin says (1859):

> "As far as mere inorganic conditions are concerned, it seems probable that a sufficient number of species would soon become adapted to all considerable diversities of heat, moisture, etc.; but I fully admit that the mutual relations of organic beings are more important; and as the number of species in any country goes on increasing, the organic conditions of life must become more and more complex".

Another, more recent view on evolution towards higher complexity is given by Murray Gell-Mann in his book [18]:

> "The movement toward higher maximum complexity can proceed in a manner that resembles diffusion, especially in nonadaptive systems like galaxies. In complex adaptive systems such as biological evolution, though, it often happens that selection pressures favor higher complexity in certain situations. The distribution of complexities in such cases will differ, in shape as a function of time, from the result of a random walk. Although there is still no reason to believe in a steady drive toward more complex organisms, the selection pressures that favor higher complexity may often be strong. To characterize those systems and environments in which complexity is highly advantageous presents an important intellectual challenge"..

But what do we mean by complexity? There are several definitions and descriptions of complexity and complex systems. For example, complexity can be studied and described in terms of structure, dynamics or function/behaviour.

Structural complexity can be seen as the interplay of many elements in a varied fashion due to many variables and structural constrains. However, the word "complex", as it may be applied to many different systems, could have many meanings. Briefly, it can be said that the degree of complexity can be measured in terms of information: "the more information that is needed to describe a system, the more complex it is". This is sometimes referred to as *algorithmic* complexity.

Another type of complexity is called *effective* complexity, which can be defined relative to a complex adaptive system that is observing it, as the length of the schema used to describe its regularities. A complex adaptive system separates regularities from randomness. The effective complexity of a bacterium, in the sense of the length of a schema, is thus related to the length of the genome.

Characteristic features of complex systems have to do with the web of (often non-linear) interrelations between variables. This setting introduces thresholds, lags and discontinuities. The feedback and feedforward loops open up for surprises and non-intuitive behaviors of the systems. In addition to nonlinearities, complex systems are often characterized by self-organization and emergence, where the behavior is often unpredictable. Complex *global* patterns can sometimes be generated by repeatedly applying simple *local* procedures or rules. Above a certain threshold of complexity, new qualities may arise (emerge).

Complex systems are often, due to their non-linear nature with feedback loops and time delays, prone to chaotic behavior. Characteristic of such behaviors include sensitivity to "initial conditions", and unpredictability at a longer time scale, while the behavior may be predictable at a shorter time scale. Thus, this irregular, or random behavior is more structured than "pure noise", which is unpredictable at all scales. Examples of systems, which have been described as chaotic include turbulence (e.g. smoke, cream in coffee), the weather, brain-waves, certain ecosystems, and the stock market. Actually, many systems with three or more coupled components can display chaotic behavior under certain circumstances (with a certain set of parameters). Although complex (chaotic) systems could be very simple, such as a double pendulum, or a chemical reaction with three substances, they often consist of a large number of components, such as in the case of a biological cell, a brain, or an ecosystem.

Many complex systems are characterized by a hierarchy of subsystems and different levels, both in time and space. In relative terms, these levels can be viewed as *micro, meso*, and *macro* levels, often equally applicable to temporal, as well as to spatial scales. In fact, often the temporal and spatial scales are correlated, so that spatially "microscopic" systems and structures are characterized by a "microscopic" time scale, i.e. "the smaller the system, the faster it moves". Similarly, larger systems are typically characterized by longer time scales (slower motions). For example, the characteristic time scale of molecules is at fractions of seconds, whereas the time scales of mesoscopic systems, such as "touchable objects", or macroscopic systems, such as solar systems, are much larger, maybe seconds to years.

In the following, we will focus on biological systems, primarily neural systems, which have resulted from a long biological evolution, with a time scale of millions of years. While this time scale is always relevant for any biological problem, often related to genetic information processing and transfer, there are much shorter time scales that should be considered, depending on the specific problem addressed. If we are interested in neural information processing associated with cognition and consciousness, the relevant temporal scales may range from milliseconds to minutes.

In general, there are many different temporal and spatial scales, ranging several orders of magnitudes, that are interlinked and interdependent in any biological system. These different scales constitute in themselves a great complexity of structures and processes. Can all this complexity be understood by chance and natural selection alone, or do we need additional laws of nature, not yet comprehended or described? Stuart Kaufmann [28] has called for such principles of organization, as some aspect of complexity:

> "Natural selection is important, but it has not labored alone to craft the fine architectures of the biosphere, from cell to organism to ecosystem. Another source – self-organization – is the root source of order. The order of the biological world, I have come to believe, is not merely tinkered, but arises naturally and spontaneously because of these principles of self-organization – laws of complexity that we are just beginning to uncover and understand".

So, how do we approach the great complexity of nature, in particular of biological systems? One way is by mathematical and computational models. This approach is today also acceptable to, or even desired by, biologists who normally shun the abstractions and simplifications that by necessity come with any mathematical treatment. The renowned biologist E.O. Wilson proclaims [53]:

> "The greatest challenge today, not just in cell biology and ecology but in all of science, is the accurate and complete *description of complex systems.* Scientists have broken down many kinds of systems. They think they know most of the elements and forces. The next task is to reassemble them, at least in mathematical models that capture the key properties of the entire ensembles."

What this kind of models can hope to achieve is primarily to find possible relations between structure, dynamics, and function. One of the toughest challenges is to relate structures and processes at different spatial and temporal scales. This is of paramount importance if we are to understand, for example, the relation between ion channel kinetics, neuronal spiking, and cortical network dynamics, associated with mental processes.

One problem is to find an appropriate measure of complexity, that can be used in the description, comparison and modelling of complex systems. There are a number of complexity measures that have been applied, also to biological/neural systems. These include various kinds of entropy measures for describing the complexity or irregularity of a system [13, 19], or more elaborate versions of multi-scale methods, such as sample entropy [30], or intrinsic mode entropy [1].

Tononi & Edelman [51] have suggested a specific complexity measure relating to cognition and consciousness, using concepts drawn from information theory. These concepts express the degree of interactions between elements of a neuronal population. The complexity of a group of neurons should be low if they fire independently (although total entropy will be high) or if they are all strongly correlated. Complexity will be high if a large number of subassemblies of varied sizes can be formed within the population.

A large cluster of neuronal groups that together constitute, on a time scale of hundreds of milliseconds, a unified neural process of high complexity is termed the dynamic core, in order to emphasize both its integration and its constantly changing activity patterns. The dynamic core is a functional cluster: its participating neuronal groups are much more strongly interactive among themselves than with the rest of the brain. The dynamic core must also have a high complexity, and its global activity patterns must be selected within less than a second out of a very large repertoire.

A part of the complexity of biological systems concerns the relation between order and disorder. For example, order could be related to regular structures or oscillatory dynamics, whereas disorder could be related to random connectivity and stochastic or chaotic dynamics.

The question of order and disorder for neural systems, also involves relations between neural activity at various scales. Traditionally, microscopic fluctuations have been regarded as detrimental to processes and functions of biological systems, and noise assumed to only corrupt information. However, there is an increasing awareness that the presumed order of biological systems has to be balanced by disorder of various kinds, and that noise and chaotic behaviors can make such systems more efficient.

It is quite clear that amplified irregular events at a microscopic level can be essential in some biological processes, such as evolution, where genetic changes are fundamental. This insight is now being recognized also in neuroscience. Internally generated fluctuations, either amplified micro events or system generated macro events may have a functional value. This also links to the relation between stability and flexibility of neural systems, where the optimal balance should be regulated according to internal and external demand.

The role of noise in neural systems was perhaps first brought to attention by Schrödinger [45] in his influential book *What is Life?*. Following his arguments, our sense organs would be useless if they were too sensitive and reacted to microscopic events, such as single action potentials. While the activity of single cells appears to be rather unpredictable and noisy, large populations of cells cooperate to produce a coherent pattern. The mass action of millions of partly noisy cells apparently makes up the orderly dynamics necessary for cognitive functions [15].

Yet, there may be even more constructive roles of noise in neural information processing. In particular, it has been found that certain sensory cells function more efficiently in the presence of noise, as a result of stochastic resonance [2, 5, 7, 44, 52].

Our own studies focus on the effects of noise (and chaos) at a cortical network level, and we have developed and used computational methods and mathematical analysis in these studies [4, 32, 33, 36, 38]. We believe such methods can be helpful in unravelling some of the intricate relations between structure, dynamics and function of neural systems. In particular, they can be used for investigating how ordered and disordered processes at various organizational levels may be linked, or how stochastic events at the microscopic level may affect the more or less orderly processes at higher levels of the nervous system. This type of investigation is essential for the understanding of a wide range of brain phenomena, including those of cognition and consciousness.

Whereas biological systems are complex and driven far away from equilibrium, most systems described mathematically in e.g. physics and chemistry are relatively simple and at equilibrium. In contrast to systems at equilibrium, complex systems far from equilibrium can exhibit a very rich dynamics, which to a large extent is unpredictable. The neurodynamics of cortical brain structures can partly be studied with experimental techniques, such as electroencephalography (EEG) and magnetoencephalography (MEG). These methods reveal a very complex dynamics, with oscillations in certain frequency bands interspersed with aperiodic, chaotic-like behavior [15, 17].

What is the significance of this dynamics? How is it related to computation and information processing in the brain, and in particular, to cognitive functions, and maybe to consciousness? It would seem that cognition requires a rich dynamics, but the relationship between neural and mental processes is far from clear. Although neuronal oscillations and chaos could be mere epiphenomena, as an inevitable consequence of the neural circuitry and without any specific biological role, these phenomena have been suggested to have functional significance.

The central idea here is that the complex dynamics of the brain, in particular at the mesoscopic scale of cortical neural networks, is a consequence of an evolutionary pressure to cope with a complex and rapidly changing environment. The internal processes of the brain presumably reflect the external processes with which they interacts.

2. Evolving Cognition

The nervous system informs the organism about the environment in which it lives and moves, processes the sensory data, relates it to previous experience, and transforms it into appropriate actions or memories. With the introduction of a nervous system, there is a tremendous increase in the way an organism can interact with the environment. Already at early stages in nervous system evolution the processing of sensory information can involve discrimination and categorization, possibly with the aid of some rudimentary learning and memory capacity. Even relatively simple organisms, such as worms, slugs, and insects are able to learn and store information in their nervous systems. In fact, much of what is known about simple learning processes and memory at a molecular level is obtained from studies of a sea slug, *Aplysia californica* [29].

For a simple nervous system, such as that of an insect or a slug, the capacity for storing new information is low. Most of its wiring would be determined genetically, which is possible for such simple but life sustaining actions as feeding, mating etc. The action control program can be specified by a small number of instructions. For example, a single specific molecule, such as a sex pheromone, can trigger the flight instructions and the mating behavior of an insect. In the more complex nervous system of, for example, a mammal, there is no way to genetically encode all the possible behaviors in the natural environment of such an animal. Instead, there must be a great capacity for learning and adaptation during the lifetime of an individual. An increase in the number of cells and synapses, and hence in the number of processing steps, will presumably make the system slower, but will also allow for a richer and more robust behavior.

At later stages in evolution, the nervous system is complex enough to allow for "Gestalt" perception, i.e. the ability to "see" an object as a whole. This ability should involve the more or less simultaneous *spatial binding* of neural activity in different parts of a cortical area [9, 10]. Although such binding has been discussed primarily with regard to vision, all sensory modalities of higher animals can be considered to function with perceptions based on patterns of nerve activities. The ability to form "internal objects" or percepts of neural activity patterns has apparently increased tremendously during evolution from reptile to mammal. Through such perceptual binding, internal representations of the external world are formed, resulting in an overall *model* of the world. An increased complexity of the evolved nervous system also enables the organism to make better predictions about the future environment, based on previous and present experience. The success of these predictions depends to a large extent on the learning and memory capacity of the animal.

At still higher evolutionary levels, reasoning, planning, and abstract thinking come into play. The behaviour of such an organism becomes less dependent on direct peripheral stimuli, although that will always be of great survival value. Communication and social interaction becomes important for the further development of consciousness [6].

Clearly, the electro-chemical (or other) processes that take place in the intricate neural networks of the brain can apparently lead to various internal states and mental processes of an animal. In lower animals the nervous system would allow only for "primitive mental" processes, such as perception, learning, recalling, drives, and emotions. To a large extent the animal is presumably unconscious of these internal processes, and it is only aware, or conscious, of the external environment.

As the nervous system evolves and attains a higher degree of complexity and organization, its processes become more advanced and sophisticated, including higher cognition, as an act of knowing, thinking, reasoning, believing, and willing. Eventually, (symbolic) language becomes an important and integrated part of the cognitive functions [43], facilitating reflection and communication.

In birds and mammals, and most articulate in man, the ability to "internalize" the world has come to include the formation of *temporal* patterns. That is, binding together sequences of neural events that would correspond to some external or internal sequences of events. This *temporal binding* allows us to perceive and understand relations between cause and effect, and to experience "the arrow of time".

For humans, the internal model of the world also includes other individuals and their minds. Models of other individuals are important for determining the significance of their behavior, and for predicting the "next steps" in that behavior.

Self-consciousness, a sense of an "I", may develop when experienced events are given a sequential context in the internal model of the world. It is obvious that many cognitive functions (but not consciousness in general) depend upon memory, but the "I" needs memories well structured in time, in order to be perceived. Obviously, also planning for the future, involving attention as well as intention, is an essential component in the making of an "I" [27]. Our subjective experience of a decision to act, as well as of a sensory stimulation, seems to require a time span of a few to several hundred milliseconds [9, 31, 47].

Cognitive functions are sometimes grouped into preattentive and higher cognitive functions, respectively. The latter type of functions are usually considered individual, or "private", in character, implying they are hardly describable objectively. The totality of all cognitive functions could be referred to as the mind. Mind can thus be regarded as involving a set of processes, including perception, emotion, memory, thinking, cognition, and reasoning. Some of these may be conscious, while others may not. In either case, typical processes related to mind would require a very high complexity of the nervous system (or any material system keeping such processes going), and cognitive functions could be seen as emergent phenomena, arising from the collective neural activity of the brain, or subsystems thereof [22, 25, 26, 39, 46, 48]. This would imply that these phenomena depend upon, but are not reducible to events at the single neuron level.

From an evolutionary point of view, it seems that cognitive functions would appear as a result of the increased complexity of the nervous system [21] (see Fig. 1). Even if one adopts the view that there is some degree of awareness (perception) in unicellular organisms, the kind of "higher" cognitive functions we are discussing here, should depend on the cooperative activity of a large number of interconnected neurons. It is also reasonable to assume that the relevant dynamics of the neural structures have evolved to cope with events and changes in the environment of the organism. Finally, since such functions as perception and associative memory supposedly involve neural activity of rather large cortical areas, corresponding to millions of neurons, it can be concluded that one should look for the material basis of cognitive functions at the level of neural networks. Networks must be crucial if mental processes are associated with patterns of neural activity.

Brain structures are characterized by their complexity in terms of organization and dynamics. This complexity appears at many different spatial and temporal scales, which in relative terms can be considered as micro, meso, and macro scales. The corresponding dynamics may range from ion channel kinetics, to spike trains of single neurons, to the neurodynamics of cortical networks and areas. The high complexity of this neurodynamics is partly a result of the web of non-linear interrelations between levels and parts, including positive and negative feedback loops [3, 4, 17].

In particular, cognitive functions such as perception, learning and memory, appear to depend on the neurodynamics of large cortical areas. These phenomena are associated with patterns of neural activity, presumably corresponding to internal representations of the environment and of the self. The spatial scale supposed to be relevant for these cognitive functions is in the order of millimeters or larger, and the corresponding temporal scale on the order of one to a hundred milliseconds. All this seems to be a result of evolutionary pressure on the nervous system to efficiently

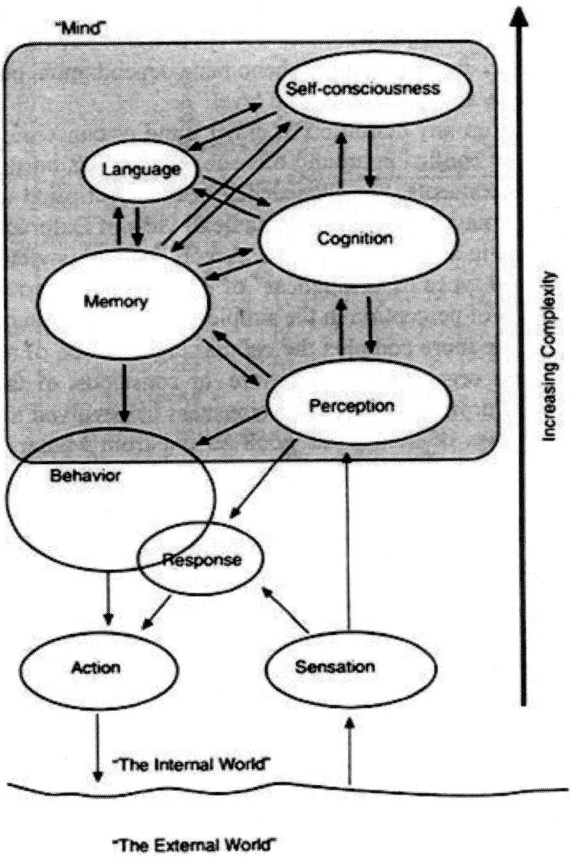

Fig. 1: A crude attempt to map various cognitive functions and relate them to each other on an "evolutionary scale" that is at the same time a measure of the degree of complexity. An increasing complexity of the biological structures allow for an increasing complexity in interactions within an organism and between its "internal" and "external" worlds. (Adopted from Liljenström, 1997).

deal with and adapt to rapid environmental changes and events.

All forms of knowledge are used by the organism in its interaction with the environment. There is usually some knowledge processing before an event in the environment (a stimulus) will result in an adequate response behaviour. Different strategies have been used during evolution to optimize the response patterns. Different requirements are imposed concerning the rate of the response, the energy used, or concerning the balance between flexibility and stability, not all easily compatible [34, 35].

Also for the central nervous system there are certain optimization problems to be solved. Presumably, the nervous system has evolved to process and store information in an efficient way, i.e. it should be optimized for (i) maximal processing rate and (ii) maximal information storage capacity. It seems likely that the nervous system has evolved primarily for a fast information transfer, from sensation (of the environment) to action. A high information storage capacity would supposedly be of a secondary, and phylogenetically younger origin, evolved for more advanced behavior in a more complex interaction with the environment.

The word "information" should be used with caution. Often, when discussing the brain's capacity for signal/information processing, one treats the environmental input as if there were many signals (implying information) with noise superimposed. Then, one discusses the capability for increasing the signal to noise ratio. However, the information of the "environmental signal" only becomes apparent, in some sense is created, in the brain. Much of the sensory input is "nonsense", until it reaches higher cortical areas where it is attaining meaning. What the brain does is to make meaning out of sensory input [16, 41]. This process involves a lot of filtering, clustering, separation, association, etc. While information alone can never produce mind, the organized amount of information processed per unit time seems to be important. A high flow rate of information may not be sufficient for mind, but it is very likely a prerequisite for it. Also the (dynamical) state of the brain is of great importance for how the information is processed, and what the result of this processing will be.

We cannot easily study evolution, let alone different options that evolution might have had during the history of life. However, with computer models we can simulate various evolutionary steps and study the effects of altered complexity and organization of the neural structures and processes. Computer models can help us understand the relationship between structure, dynamics, and function, and allow us to test different possibilities. However, it is important to bear in mind that all variables and parameters introduced in a model should have a counterpart in the real system. With appropriate computer models we can make simulation "experiments" that would be inconceivable, or would take an unrealistic long time to carry out with the real system. For example, if a model parameter can be related to the concentration of a certain transmittor substance or neuromodulator, its effects on the dynamics and functions of the system can be investigated in detail using computer simulations.

With the type of computer models we have developed and reported elsewhere [32, 33, 36, 37], attractor neural networks with biologically inspired connectivity, it is possible to capture features of real neural systems and get simulation results that in many respects closely resemble experimental results. This kind of models can be useful for finding out which neural properties and network architecture could be responsible for the characteristic dynamics of the real system. By exploring the parameter space for e.g. connection strengths, delays, gains, etc. we can point at plausible solutions to neurobiological problems, and put constraints on the possibilities of the natural system. Fig. 2 shows a computer simulation of a pattern recognition situation, where the system continuously learns and recalls patterns corresponding to different sensory inputs.

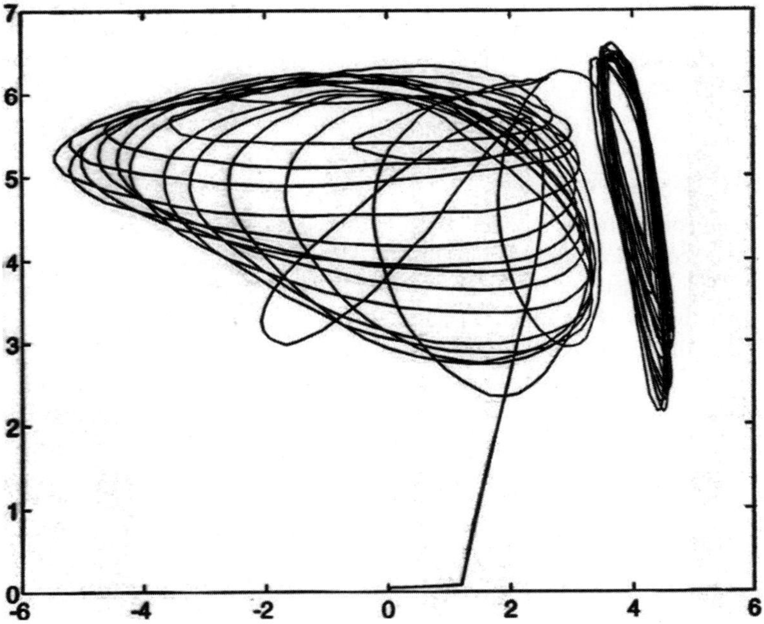

Fig. 2: Computer simulation of continuous learning and recall. The two separate "limit cycle attractors" correspond to two separate memory states; the left one to a previously stored pattern and the right one to a new memory state, which results from learning an unrecognized input pattern. The trajectory corresponds to the output signals for two arbitrary chosen network nodes, plotted against each other. Scales are arbitrary.

Although we cannot make any conclusive statements based on the models alone, theories for certain cognitive functions, such as perception, learning and memory can be tested, and critical experiments can be suggested. In our previous work, we have pointed at the relationship between the network structure and the complex dynamics it displays, using neural network models of the different cortical structures [21, 23, 32, 37]. Furthermore, we have linked the dynamics of the system to its function, and shown that the complex behavior can enhance the performance, by increasing fidelity and rate of information processing [3, 34, 38].

We do not believe that we are able, maybe ever, to model or understand all aspects of conscious mind, such as subjectivity, or understanding itself, but certain mental processes might lend themselves to mathematical modeling and computer simulation. These would primarily be cognitive functions based on neural computations, that could be mimicked in an artificial system. Instead of arguing about which cognitive functions might be based on computable processes, and which might not be, we use computational modeling to explore how far such models may be carry us in the direction towards an understanding of the brain-mind relation(s). Perception, learning, associative memory, prediction, and even attention are fundamental cognitive functions that people try to describe and simulate with computer models [9, 14, 21, 22, 26].

3. Evolving Consciousness

There is no consensus on any definition of neither mind nor consciousness. Indeed, much of the confusion around this subject depends on the disparate usage of these concepts. For consistency in the description of biological processes, one may adopt the (rather radical) view of Max Delbrück [12], and use the terms in a very broad sense. According to this view, consciousness includes all types of "awareness" of external and internal events, from a crude form of perception in the simplest organisms to self-consciousness in Man (c.f. Fig. 1). The more complex the cellular organization of an organism is the more this organism can be aware, or conscious, of the external and internal world. In this sense, consciousness has evolved together with other properties of life. In his book [12], *Mind from Matter?*, Delbrü regards perception as the basic form of consciousness and he states (p. 43) that,

> "The unity and continuity of life is equally manifest in its psychic aspects. Perception in plants and animals is a familiar phenomenon, but the beginnings of perception are also clearly present in microorganisms, in which adaptive behavior demonstrates that they can detect and evaluate signals from the environment and respond appropriately". This is not a dominating view in science today, but it is appealing in the sense that it relates processes in simple organisms with those in higher, and provides a continuity in the evolution of consciousness. Yet, we find it reasonable to believe that the emergence of conscious cognition was a major transition in the evolution of life. Increasing levels of consciousness, we believe, are associated with increasing complexity, and conscious cognition would require a rather complex nervous system, not found in any primitive animal.

We assume there has been an evolution of consciousness in smaller or larger steps, but in parallel and in interaction with the evolution of the nervous system. At present, we cannot know at what stage in evolution, with what organism, the first signs of conscious cognitive processes appeared, but it can be argued that mammals, and possibly birds, possess this quality. These animals are believed to have Gestalt perception of objects and are able to think in abstract symbols, to a lesser or higher degree (for example, a chimpanzee can do this to a much higher degree than a mouse). They can be assumed to have subjective experiences, although not necessarily be aware of themselves as individuals [20].

Cognitive processes are clearly multi-levelled, and in humans these would range from unconscious to highly conscious cognitive activities. All cognitive processes are clearly not conscious. On the contrary, most cognitive processes are unconscious, but at some degree of complexity and arousal some cognitive processes become conscious. Furthermore, it is apparent that consciousness appears at different levels of complexity. There are reasons to believe that full human consciousness shows more levels than that of any other species. For instance, there is no clear evidence that other species possess more than rudimentary forms of self-consciousness, i.e. consciousness/awareness of a self and the capability to reflect about it. The emergence of self-consciousness and human language must have constituted another major transition in evolution. It implied a critical, goal-directed thinking which has changed the world dramatically.

We think consciousness is central for higher cognitive functions, in that conscious cognition would provide prediction, expectation, volition, plans, goals, hopes etc. beyond the immediate perception. It would have consequences for choosing mating partners, for securing good living conditions for offsprings etc. We conjecture it would be far more efficient than unconscious "planning". Conscious cognitive functions may also permit a more complex interaction with other individuals and with the environment.

Predictions and expectations that bring the cognitive processes into a (distant) future also open up for a more advanced planning and setting up of goals. Free will can be developed out of desires, and this should constitute a strong driving force in nature. Thus, these conscious functions can indeed interact with the complexity of the nervous system, leading to further complexity by speeding up evolution towards higher complexity.

The complex neurodynamics, primarily at a mesoscopic level of the brain, seem significant for cognitive functions and conscious activity. It has been related to perception, attention and associative memory, but also to volition and activity in the sensory and motor areas of the brain. Even though many details are still unknown, it is obvious that there is an interplay between the neurodynamics of the sensory and motor pathways, essential for the interaction with our environment [8, 42].

We explore our world in a perception-action cycle [17]. The development of our cognitive and conscious abilities depend on an appropriate interaction with the complex and changing environment, in which we are embedded. Our perceptions and actions develop and are refined to effectively deal with our world.

Intention can be viewed as a precursor to our (free) will, as an "urge" or "desire" to act in a certain direction, to attain a certain goal. Voluntary movement, or more generally, behaviour, is based on perception and past experience (memory), which are required for prediction of internal and external interactions. *Attention* may provide information about the internal and external worlds, but intention guides our actions.

Attention and intention could be seen as complementary aspects of consciousness, with their effective neural counterparts of perception and action, respectively, when implemented through the interactions within sensory and motor hierarchies of the nervous system. They correspond to an "inward" and an "outward" going activity of consciousness (see Fig. 2). Both of these aspects seem essential for exploring our external, as well as internal world.

Attention and intention, as complementary aspects of consciousness, are presumably not only part of the human experience, but could be much more universal. For example, (visual) attention has been studied in insects, such as fruit-flies [50] and honey bees [49], and can be expected to be found also in other invertebrates. Likewise, intentional behaviour can be attributed to many types of (in particular social) animals, including crustaceans [24, 40].

Indeed, these aspects of consciousness might be intimately linked to life at all levels, albeit not in the same sense as we use the terms for humans (or other mammals). Naturally, in organisms without a nervous system, attention and intention could never be implemented through any sensory or motor hierarchy of such a system, but would rather have to be implemented through other molecular or cellular structures.

While intention and attention may be ubiquitous to all life, higher conscious cognition, and in particular the awareness and reflection of a self, seems to appear only with human beings, resulting in the notion of a free will and responsibility for our actions. As far as we can tell, no other species would have such considerations, and also no sense of a self with a beginning (birth) and an ending (death), which eventually may lead to the belief in an immortal soul with a purpose in life, as fundamental aspects of any religion.

4. Evolving Conclusions

Although we regard consciousness, in its broadest sense, to be part of an evolutionary process, it is possible that its fundamental aspects, such as subjectivity and sense of free will, may not have any survival value. They could just be a consequence of the complexity of the human brain. Yet, there are many reasons to believe that (the evolution of) a conscious mind actually does have a functional role. For example, it could give an additional motivation to stay alive, to "keep the self intact". By perceiving a self that should live on for many years, there can be better planning both for the individual and for coming generations. It also enables the individual to endure hardships for rewards that may come later in life, hardships that perhaps would be avoided if a future life were not seen. It enables man to behave against "the law of least resistance". Thus, the capabilities to predict future events have been extended to the anticipation and expectation of a continuous individual life. Also, if one has a (good) model of oneself one can infer one's own thinking, feelings and behaviors also to others, thus allowing for a more efficient interaction with other individuals, and better prediction of their behavior. The ultimate pattern in our understanding of the world results in an experience of *meaning*, a purpose and a context in which all (or most) events can be ordered and related to each other.

This kind of reasoning leads us to a view that knowledge is (at least) as old as life itself, and that knowledge acquisition was essential already for the first organisms, to increase their survival probability. It leads us to see knowledge acquisition and learning as biological processes that have existed and evolved gradually with the life forms. A major step was taken with the introduction of a nervous system, which enabled a tremendous increase in the way an organism could interact with the environment. In summary, an evolutionary perspective leads us to the view that adaptation and (phylogenetic and ontogenetic) learning is a widespread and old property of living organisms. It was an integral and essential part of early life forms, that appeared about 3.5 billion years ago. We suggest that knowledge processing mediated by a centralized nervous system, that is cognition, shows the same principal features as non-neural adaptive processes. Similarly, consciousness can be said to appear, to different degrees, at different stages in evolution. Both cognition and consciousness, and in particular conscious cognition, depends heavily on the organization and complexity of the organism, primarily with regard to the nervous system.

Consciousness and cognition apparently evolve through interaction with the environment, where the organism is embedded. Such exploration of the environment requires both attention and intention. The evolution of consciousness presumably occurs through larger leaps when new species appear, and through smaller steps within a species and even smaller steps within an organism. A great leap of consciousness must have happened when we became aware of ourselves and our own cognitive activity. That leap should also include an awareness of our intentions, a sense of a free will. This more advanced consciousness is presumably experienced, or at least reflected on, by humans alone.

If intentionality is a fundamental aspect of consciousness, and if an extended form of intentionality corresponds to our sense of free will, it is difficult to fit it within the framework of current science, which is based on deterministic laws and chance. In our view, consciousness provides a freedom beyond the indeterminism of quantum physics or the determinism of classical physics.

The effects of consciousness is most clearly linked to intention, as its more active aspect, which has to be considered if we are to understand its role in evolution and in the universe. Exploration, through attention-perception and intention-action, is fundamental to all levels of existence. In this sense, consciousness is inseparable from life, expanding and evolving together. It is expressed and evolving through inter-relations (attention) and inter-actions (intention). Hence, consciousness can be regarded as a driving force in nature. (Fig. 3 summarizes this view).

To conclude, while cognition and consciousness may be different in nature, their evolution seems to depend on a growing complexity of the correlated structures and processes. Such complexity can be expressed through relations of various kinds. Life, cognition and consciousness concern and evolve through relations,: between different parts or 'aspects' of an individual (e.g. attention/perception and intention/action), as well as between individuals and their environments.

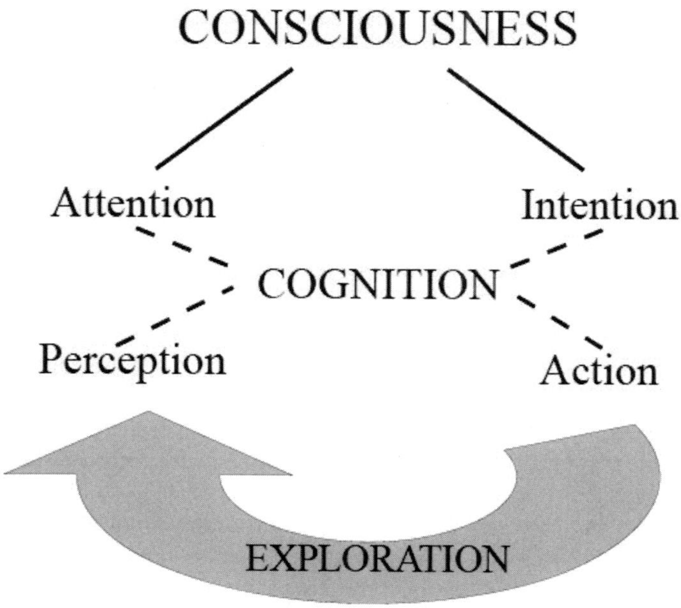

Fig. 3: The relation between the "implicate" order of consciousness, as in the dual aspect of attention and intention, and the "explicate order" of exploration, as in the perception-action cycle.

As the biological structure, notably the neural system, grows and becomes more complex, higher and higher levels of consciousness can be expressed, where the organism becomes increasingly aware of its external and internal environment, leading eventually to human self-awareness. Further evolution may lead to conscious choices and insights of the consequences of thoese choices, which in turn may result in reflections on responsibility and moral behavior.

With such an insight, we may realize that we are living in a world of emergent evolution, entangled in a web of life, with the future of our own and other species in our hands. Let us carefully explore our outer and inner universes, cultivating our relations in all directions and all levels. This may be our only chance!

References

[1] H. Amoud *et al.*, *J Neuroengineering Rehabil* **4** (12), (2007).
[2] V. S. Anishchenko, A. B. Neiman, M. A. Safanova, *J. Stat. Phys.,* **70**, 183–196, (1993).
[3] P. Århem, H. Braun, M. Huber and H. Liljenström, Non-Linear state transitions in neural systems: From ion channels to networks. In: H. Liljenström & U. Svedin

(Eds.) *Micro - Meso - Macro: Addressing Complex Systems Couplings*, 37–72 (World Scientific, 2005).
[4] P. Århem, H. Liljenström, Fluctuations in neural systems: From subcellular to network levels. In: F. Moss & S. Gielen (Eds.) *Handbook of Biological Physics,* **4** Neuroinformatics, Neural Modelling, 83–125 (Elsevier, 2001), .
[5] G. Balazsi, L.B. Kish, F. Moss, *Chaos* **11**, 563–569 (2001).
[6] H. Barlow, The biological role of consciousness. in *Mindwaves*, 361–374 (Blackwell 1987).
[7] A. Bulsara, E. W. Jacobs, T. Zhou, F. Moss, and L. Kiss, *J. theor. Biol.,* **152**, 531–555 (1991).
[8] R. Cotterill, *Enchanted Looms – Conscious Networks in Brains and Computers*, (Cambridge University Press, 1998).
[9] F. Crick, *The Astonishing Hypothesis - the scientific seardh for the soul*, (Simon & Schuster, 1994).
[10] F. Crick and C. Koch, Towards a neurobiological theory of consciousness, *Seminars Neurosci.* **2**, 263–275 (1990).
[11] C. Darwin, *On the Origin of Species by Means of Natural Selection* (John Murray, 1859).
[12] M. Delbrück, *Mind from Matter?* (Blackwell Scientific Publ, 1986).
[13] J.-P. Eckmann, D. Ruelle, *Reviews of Modern Physics* **57**, 617–656, (1985).
[14] G. Edelmann, *The Remembered Present –A Biological Theory of Conciousness*, (New York: Basic Books, 1991).
[15] W.J. Freeman, *Mass Action in the Nervous System* (New York: Academic Press, 1975).
[16] W.J. Freeman, *Sci. Am.* **264**, 78-85 (1991).
[17] W.J. Freeman, *Neurodynamics: An Exploration in Mesoscopic Brain Dynamics* (London: Springer, 2000).
[18] M. Gell-Mann, *The Quark and the Jaguar: Adventures in the Simple and the Complex*, (New York: W. H. Freeman & Co. 1994).
[19] P. Grassberg, I. Procaccia, *Phys. Rev. Lett.* **50**, 346-349 (1983).
[20] D.R. Griffin, *Animal Minds*, (University of Chicago Press, 1992).
[21] Y. Gu, H. Liljenström,*Cognitive Neurodynamics* **1**, 275–285, (2007).
[22] H. Haken, *Synergetic Computers and Cogntion* (Springer, 1991).
[23] G. Halnes, H. Liljenström, P.Århem, *Biosystems* **89**, 126–134 (2007).
[24] M.D. Hauser, D.A. Nelson, *TREE* **6**, 186–189 (1991).
[25] W. Heiligenberg, *Ann. Rev. Neurosci.* **14**, 247–267 (1991).
[26] J. Hopfield, Neural networks and physical systems with emergent collective computational abilities, in *Proc. Natl. Acad. Sci. USA* **79**, 2554–2558 (1982).
[27] D.H. Ingvar, The will of the brain: Cerebral correlates of willful acts, *J. theor. Biol* **171**, 7–12 (1994).
[28] S. Kaufmann, *At Home in the Universe – The Search of the Laws of Self-Organization and Complexity* (Oxford University Press, 1995)
[29] E.R. Kandel, *Cellular Basis of Behavior, an introduction to behavioral neurobiology* (W. H. Freeman and Company 1976).
[30] D.E. Lake, J.S. Richman, M.P. Griffin, J.R. Moorman, *Am J Physiol Regul Integr Comp Physiol* **283**, R789–97, (2002).
[31] B. Libet, *Behav. Brain Sc.* **8**, 529–566 (1985).
[32] H. Liljenström, *Int. J. Neural Syst.* **2**, 1–15 (1991).
[33] H. Liljenström, *Int. J. Intell. Syst.* **10**, 119–153 (1995).
[34] H. Liljenström, *Intl. J. Neural Systems* **7**, 497–505 (1996).

[35] H. Liljenström, *Neuropsychopharmacology* **28**, S64-S73 (2003).
[36] H. Liljenström, Inducing transitions in mesoscopic brain dynamics. In: D. A. Steyn-Ross, M. Steyn-Ross (Eds.), *Modeling Phase Transitions in the Brain*, **149**, 149–179 (Springer Series in Computational Neuroscience 4, 2010).
[37] H. Liljenström, M. Hasselmo, *J. Neurophysiol*, **74**, 288–297 (1995).
[38] H. Liljenström, X. Wu, *Intl. J. Neural Systems*, **6** 19–29 (1995).
[39] D. MacKay, *Brains, Machines & Persons* (London: W. Collins Sons & Co. 1980)
[40] J. Mather, *Journal of Cosmology*, **14** (2011).
[41] H.R. Maturana, F.J. Varela, *The Tree of Knowledge* (Boston: Shambala, 1992).
[42] B. Merker, Nested ontology and causal options: A paradigm for consciousness (2010).
[43] P. Munz, The Phenomenon of Consciousness from a Popperian Perspective. In: H. Liljenström P. Århem (eds.), *Consciousness Transitions- Phylogenetic, Ontogenetic and Physiological Aspects.* (Elsevier 2008).
[44] D. Russell, L. Wilkens and F. Moss, *Nature* **402**, 219–223 (1999).
[45] E. Schrödinger, *What is Life?* (Cambridge University Press, 1943).
[46] C. Skarda, J. Freeman, *Brain and Behavioral Sciences* **10**, 161–195 (1987).
[47] C.S. Soon, M. Brass, H-J. Heinze, J-D. Haynes, *Nature Neuroscience* **11**, 543–545 (2008).
[48] R.W. Sperry, *Science and Moral Priority,* (New York: Columbia University Press, 1983).
[49] M.V. Srinavasan, *Annu. Rev. Entomol* **55**, 267–284 (2009).
[50] B. van Swinderen, The remote roots of consciousness in fruit-fly selective attention, In: H Liljenström, & P. Århem (Eds.), *Consciousness Transitions - Phylogenetic, Ontogenetic and Physiological Aspects.*, 27–44 (Elsevier, 2008)
[51] G. Tononi & G. M. Edelman, *Science* **282**, 1846-51 (1998).
[52] K. Wiesenfeld and F. Moss, *Nature* **373**, 33-36 (1995).
[53] E.O. Wilson, *Concilience* (New York: Knopf 1998).

SELF-ASSEMBLY, MODULARITY AND PHYSICAL COMPLEXITY

S.E. AHNERT

TCM, Cavendish Laboratory, University of Cambridge
JJ Thomson Avenue, Cambridge CB3 0HE, United Kingdom
E-mail: sea31@cam.ac.uk

Self-assembly is ubiquitous in physics, chemistry and biology, and has many applications in materials science and engineering. Here we present a general approach for finding the simplest set of building blocks that will assemble into a given physical structure. Our procedure can be adapted to any given geometry, and thus to any given type of physical system. The amount of information required to describe this simplest set of building blocks provides a quantitative measure of the structure's physical complexity, which is capable of detecting any symmetry or modularity in the underlying structure. We also introduce the notions of joint, mutual and conditional complexity for self-assembling structures. We illustrate our approach using self-assembling polyominoes, and demonstrate the breadth of its potential applications by using it to quantify the physical complexity of protein complexes.

Keywords: Self-assembly; Algorithmic Complexity; Polyominoes

1. Introduction

1.1. *Algorithmic Complexity*

Complexity is a widely used term, with many definitions. One of the most rigorous of these is the notion of *algorithmic complexity*, a concept discovered independently by Kolmogorov [1] and Chaitin [2]. For a given bit string the algorithmic complexity is the length of the shortest program on a universal Turing machine that will produce the string in question. This quantity is uncomputable, but the length of any algorithmic description of a given piece of information forms an upper bound on its Kolmogorov complexity. Given a particular algorithmic language we can therefore obtain useful estimates of the relative complexity of different data sets. This approach has been successfully applied to microarray data analysis, for example in 3. Here we will address the question whether it is possible to use the same framework to measure the physical complexity of a structure. For this to be possible we need to first define an equivalent of the programming language for a structure. We will use self-assembly rules as our algorithmic framework.

1.2. *Self-assembly*

Self-assembly is a widespread process in physics, biology, chemistry and materials science. Self-assembling structures include micelles [4], DNA [5], protein complexes [6], and viruses [7]. The building blocks of self-assembling systems contain the information about the final structure within them, and do not require an external agent to drive or direct the assembly process. While the framework we introduce is completely general, we will focus our attention mostly on lattice structures known as polyominoes.

2. Polyomino Self-assembly

Historically, polyominoes have been studied extensively in the context of combinatorics [8], but have recently also attracted the attention of researchers interested in structural complexity [9]. In 2D, polyominoes are connected sites on a two-dimensional square lattice. We will view the sites as square tiles, and each of the four sides of a tile has a 'colour', which can attract one or more other colours. These attractions are defined by an interaction matrix between colours. A given structure can usually be constructed using a variety of building block sets. A set of building blocks, together with the interactions between colours, constitutes a *self-assembly kit.* In our model a seed tile initiates the self-assembly and then tiles are randomly selected from the building block set and placed randomly next to existing bound tiles (initially only the seed tile). If attracting colours are adjacent, the tile becomes bound. Given a particular structure A we can minimize the information $I(A)$ required to specify the building block set that gives rise to this structure. This particular building block set we will call a *minimum self-assembly kit* of A and the minimum of $I(A)$ over all kits that build the same structure A is $K(A)$, the *complexity* of A. The quantity $I(A)$ is given by:

$$I(A) = \log_2(c_s + 1) + \sum_j c_j \log_2 c + \log_2 N'(c_j)$$

where c_s is the number of self-interacting colours, c is the total number of colours, and c_j is the number of colours on block j. $N'(c_j)$ is a combinatorial factor which enumerates the ways of distributing exactly c_j colours on the four faces of the building block, taking into account rotational symmetry. This term is related to the theory of necklaces in combinatorics. For a more detailed explanation of the terms in $I(A)$, see 10.

For a tree-like polyomino (see 1) we can find the minimum kit by iteratively labeling those blocks in the structure that have one unlabelled neighbour, starting at the leaves of the tree. The labels at each stage distinguish differing topological neighbourhoods. For non-tree-like polyominoes we can search across all spanning trees of the contact graph of the blocks. However, for certain non-tree-like polyominoes, particularly ring structures, this will not find the minimum kit. For such structures we can employ a second type of iterative labeling algorithm in which we label all blocks simultaneously, in every iteration, according to the labellings of their neighbourhood. To find the minimum kit for a given arbitrary structure we must therefore now search across all subgraphs of the contact graph, and apply the two iterative labeling algorithms to the tree-like and ring-like parts respectively. This search is computationally expensive, but when dealing with real world structures we can often make educated guesses or assumptions that will drastically cut down the number of possible subgraphs.

3. Studying Evolution with Polyomino Self-assembly

Genetic algorithms are optimization algorithms that employ an evolutionary principle. A population of potential solutions to a given problem, encoded in some suitable 'genomic' form, reproduces asexually or sexually while being subject to mutations as well as selection pressures implemented by a fitness map. The fitness criterion drives the population towards the best solution to the problem. By viewing the colouring of building blocks as the 'genotype' and the final structure as the 'phenotype', we can evolve towards a given target structure, or a given desired property of the target structure. Fig. 2 shows the distribution of complexity for polyominoes with target size eight (so-called octominoes). What is interesting is that the symmetric and modular structures are much more frequent that asymmetric and non-modular structures.

4. Protein Complex Self-assembly

The above approach for measuring the complexity of a given physical structure employs self-assembly as a *language* for describing a structure. When using this approach to measure the complexity of a real structure it therefore does not matter whether this structure does in fact form through self-assembly in the physical

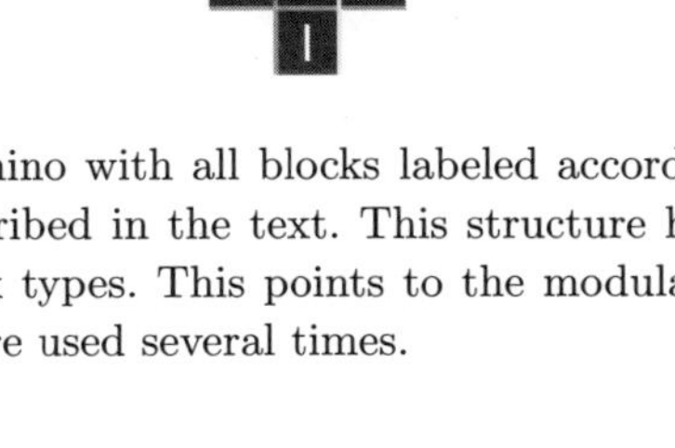

Fig. 1: Tree-like polyomino with all blocks labeled according to the first iterative labeling algorithm described in the text. This structure has 23 blocks and 9 self-assembly building block types. This points to the modularity of the structure, as many building blocks are used several times.

world. If it does however, the minimum assembly kit helps to identify symmetry and modularity in the structure, since these features need to be exploited in order to describe the structure in the most efficient way, leading to the repeated use of the same building block types. Symmetry and modularity play an important role in many real physical or biological self-assembly systems.

When applying this algorithm to real self-assembling structures we use a similar approach to the polyomino system with the main difference lying in the geometry and topology of the building blocks. These depend on the physical system in question. The particular example we will look at for the rest of this paper is that of protein complexes. Proteins bind together to form these quaternary structures, which fulfill many important functions in biological organisms. The structural data we use comes in the form of weighted contact graphs between the proteins in a complex. The weights specify the number of residues in contact between a pair of proteins. We obtain this data from the 3DComplex database [11], a comprehensive library of protein quaternary structure.

We run a version of the minimum assembly kit labeling algorithm on these structures, with two notable differences to the polyomino case: Firstly we assume that ring structures are built by self-limiting growth, which means that from the

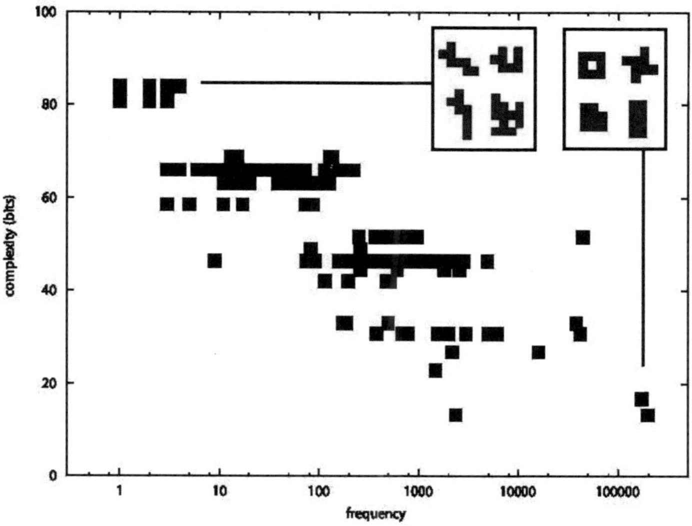

Fig. 2: Distribution of octominos evolved using a genetic algorithm. The most frequent structures are also those with the lowest complexity – in other words the most symmetric and modular.

perspective of the assembly block it is forming an infinite chain. Only the geometry of the building block turns the infinite chain into a closed loop. In other words, if proteins form a ring of size six they are not 'counting' to six (or three, exploiting symmetry). This is a reasonable assumption for protein complexes. Secondly, we do not search across all subgraphs of the complex but instead calculate the complexity for the series of subgraphs formed by successively removing interfaces, from the weakest to the strongest. The thought behind this is that the interfaces of a contact graph in a self-assembling structure can usually be divided into circumstantial and essential self-assembly interfaces, and that the weaker interfaces are much more likely to be circumstantial. Many different structures in the 3DComplex database, with different numbers of edges, turn out to have the same minimum self-assembly kit. This allows protein complexes to be classified in a more systematic way. It is also known that complexes assemble by joining the subunits in order of decreasing interface size. By considering the different interactions between the different building block types, and by switching them on in order of decreasing average interface size, we can therefore predict the sequence in which these complexes assemble. Furthermore we can search the 3DComplex database for complexes that match the assembly intermediates of this complex, pointing us to its potential evolutionary precursors. The relationship between evolution and assembly of protein complexes is discussed in great detail in 6.

5. Joint, Mutual and Conditional Complexity

The language of self-assembly allows for the definition of distance measures between structures by considering the notion of a *joint assembly kit* of two structures A and B. Appropriate subsets of this joint kit will assemble into A or B respectively, and the least amount of information required to describe such a joint kit is the *joint complexity* of A and B. The amount of information required to describe the overlap between the two subsets is the *mutual complexity*. Similarly, the additional amount of information required to build structure B given the set for structure A is the *conditional complexity* of B given A. These measures can serve to measure the distance between structures. Two structures can look very different and yet be closely related in terms of these distances, which makes this approach a useful tool for the systematic analysis and classification of structural data.

6. Conclusion

We have introduced self-assembly as a language for describing physical structures, thereby providing a quantitative measure of their physical complexity. This approach can be applied to polyominoes, which also serve as a model for studying evolutionary questions such as the emergence of symmetry and modularity, and the increase in complexity throughout evolution. The self-assembly framework can also be applied to protein complexes in order to classify them and make predictions about their likely assembly pathway. Finally this approach can also be used to construct distance measures between physical structures through the concepts of joint, mutual and conditional complexity. A much more detailed account of this work can be found in 10 and 12.

7. Acknowledgements

The author is supported by the Royal Society, UK.

References

[1] A. N. Kolmogorov, *Problems in Information Transmission* **1**, 1 (1965).
[2] G. J. Chaitin, *Journal of the ACM* **13**, 547 (1966).
[3] S. E. Ahnert, K. Willbrand, F. C. S. Brown and T. M. A. Fink, *Bioinformatics* **22**, 1471 (2006).
[4] J. Israelachvili, *Langmuir* **10**, 3774 (1994).
[5] E. Winfree, F. Liu, L. A. Wenzler and N. C. Seeman, *Nature* **394**, 539 (1998).
[6] E. D. Levy, E. B. Erba, C. V. Robinson and S. A. Teichmann, *Nature* **453**, 1262 (2008).
[7] H. Fraenkel-Conrat and R. C. Williams, *Proceedings of the National Academy of Sciences of the United States of America* **41**, 690 (1955).
[8] S. W. Golomb, *Polyominoes: Puzzles, Patterns, Problems, and Packings*, 2nd edn. (Princeton University Press, 1991).
[9] D. Soloveichik and E. Winfree, *SIAM J. Comput.* **36**, 1544 (2007).
[10] S. E. Ahnert, I. G. Johnston, T. M. A. Fink, J. P. K. Doye and A. A. Louis, *Phys. Rev. E* **82**, p. 026117 (2010).
[11] E. D. Levy, J. B. Pereira-Leal, C. Chothia and S. A. Teichmann, *PLoS Comput Biol* **2**, p. e155 (2006).
[12] D. J. P. K. Johnston Iain G., Ahnert Sebastian E. and L. A. A., *Submitted to Physical Review E* (2011).

THE CATEGORY OF TOPOLOGICAL THERMODYNAMICS

R.M. KIEHN
University of Houston, Texas, USA
E-mail: RKiehn2352@aol.com
http://www.cartan.pair.com

The Category theory of Topological Thermodynamics and Continuous Topological Evolution can be used as an abstract, but universal, foundation for understanding, among other things, scale invariance, chaos, emergence, self-similarity, dynamical systems, irreversible processes, and other non-linear, and even statistical, phenomena. The abstract Category theory, as an exterior differential system, has solutions which can be put into correspondence with numerous physical experiments in different disciplines.

1. Introduction

Mathematicians consider a Category Theory, C, to be a collection of abstract objects, $ob(C)$, constructed in terms of ingredients, with morphisms that continuously map an object X to another object Y. Herein, the interest is focused on the abstract Category of objects and morphisms that represent Topological Thermodynamics and Continuous Topological Evolution. The objects are a collection of finite topological spaces of sets whose ingredients are exterior differential forms and/or differential form densities. The morphisms between sets are constrained to be homotopic classes of topologically continuous but possibly non-invertible maps. The coefficient functions of the exterior differential forms define abstract, and universal, thermodynamic systems. The homotopic, topologically continuous, morphisms define abstract thermodynamic processes. These morphisms include diffeomorphic processes that preserve geometry and metric scales, homeomorphic processes that preserve deformable topological features, which are scale invariant, and irreversible processes that permit topological change.

2. Topological Spaces of Exterior Differential Forms

As all physical measurements are over finite space-time intervals, a simplified description of thermodynamic topological systems can be developed in terms of 3 finite topologies. The first topology of interest is the disconnected, metrizable, Hausdorff $T2$ discrete topology suitable for the study of equilibrium systems of distinguishable particles (1964). The second topology of interest is the disconnected, not metrizable, Kolmogorov $T0$ (poset 3) topology suitable for the study of NON-equilibrium systems of distinguishable particles (1987). The third topology of interest is the Not-$T0$

topology which is without separation axioms, so that some - if not all - of the topological lattice closures will be identical, and therefor indistinguishable. Such Not-$T0$ topologies are suitable for the study of statistical properties of thermodynamic systems of complex wave-diffusion ensembles, including indistinguishable Bosons and Fermions. Such Non-$T0$ topologies should not be used to define particles, which require distinguishable closures.

3. Particle Thermodynamic Systems (1962–2009)

Any exterior differential 1-form of Action $A = A_k(x, y, z, t)dx^k$, can be used to generate the topological structure suitable to describe thermodynamic systems of distinguishable particles. Consider the Pfaff Sequence which consists of the p-forms that can be generated by algebraic operators of the exterior differential d, and the inner product, $\hat{}$:

$$PS = [A, dA, A\hat{}dA, dA\hat{}dA] = [A, F, H, K].$$

It is convenient to give suggestive names to the 4 elements of the Pfaff Sequence.

$$A = A = \text{Topological Action 1-form} \tag{1}$$

$$F = dA = \text{Topological Vorticity 2-form} \tag{2}$$

$$H = A\hat{}dA = \text{Topological Torsion 3-form} \tag{3}$$

$$K = dA\hat{}dA = \text{Topological Parity 4-form} \tag{4}$$

The number of non-zero elements in the PS defines the Pfaff Topological Dimension (or irreducible class) of the Topology generated by A. Thermodynamic systems of particles are in 1 to 1 correspondence with the Pfaff Topological Dimension:

$$PTD(A) \text{ and Thermodynamic Systems}$$

$$\{A, 0, 0, 0\}PTD(A) = 1 \text{ Equilibrium} \tag{5}$$

$$\{A, dA, 0, 0\}PTD(A) = 2 \text{ “isolated” Equilibrium} \tag{6}$$

$$\{A, dA, A\hat{}dA, 0\}PTD(A) = 3 \text{ Closed non-Equilibrium} \tag{7}$$

$$\{A, dA, A\hat{}dA, dA\hat{}dA\}PTD(A) = 4 \text{ Open non-Equilibrium} \tag{8}$$

Using the fact that the exterior differential, d, and the Identity, I, define a Kuratowski closure operator, $Kcl = I \cup d$, it becomes apparent that the exterior differential of a p-form, Σ, defines the limit set, $d\Sigma$, as a $p + 1$ form. The closure of Σ is equal to, $Kcl(\Sigma) = \Sigma \cup d\Sigma$. It follows that the complete topological structure can be constructed from a given 1-form of Action, A. The basis for the topology defined by the (arbitrary) 1-form, A, is composed of the elements in the Pfaff Sequence.

$$Basis : \{A, Kcl(A), A\hat{}dA, Kcl(A\hat{}dA)\} = \{A, A \cup F, H, H \cup K\}.$$

The Open sets of the Cartan - Kuratowski topology are generated by composing all unions of the Basis sets. For 4 ingredients, the Cartan-Kuratowski topology becomes recognizable as the equivalent of the disconnected Kolmogorov $T0$ (poset 3) topology. For two ingredients the topology becomes recognizable as the metrizable Hausdorff $T2$ topology, which acts as the foundation of most metric based geometries.

The correlation matrix, $[\partial A_k/\partial x^j]$, constructed from the system 1-form of Action, A, defines a Caley-Hamilton characteristic polynomial, Φ, of 4th degree in terms of the matrix similarity invariants, $\{Xm, Yg, Za, Tk\}$. The hypersurface

$$\Phi = \rho^4 - Xm\rho^3 + Yg\rho^2 - Za\rho + Tk = 0,$$

can be interpreted as a thermodynamic phase function in the symplectic topological space $PTD(A) = 4$, and has the ubiquitous (homeomorphic) properties of a van der Walls gas in the subset contact topological space, where $PTD(A) = 3$. Complex eigenvectors of the Jacobian matrix represent anti-symmetric features of the Jacobian matrix for the universal van der Waals gas. Note that two of the eigenvectors above the critical isotherm of the van der Waals gas are complex Macroscopic Isotropic Spinors of zero norm. Such direction fields can be put into correspondence with Majorana Spinors.

Based on the Pfaff Topological Dimension of the work 1-form, W, thermodynamic reversible processes imply that the Heat 1-form, Q is integrable. $Q\hat{}dQ = 0$. Consider the three cases of the classic Extremal, Bernoulli-Casimir, and Helmholtz processes:

Extremal: $PTD(W) = 0, W = 0$,
(Hamiltonian)
Bernoulli-Casimir: $PTD(W) = 1$, W exact
(Hamiltonian)
Helmholtz: $PTD(W) = 1$, W closed
(Conservation of Vorticity)

Each of these evolutionary flows are thermodynamically reversible, as both the Work and Heat 1-forms are topologically closed: $dW = 0 = dQ$, implies $Q\hat{}dQ = 0$. The 1-form A is of Pfaff Topological dimension < 3, hence the system is an equilibrium or isolated-equilibrium system.

These particle aspects have been well documented with many examples in numerous publications over the years, starting in 1962. This work has been collected into 6 monographs, with extensive references [1].

4. Process Morphisms for Particle Systems

Following E. Cartan's method (called Cartan's magic formula by Marsden), the "laws of motion" that describe the evolution of the $T2$ or $T0$ "particle" topologies will be generated by applying the homotopic Lie differential (not derivative), $L(V4)$, to the exterior differential 1-form of Action, A, that encodes the system:

$$L(V4)A = i(V4)dA + d(i(V4)A) \Rightarrow Q.$$

With a change of notation,

$$L(V4)A = W + dU \Rightarrow Q,$$

and Cartan's Magic formula becomes recognizable as a topological, universal, expression of the First Law of Thermodynamics.

The direction field, $V4$, representing a process germ, can be an element of a differential semi-group, or an element of a complex spinor space. In such cases, when $Q\hat{}dQ \neq 0$, the processes are not uniquely reversible. The expression $Q - W = dU$ should be recognized as an element of a deRham cohomology, when the Heat and Work 1-forms are closed (equilibrium). However, the derived expression for the topological first law is valid even when the Heat and Work 1-forms are not integrable (non-equilibrium). The concept of a deRham cohomology can be extended to 3-forms and non-equilibrium systems.

The Cartan homotopy operator, $L(V4)A$, operating on exterior differential forms that are $C2$ differentiable, is topologically continuous. The 4 components of $V4$ can consist of both vector/spinor direction fields (˜germs) and can be used to define a 3-form Current density. The Cartan homotopy maps the Limit points, d, of the ingredients, Σ, of the initial topology into the Closure, $\Sigma \cup d\Sigma$, of the ingredients of the final topology. For 1-forms, the initial state topology generated by A is not necessarily the same as the final state topology generated by Q. In fact, Topological Evolution is a necessary requirement for process irreversibility. Even though the topology changes, the Cartan homotopic process is topologically continuous.

5. Topological Torsion

It is remarkable that any 1-form of action, A, can be used to define a topology, and an equation of state via the Jacobian matrix of the coefficients. Perhaps even more remarkable is the fact that the system 1-form, A, also can be used to define a unique non-zero 3-form current, $H = A\hat{}dA$, associated only with non-equilibrium systems. This current 3-form is defined as the topological torsion 3-form and does not exist in equilibrium systems. The Topological Torsion vector, $T4$, (not $V4$) is a 4 component direction field that generates the topological torsion 3-form by contraction with the space-time volume element:

$$i(T4)\Omega(dx^k) = i(T4)dx\hat{}dy\hat{}dz\hat{}dt = A\hat{}dA.$$

The Topological Torsion vector is an associated vector with respect to 1-forms of Action, A, Work, W, and the Heat, Q. That is $i(T4)A = 0$, $i(T4)W = 0$, $i(T4)Q = 0$, hence the process, $T4$, is transversely adiabatic. The equation of homotopic evolution relative to $T4$ becomes:

$$L(T4)A = W + 0 = \sigma A = Q.$$

The 4 divergence of the Topological Torsion vector, $T4$ (which is equivalent to the Kuratowski closure of the 3-form, $A\hat{}dA$) determines the Space-Time Irreversible

Dissipation Coefficient σ.

$$d(A\hat{}dA) = dA\hat{}dA = \sigma(dx\hat{}dy\hat{}dz\hat{}dt).$$

On regions where $\sigma \neq 0$, the $PTD(A)$ is 4. The thermodynamic system is not in equilibrium, as

$$L(T4)A\hat{}L(T4)dA = Q\hat{}dQ = (-\sigma) = (\sigma)^2 A\hat{}dA \neq 0.$$

Then it seems reasonable to define

$$(\sigma)^2 \simeq EntropyProductionRate,$$

and the process in the direction of the Torsion vector $T4$ is thermodynamically irreversible.

Examples demonstrate that to reduce irreversible dissipation and turbulence in a Plasma, minimize $\sigma = (E \circ B)$. To reduce irreversible dissipation and turbulence in a Navier Stokes fluid minimize $\sigma = (a \circ \varpi)$, where

$$(a \circ \varpi) = \{\text{grad}P/\rho + \mu_B \text{ grad div}V - \nu V x \nabla^2 V\} \circ \varpi$$

and μ_B = bulk viscosity, ν = shear viscosity, a = acceleration, ϖ = vorticity.

Note that turbulent solutions imply $\sigma = (a \circ \varpi) \neq 0$, thereby proving that there exist solutions to the Navier-Stokes equations that are thermodynamically irreversible, and thereby turbulent.

The fact that any differential 1-form can be used to define a topological space came to my attention about 1987, after some 20 years of studying Cartan's utilization of differential forms and their application to the theory of integral invariants. I defined this topological space as the disconnected Cartan topology of exterior differential forms. However, it was not until 2009, when I attended my first conference for professional topologists at Haceteppe, Turkey, that I learned that the Cartan topology of 4 differential form ingredients was exactly equivalent to a Kuratowski representation of the disconnected Kolmogorov $T0$ (poset 3) topology, based upon the power set of 4 exterior differential forms as ingredients.

6. Emergence of Topological Defects

The Category theory of Topological Thermodynamics yields examples explaining how continuous irreversible dissipative processes can form emergent Closed topological defect structures, far from equilibrium, in an Open thermodynamic environment of $PTD = 4$. These examples give formal justification to Progogine's conjectures of dissipative emergence. Continuous topological evolution can describe the irreversible evolution on an "Open" symplectic non-equilibrium domain of Pfaff dimension 4, with evolutionary irreversible orbits being attracted to a contact "Closed" non-equilibrium domain of Pfaff dimension 3, with an ultimate decay to the "Isolated-Equilibrium" domain of Pfaff dimension 2 or less (an integrable Caratheodory surface). The experimental creation of Falaco Solitons in a swimming pool are visual examples of the thermodynamic emergence concept [2].

7. Statistical Thermodynamic Systems (2009–2011)

The third partition utilized by the Category Theory of Topological Thermodynamics involves Not-$T0$ topologies which are not constrained by separation axioms. Some, if not all, of the closed singleton subsets of the Not-$T0$ topologies are indistinguishable. An example is given by the indiscrete topology of two ingredients: the null set and the top Pfaffian 4-form density μ. The Non-$T0$ topologies are useful for describing differential form densities which are homogeneous, and have connected and/or not simply connected components. Note that the particle topologies have disconnected components, while the statistical topologies have connected components. Such Not-$T0$ partitions are applicable to the understanding of complex wave, diffusion, and statistical systems, which are based upon indistinguishable ingredients, herein defined as "complex wavelets", not "particles".

Equations that describe continuous topological evolution of the Not-$T0$ topologies are constructed in terms of Cartan's homotopic operator, the Lie Differential, acting on the sets of exterior differential form densities. The fundamental equation of continuous topological evolution of densities,

$$\mu = \rho(x^k)\Omega(dx^k) = \rho(x^k)\{dx\hat{\ }dy\hat{\ }dz\hat{\ }dt\},$$

is defined in terms of $4D$ Currents, $\mathbf{J}$, and the Lie differential:

$$L(\mathbf{J})\mu = i(\mathbf{J})d\mu + d[i(\mathbf{J})\mu] = 0 + d[i(\mathbf{J})\mu] \Rightarrow \kappa(x^k)\rho(x^k)\Omega(dx^k).$$

This coefficient expression, $\kappa(x^k)$, can be interpreted as the "bulk-dissipation, or self-similarity, or homogeneity" coefficient. If the function, $\kappa(x^k)$ vanishes, the density, μ, is a local evolutionary 4volume invariant. If $\kappa = +1$, the evolution is homogenous of degree 1, a necessary requirement for positive definite additivity of indistinguishable Bosons. If $\kappa = -1$, it is conjectured that the system describes indistinguishable (mod Spin) Fermions.

Factorization of the formula of continuous topological evolution, above, yields a necessary PDE equation that describes a 4D divergence relation with inhomogeneous source terms:

$$\text{div4}[\mathbf{J}^j(x^k)] = \kappa(x^k) - \{\mathbf{J}^j(x^k)(\partial ln\rho(x^k))/\partial x^j\}.$$

Realizations of this ubiquitous partial differential system of homotopic topological evolution can be recognized as the inhomogeneous wave equation or diffusion equation, the fractal minimal surface equation, the Klein-Gordan equation, the Schroedinger equation, the Hopf conjugate Spinor equation, the Landau-Ginsburg equation of superconductivity, and the Petrov blackhole equations of relativity theory. The term in brackets, $\{\}$, represents a generalization of the non-linear Eikonal expression, whose zero sets represent propagating singularities and discontinuities in the Maxwell theory of electromagnetic waves, the Raleigh-Taylor wavecrests so cherished by surfers, and the von Karmon mushroom cloud singularities seen in liquid and plasma fluids.

For example, consider the trial solution

$$\mathbf{J} = [\text{grad}\Psi, -D\Psi]$$

$$\text{div4}\mathbf{J} = \nabla^2\Psi - D\partial\Psi/\partial t$$

substituted into the 2nd Cartan homotopic evolution formula,

$$\text{div4}\mathbf{J} = \mu - \mathbf{J}^k \partial ln\rho/\partial x^k,$$

to yield

$$\nabla^2\Psi - D\partial\Psi/\partial t = \{\kappa - \mathbf{J}^k \partial ln\rho/\partial x^k\}.$$

The equation is to be recognized as the inhomogeneous Diffusion equation where the terms on the right hand side can be interpreted as interactions, or source terms. The solutions to the diffusion equation consists of indistinguishable complex distributions (wavelets). If the diffusion coefficient is imaginary, $D \Rightarrow h/2\pi i$, then the equation of homotopic evolution has the format of the Schroedinger equation.

Many examples can be found monograph Vol. 5 at http://www.lulu.com/kiehn.

8. Significant Results

As stated above, the Category of Topological Thermodynamics and Continuous Topological Evolution can be used as an abstract, but universal, method for understanding, among other things, scale invariance, chaos, emergence, self-similarity, dynamical systems, other non-linear, and even statistical, phenomena. The abstract categorical foundation, as an exterior differential system, admits solutions which can be put into correspondence with numerous physical experiments, in different physical disciplines.

- The method explains topological distinctions between equilibrium $PTD(A) < 3$ (integrable) and non-equilibrium $PTD(A) > 2$ (non-integrable) thermodynamic systems of distinguishable particles, based upon the concept of Pfaff Topological Dimension.
- The method explains topological distinctions between thermodynamically reversible, integrable processes ($Q\hat{}dQ = 0$) and irreversible non-integrable ($Q\hat{}dQ \neq 0$) processes.
- The method explains topological distinctions between thermodynamic systems based on distinguishable sets ($T0$ or $T2$ topologies), and thermodynamic systems based upon statistical distributions of indistinguishable sets (Not-$T0$ topologies).
- The method explains that Entropy can consist of two simultaneous components: one component is based on distinguishable sets (massive particles) and the other component is based upon indistinguishable sets (radiation).
- The method demonstrates that the First Law of Thermodynamics is a statement of cohomology, where the difference of the two (not necessarily integrable) 1-forms of Heat Q and Work W are equal to an exact differential of internal energy, dU.

• The method demonstrates that the decay of turbulence can be described in terms of a topologically continuous process, but the creation of turbulence can not.

• The method indicates that the top down nested sequence of Open, Closed, Isolated, and Equilibrium thermodynamic domains is in 1-1 correspondence with the Pfaff Topological Dimension $[4, 3, 2, 1]$ of the 1-form, A, used to create the Cartan topological structure.

• The method suggests that the environment of the universe is not a vacuum (a void empty set $PTD(A) = 0$), but is best described as the non-integrable Open system of $PTD(A) = 4$, in which domains of $PTD(A) < 4$ are distinguishable as topological defects in the environment.

• Non-Equilibrium open systems of Pfaff Topological Dimension 4 ($PTD(A) = 4$) can decay (in finite time) to Closed systems of Pfaff Topological Dimension 3 ($PTD(A) = 3$), thereby justifying Prigogine's conjecture of emergence by means of dissipative, but topologically continuous, processes. These metastable $PTD(A) = 3$ Closed states are topological defects in the Open systems of $PTD(A) = 4$.

• Topological change is a necessary condition for thermodynamic irreversibility.

• The constraints of topological continuity and non-homeomorphic processes establish a logical basis for thermodynamic irreversibility and the arrow of time.

• Any synergistic system of parts defines a topology such that the Category of Topological Thermodynamics is applicable to economic systems, political systems, as well as to biological systems. Such systems admit long-lived states far from equilibrium (adulthood). However, the theory can be used to design subtle perturbations to such systems such that their metastable lifetimes can be extended or destroyed.

• $C2$ Continuous Topological Evolution permits irreversible processes, for which, $Q\hat{}dQ \neq 0$. Segmented $C1$ processes approximating smooth $C2$ processes can be reversible, $Q\hat{}dQ = 0$.

• On odd-dimensional spaces, sequential $C1$ (translational) processes can be thermodynamically reversible, while intransitive $C2$ processes (rotation and expansion with a fixed point) can be thermodynamically irreversible.

• The relativistic Twin Paradox is resolved if the evolutionary paths of each twin are defined by processes that cause (unequal) topological change. If there is no topological change, or if each twin suffers the same topological change, there is no disparate biological aging.

• Adiabatic processes are transverse to the heat 1-form, $(i(\rho V)Q) = 0$. Adiabatic processes need not be quasi-static, and can be reversible or irreversible. Topological Transitions can occur in finite (often short) time.

• The work 1-form, W, is always transverse to the $4D$ process direction field, $\rho V4$, but the heat 1-form, Q, may or may not be transverse, The heat 1-form, Q can have longitudinal components in the direction of the process, corresponding to irreversible dissipation.

• Engineers should be guided by the universal concept of minimizing longitudinal heat in order to improve efficiency. For fluids, this idea can be translated to reducing vorticity components in the direction of fluid accelerations (by using wing tip tabs on commercial jets).

• For non-equilibrium systems, the 3-form of Topological Torsion (an $N-1$ form current) is not zero: $A\hat{}dA = i(T4)dx\hat{}dy\hat{}dz\hat{}dt \neq 0$. The Topological Torsion vector, $T4$, is deduced intrinsically from the 1-form A that encodes the thermodynamicsystem; $\rho T4$ can be used as a direction field that describes a unique process current density.

• For $PTD(A) = 3$ "closed" thermodynamic systems, the process current has zero divergence, and the $4D$ volume element is a conformal invariant (any ρ). This result is the space-time extension of the Liouville theorem that preserves the phase-space volume element in classical theory.

• For a $PTD(A) = 4$ "open" thermodynamic systems, the Topological Torsion vector does not have zero divergence, and so the process current $\rho T4$ may not be closed for arbitrary ρ (that is, the divergence of the process current is not zero).

• The combination of continuity and non-equilibrium, $PTD(Q) > 2$, requires a causal direction to the arrow or time.

• $PTD(A) = 3$ domains can have a local basis in terms of one complex Spinor pair with complex conjugate eigenvalues, and one real vector with eigenvalue zero. The eigenvalue 0 state can represent metastable (long-lived) configurations far from equilibrium. Such domains are locally "contact" manifolds.

• A key artifact of non-equilibrium is the existence of Topological Torsion current 3-forms, Topological Spin current 3-forms and Topological Adjoint current 3-forms. All of these current 3-forms are similar to the Amperian charge current 3-form of electromagnetic theory, but are related to different species of dissipative phenomena, which occur only in non-equilibrium systems. The dissipation coefficients are related to the non-zero divergences of the vector coefficients of the various 3-forms. For example, in electromagnetic systems, the dissipation coefficient is proportional to $E \circ B$ [3]; in hydrodynamics, the dissipation coefficient is called "Bulk Viscosity" [4].

• $PTD(W) = 4$ domains can have a local basis in terms of two complex Spinor pairs. Locally, such domains are symplectic manifolds. In such symplectic domains there exist local density distributions (sub domains), ρ, such that the divergence of any process current is zero in that sub domain. Such sub domains are metastable contact manifolds. In other words, the symplectic domain can contain defect structures in the form of contact sub domains. It can be demonstrated in terms of continuous topological evolution that such local density distributions, which define a "stationary" state, can emerge as a topological defect in a $PTD(W) = 4$ system, by means of a dissipative processes [5].

• The Adjoint Current, $J_{adjoint}$, like the Torsion Current, can be constructed entirely from the topological features of the thermodynamic system as determined by the functional coefficients of the 1-form of Action, A. The Adjoint Current also admits an infinity of integrating factors. However, when the 1-form coefficients, A_k are divided by a Holder Norm, λ, of homogeneity index one (known as the Gauss map), then the components of the Adjoint Current are defined as $|J_{adjoint}\rangle = [Jacobian(A_k/\lambda)] \circ |(A_k/\lambda)\rangle$. The 3-form of the adjoint current is defined as $J_{adjoint} = i(J_{adjoint})dx\hat{}dy\hat{}dz\hat{}dt$, and has zero divergence globally, and a pre-image 2-form that is not closed, $G_{adjoint}$, such that $dG_{adjoint} = J_{adjoint}$. It can be demonstrated that the adjoint current is related to the cubic curvature of the shape matrix, leading to the idea that the source of electromagnetic charge is related to cubic curvatures, similar to the idea that mass is related to quadratic curvatures.

• Examples of thermodynamic systems can be given to demonstrate that the conjectured format of the London Current of superconductivity, where $J = \chi\ A$, can be deduced as an emergent consequence of the Topological Theory of Thermodynamics [6].

• Examples can generate a Spin current 3-form, $S = A\hat{}G$, where, formally, the Spin current is proportional to the Lorentz force (the space-time components of the Work 1-form, W). This is a new interpretation of an old result, $J = \sigma(E + V \times B)$, which is Ohm's law. The new part is due to the idea that the dissipation is due to Spin Currents and the transport of collective spins, $A\hat{}G$.

• The topological structure of wave density domains of Pfaff dimension 2 or less creates a connected, but not necessarily simply connected topology. Evolutionary predictive solution uniqueness is possible.

• The topological structure of particle domains of Pfaff dimension 3, or more, creates a disconnected topology of multiple components. If solutions to a particular evolutionary problem exist, then the solutions are not unique. Envelope solutions, such as Huygen wavelets and propagating tangential discontinuities (called signals, or wakes) are classic examples of solution non-uniqueness. Topological Torsion is an artifact of non-uniqueness, and must be non-zero if a hydrodynamic system exhibits turbulence.

• All Hamiltonian, Symplectic-Bernoulli and Helmholtz processes are thermodynamically reversible. In particular, the work 1-form, W, created by Hamiltonian processes is of Pfaff Topological Dimension 1 or less. In all reversible cases the Work 1-form is closed, $dW = 0$. The evolutionary equations in such cases are time reversal invariant.

• The assumption of uniqueness of evolutionary particel solutions (which implies the Pfaff Topological Dimension of the thermodynamic system be equal to 2 or less), or the assumption of homeomorphic evolution, have imposed constraints upon classical mechanics that eliminate any time asymmetry, and the existence of isotropic spinors.

• The Lie differential acting on differential forms is not necessarily the same as a linear affine covariant differential acting on differential forms. The covariant differential always defines an adiabatic process, where the Lie differential does not.

• The particle view is based upon the 1-form of action whose coefficients, $[A_m]$, admit a correlation Jacobian matrix $[\partial A_m/\partial x^n]$.

• The statistical point of view is based upon a current N-1 form whose coefficients, $[C^m]$, admit a collineation Jacobian matrix, $[\partial C^m/\partial x^n]$, with a trace equal to a divergence of the Current.

• On spaces of $PTD(A) = 4$, the correlation Jacobian matrix has a characteristic Cayley-Hamilton polynomial that defines an quartic equation of state in terms of similarity invariants. The characteristic polynomial produces an implicit hypersurface function that can have envelopes and edges of regression in the format of the Gibbs function for a (universal and deformable) van der Waals gas (See image in 7).

• The correlation can be mapped into a reduced characteristic polynomial representing a quartic Universal Thermodynamic Phase function, with an envelope, which, below the critical point, has features of a Higgs potential (See image in 8).

• The reduced polynomial yields universal analytic expressions for the critical point, and the binodal and spinodal lines, in terms of the similarity invariants. The same technique can be applied to dynamical systems (See image in 9).

• The collineation can have complex eigenvalues and eigenvectors, even though the maximal rank matrix elements are real. These complex eigenvectors have a zero norm, and are generators of conjugate minimal surfaces, that have both statistical and fractal components.

• The Jacobian matrix can have complex eigenvalues and eigenvectors, even though the matrix elements are real. These complex eigenvectors have a zero norm, and are generators of conjugate minimal surfaces. Cartan defined such sets as Isotropic Spinors (Majorana Spinors, not Dirac Spinors). Pairs of non-co-linear Spinors define an area, but the "norm" of each Spinor is Zero!

• The hypersurface minimal surface can be generated by a holomorphic function that includes both the Gibbs entropy and a Mandelbrot fractal germ, $\Theta = (zlnz - z) + (a - bz^2)$. The third partial derivative of Θ leads to conjugate pairs of minimal surfaces. and the Mandelbrot germ vanishes. All functional iterates remain holomorphic and hence generate minimal surfaces with fractal boundaries.

• Topological fluctuations can be induced by processes that have components in terms of thermodynamic macroscopic Spinors. Thermodynamic Macroscopic Spinors are non-zero complex eigenvectors with complex eigenvalues (and zero quadratic form) of the antisymmetric 2-form (or matrix) representing the "Limit Points", dA, of the 1-form of Action, A. Such Macroscopic Spinors are capable of representing minimal surface conjugate pairs.

• Topological Insulators correspond to Impedances defined in terms of quantized Topological Spin, and Spin pairs coupled via Falaco Solitons. If the divergence of the Spin 4-vector vanishes, the Spin Current is time reversal invariant (See image in 10).

• The thermodynamic processes that lead to self-similarity of a Current 3-form $L(J)C = \sigma C$ can generate fractals and holographic effects where small neighborhoods replicate the whole, approximately. The homogeneity coefficient is the trace of the Jacobian Collineation: $\sigma = Trace[\partial C^m/\partial x^n]$, or the divergence of the Process vector field.

• **Following Landau,** it is conjectured that a turbulent non-equilibrium thermodynamic cosmology can be constructed in terms of a dilute van der Waals gas near its critical point. The conjecture yields an explanation for:

a The granularity of the night sky as exhibited by stars and galaxies is due to density fluctuations near the critical point, and the Newtonian law of gravitational attraction proportional to $1/r^2$ is due to a correlation between fluctuations (due to Landau).

b The conformal expansion of the universe as an irreversible phenomenon-associated with Quartic similarity invariants in the thermodynamic phase function, and conformally related to dissipative effects.

c The possibility of domains of negative pressure (explaining what has recently been called "dark energy") could be due to a classical "Higgs" mechanism for aggregates below the critical temperature (see image in 11).

d The possibility of domains of negative temperature (explaining what has recently been called "dark matter") could be due to macroscopic collective states of ordered spins. The conjecture is that Positive temperature radiates, Negative temperature does not. The conjecture is that black holes could be negative temperature states of collective spins.

e The possibility of domains where gravitational effects (quadratic similarity invariants, or 2nd order Gauss curvature effects) appear to be related to entropy and temperature properties of the thermodynamic system, and where cubic curvature effects could impede gravitational collapse.

f Black Holes (generated by Petrov Type D solutions in gravitational theory) are related to Minimal Surface solutions to the Universal thermodynamic 4th order Phase function.

9. Epilogue

Nowadays, the student of modern physics is trained to accept the jargon that there is a dualism between waves and particles. A precise definition of the dualism is rarely formulated in simple terms. For example, in the domain of electromagnetism, the student is taught that electromagnetic waves (which "are" solutions to the Maxwell "wave equation") consist of objects called photons (a type of Baryon that has spin zero); but, what exactly is a photon? This question continues to invoke a multitude of responses from optical engineers and physicists at the yearly SPIE conferences around the world.

The difference between waves and particles can be found in terms of topological ideas. Waves are collections of indistinguishable objects, and particles are collections of distinguishable objects. Waves transport energy and momentum without transporting mass. Particles transport energy and momentum in the form of mass. Waves are artifacts of Not-$T0$ topologies, where particles are artifacts of T0 topologies. Yet it is commonplace in the scientific community to ignore these topological distinctions.

It is conjectured that the holy grail of finding a marriage between Quantum Mechanics and Gravity will require the recognition that Quantum Mechanics (of radiation) is based upon connected Not-$T0$ topologies and that Gravity (of particles) is based upon disconnected $T0$ or $T2$ topologies.

References

[1] http://www.cartan.pair.com/ebooks.htm.
[2] http://www22.pair.com/csdc/pdf/falaco85o.pdf.
[3] http://www22.pair.com/csdc/download/ebookvol4.pdf.
[4] http://www22.pair.com/csdc/download/ebookvol3.pdf.
[5] http://www22.pair.com/csdc/pdf/falaco85o.pdf.
[6] http://www22.pair.com/csdc/download/ebookvol5.pdf.
[7] http://www22.pair.com/csdc/graphics/univgibb.jpg.
[8] http://www22.pair.com/csdc/graphics/Higgs.jpg.
[9] http://www22.pair.com/csdc/graphics/Binodal.jpg.
[10] http://www22.pair.com/csdc/graphics/SpinCurrents.jpg.
[11] http://www22.pair.com/csdc/graphics/Nonequil.jpg.

ANTI-PHASE SPIKING PATTERNS

M.P. IGAEV*, A.S. DMITRICHEV, V.I. NEKORKIN

Institute of Applied Physics of RAS,
46 Uljanov Street, 603950 Nizhny Novgorod, Russia
**E-mail: mooz@list.ru*
www.ipfran.ru

Spatio-temporal dynamics of the chain system modeling collective behavior of electrically coupled nonlinear cells is investigated. The dynamics of a local cell is described by two-component Morris–Lecar model. It is shown that such a system supports formation of two stable types of anti-phase spiking patterns (ASP). These are travelling and oscillatory type. High multistability of the travelling ASP is indicated.

Keywords: Anti-phase synchronization, pattern formation, nonlinear dynamics, nonlinear waves

1. Introduction

Having a clear view of the neuronal system dynamics is significant for an understanding of their functions and it is also irreplaceable when one endeavors to devise mathematical models of the neurons or neuronal networks. The most popular and practically important branch in such kind of researches is an investigation of small ensembles of neurons, where both neurons and their mutual coupling are described by sufficiently realistic mathematical models, based on the experimental data [1, 2]. But in spite of simplicity of such models, they possess a great number of collective behavior regimes on account of ultrahigh multistability. Certainly, some of these regimes have no definite physical interpretation, but nevertheless, it can be of service to understand the basic information transmission processes in the brain.

The functioning of such complex neuronal systems as retina and other sensory networks is one of the crucial problems of neuroscience nowadays [3–5]. Retinal structure is adequately high-ordered, which would have simplified to devise mathematical models and to simulate the networks numerically. But at the same time this results in general view of connectivity patterns and brain principles of operation at large. At present the new mathematical models of retinal-like structures are suggested and some progress in understanding basic properties of such systems is achieved. But we take the view, that the very neuron models mentioned above can be useful in investigating the ordered neuronal structures.

In Fig. 1 the most typical impulse responses of retinal ganglion cells are shown. It is important that the surround cell response is out of phase with the central one.

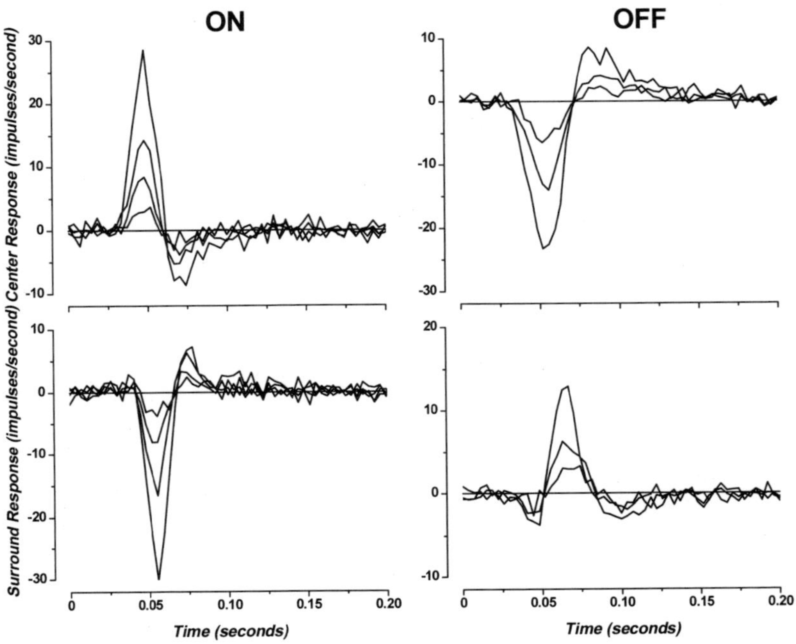

Fig. 1: The first order responses of a P–type primate ganglion cell center and surround (from Benardete and Kaplan, 2001).

This makes it possible to apply an anti-phase synchronization model of two electric-coupled neurons to account for the fact of such complicated oscillatory structures existence that will be discussed later.

In this paper we shall consider and provide the motivation for a discussion of the special class patterns formation in neuronal chain with electric coupling in terms of nonlinear dynamics. Firstly, we briefly study the most simple two-component Morris–Lecar neuron model [6], but the most convenient for analysis though. In the subsequent sections we discuss a special class of collective activity, which was named "anti-phase spiking patterns". Further the local mechanism of formation will be discussed. We shall show that exactly anti-phase synchronization regime between neighborhood elements is crucial for existence and stability of the ASP. In the concluding section we provide a brief summary of how the results of this paper help to understand the basic properties of such neuronal ensembles and give us some fundamentals for future work in other areas of the nervous system.

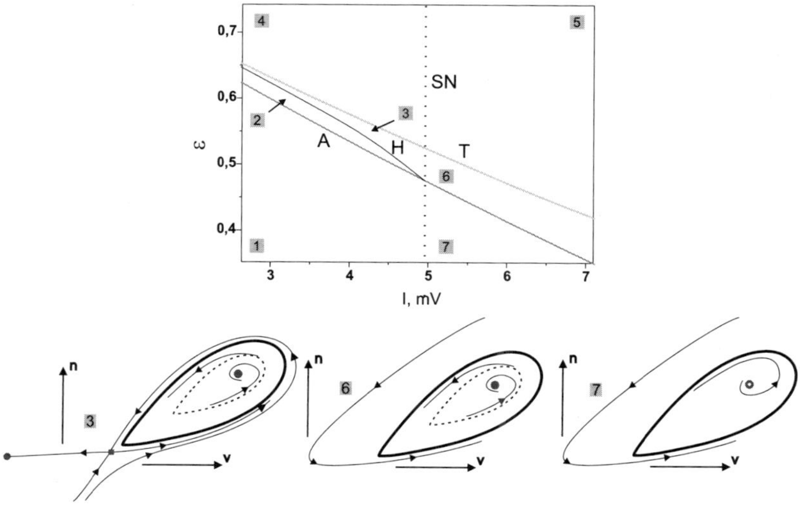

Fig. 2: Two-parametric bifurcation diagram of the system Eq. (1). Here $I = \frac{I_{ext}}{g_K}$ and $\varepsilon = \varepsilon_0 \frac{C_M}{g_K}$. See the main text for more explanation.

2. Local Cell Dynamics

In this section we discuss the dynamics of a single neuron. As it has been told in the introduction, it is described by well-known Morris–Lecar model:

$$\begin{cases} I_{ext} = C_M \frac{dv}{dt} + g_L(v - V_L) + g_{Ca}(v - V_{Ca})M_\infty + g_K(v - V_K)n, \\ \frac{dn}{dt} = \frac{N_\infty - n}{\tau_N}, \end{cases} \tag{1}$$

where functions $M_\infty = 0.5(1 + th\frac{v-V_1}{V_2})$, $N_\infty = 0.5(1 + th\frac{v-V_3}{V_4})$ and $\tau_N = \frac{1}{\varepsilon_0}ch^{-1}(\frac{v-V_3}{2V_4})$. Parameters ε_0 and I_{ext} are variable, but the rest of the parameters are taken from [7] and correspond to the I class excitability of the neuron cell. Figure 2 shows the fragmentation of the (ε_0, I_{ext})-plane into seven regions. Each region is characterized by the presence of the special behavior of the model Eq. (1).

Detailed analysis of Eq. (1) for the whole parametric set is given in [7] as well. But we briefly discuss principal bifurcational features for the (ε_0, I_{ext})-parametric plane and the key dynamical regimes required to us for describing ASP will be illustrated.

Since a special class of oscillatory movements in such neuronal media is investigated, non-oscillatory regimes 1, 2, 4, 5 are ignored and only regions 3, 6, 7 will be considered, where the stable limit cycle attractor exists. In addition, the proximity of the system Eq. (1) to the saddle-node bifurcation is important as well, since the presence of both slow and fast movements areas has an effect on anti-phase synchronization regime stability.

3. Anti-phase Spiking Patterns

Typical wave processes in neuronal systems include travelling excitation pulses (action potentials) and wave fronts (moving interfaces between different levels of neuronal activities). But in case of weak coupling the behavior of such systems has a distinction in kind. Whatever the case, the important property of such waves is that their principal characteristics (shape, velocity, timing intervals) are defined only by intrinsic properties of the medium and do not depend on initial conditions or perturbations within certain limits.

Here we focus on numerical simulation results. Let us consider a periodical chain of Morris–Lecar neurons. This model has the following form:

$$\begin{cases} I_{ext} = C_M \frac{dv_j}{dt} + I_L(v_j) + I_{Ca}(v_j, M_\infty) + \ldots \\ \qquad\qquad \ldots + I_K(v_j, n_j) + d(v_{j-1} - 2v_j + v_{j+1}), \\ \frac{dn_j}{dt} = \frac{N_\infty(v_j) - n_j}{\tau_N(v_j)}, \end{cases} \quad (2)$$

$j = \overline{1, N}, \quad v_j(t) = v_{j+N}(t).$

According to the numerical simulation of Eq. (2) it is possible to characterize two stable classes of ASP (see Fig. 3). Above we discussed some general features of the ASP while the internal chain dynamics was not taken into account. For

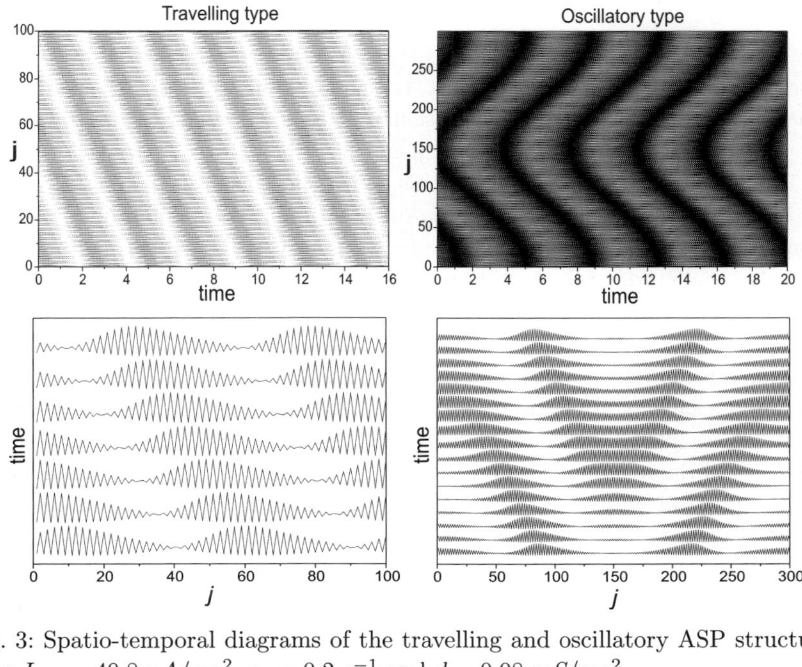

Fig. 3: Spatio-temporal diagrams of the travelling and oscillatory ASP structures. Here $I_{ext} = 40.8\ \mu A/cm^2$, $\varepsilon_0 = 0.2\ s^{-1}$ and $d = 0.08\ mS/cm^2$.

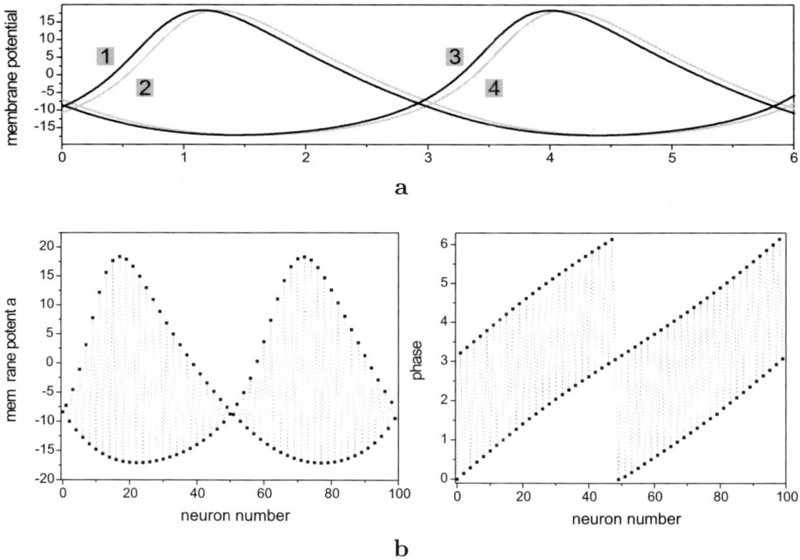

Fig. 4: (a) Waveforms of four nearest-neighbor elements. (b) Potential and phase distribution along the chain. Parameters $I_{ext} = 40.8\ \mu A/cm^2$, $\varepsilon_0 = 0.2\ s^{-1}$ and $d = 0.08\ mS/cm^2$.

the anti-phase spiking patterns to exist, the neuronal network Eq. (2) has to meet three requirements. First of all, only oscillatory-type neurons are important. Consequently, we should take the values of ε_0 and I_{ext} from 3, 6 or 7 parametric regions (see Fig. 2) to provide the spiking activity of a single neuron. Secondly, just the functioning of each cell close to the saddle-node bifurcation determines the anti-phase synchronization regime stability between two weak-coupled neurons [8, 9] and makes the interneural interaction problem more relevant. And finally, just the travelling structures are investigated, therefore, we should assume that there is a certain phase shift between oscillations along the chain. Figure 4(a) illustrates the evolution of membrane potentials of the next nearest and neighboring neurons. The neighboring elements are anti-phase synchronized while the next nearest elements oscillations are quasi-synchronous with a constant phase shift which increases under the growth of the neuron number.

The numerical simulation of the Eq. (2) makes it clear that the oscillation phase distribution is not monotonic along the chain but discontinues at transition from one element to another (see Fig. 4(b)). Summing up the results of numerical experiments we arrive at a conclusion that dependence of the oscillation phase on the neuron number meets some requirements which can be expressed in following form:

$$\varphi_{j+m} = \varphi_j + \frac{(m - \delta_m)}{2}\varphi^0 + \pi\delta_m, \tag{3}$$

where $\varphi_j = \varphi_{j+N}$ - oscillation phase of the j-th element, $j = \overline{0, N-1}$, $m = \overline{0, N - \delta_N}$, $\delta_m = 0$ if m is even, $\delta_m = 1$ if m is odd and φ^0 is a constant phase shift between every pair of elements. Without loss of generality, we can set φ_j and j equal to zero in Eq. (3) because it's possible to measure the oscillation phase from any element in the chain.

Relying on the basic phase relationships Eq. (3) it is possible to derive the expression for the ASP spatial period λ. In addition, only the integral number of ASP structures can fit into the full length of the neural chain and, in consequence of this fact, the constant phase shift φ^0 possesses the discrete values:

$$\lambda_k = \frac{2\pi(1 - \delta_{\lambda_k})}{\varphi_k^0} + \delta_{\lambda_k}, \quad \varphi_k^0 = \frac{2\pi(1 - \delta_{\lambda_k})}{\frac{N}{k} - \delta_{\lambda_k}}, \tag{4}$$

where k is mode number.It means that there is multistability in the system Eq. (2). Figure 5 shows the examples of the ASP structures with the same values of parameters ε_0, I_{ext} and d as in Fig. 4(b) and with the the same chain length $N = 100$.

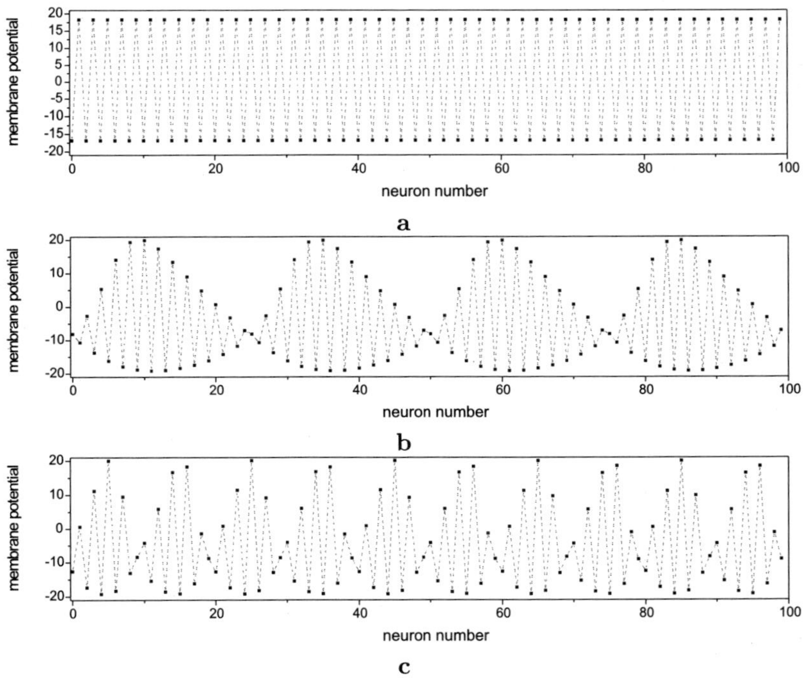

Fig. 5: Examples of travelling ASP structures: (a) $k = 0$ (b) $k = 4$ and (c) $k = 10$.

4. Conclusion

The special class of spatio-temporal dynamics of a chain system modeling electrically coupled neuron-like media has been investigated. The system Eq. (2) was considered for small values of coupling strength d when the discreteness of such media has an effect on collective behavior. We have shown that at the fulfillment of requirements listed above (see Sec. 3) the neural chain is capable to produce two types of stable ASP structures without any external perturbing actions. This provides their robustness and self-sustainability as information carriers for inter-neuron communication. In addition, the special ASP internal structure should be observed as well. All elements in the network have been found to be anti-phase synchronized in pairs and the constant phase shift along the chain provides the wave nature of neurons activity. Obviously, these results, as well as being useful for understanding communication between neurons, are applicable in synchronization problems of nonlinear elements. It should be noted that the similar phenomenon has been observed in oscillatory systems with cross diffusion (private communication, M.A. Tsyganov, 2010).

This work has been supported by the Russian Foundation for Basic Research (Grants 09–02–91061, 09–02–00719, 10–02–00643) and the Ministry of Education and Science (Governmental contracts P942, 02.740.11.5188 and 14.740.11.0348).

References

[1] G. B. Ermentrout and D. H. Terman, *Mathematical Foundations of Neuroscience* (Springer, 2010).
[2] Y. Kuramoto, *Chemical Oscillations, Waves, and Turbulence* (Mineola, NY: Dover Publications, 1984).
[3] E. Kaplan and E. Benardete, *Progress in Brain Research* **134** (2001).
[4] M. J. Berry, D. K. Warland and M. Meister, *Neural Network* **14** (2001).
[5] G. D. Field and E. J. Chichilnisky, *Annual Review of Neuroscience* **30** (2007).
[6] C. Morris and H. Lecar, *Biophysical Journal* **35** (1981).
[7] K. Tsumoto, T. Yoshinaga and H. Kawakami, Bifurcation in a Morris-Lecar Neuron Model, *in Proc. of International Symposium on Nonlinear Theory and its Applications* (2002).
[8] A. Sherman and J. Rinzel, *Proceedings of the National Academy of Sciences of the USA* **89** (1992), pp. 2471–2474.
[9] S. K. Han, C. Kurrer and Y. Kuramoto, *Physical Review Letters*, **75** (1995), pp. 3190–3193.

PART C

Data/System Representation

REALITY, MODELS AND REPRESENTATIONS: THE CASE OF GALAXIES, INTELLIGENCE AND AVATARS

Jean-Claude HEUDIN*

IIM Imedialab and Pôle Universitaire Léonard de Vinci
Paris La Défense, 92916, FRANCE
**E-mail: jean-claude.heudin@devinci.fr*

This paper discusses relationships between reality, models and representations in the study of complex systems. We present three different case studies: the modeling of colliding galaxies, the modeling and simulation of human intelligence, the creation of digital avatars. We clarify the relationships between reality, models and representations using a semiotic approach. We show that visual representation is as important as the modeling and that there is no possible equivalence between the real phenomena and its model and representation.

1. Introduction

Simplicity is not the rule: a majority of systems are complex rather than simple. Modeling complex systems is therefore crucial for many domains in science. A complex system can be basically defined as composed of a large number of interacting elements showing collective patterns or global behaviors. Because they are generally the result of a high number of non-linear interactions, such global properties cannot be captured by only considering the properties of individual elements.

No consensus exists on how to model complex natural or artificial phenomena. The classical approach tries to reduce the system to its underlying fundamental parts and processes. A dynamical model is then formulated to predict the evolution of the system by decreasing the high number of operative degrees of freedom, simplifying the underlying processes using approximations, averaging or smoothing parameters. However, when simplifications are too strong, results become inadequate and do not reflect the real behaviors. In most cases, global properties simply disappear.

Recently, the study of complex systems in a unified framework has become recognized as a new interdisciplinary scientific field [1]. The approach defines complexity levels and their related components. Then a simulation attempts to capture the behavioral essence of the complex phenomena viewing it as a large population of simple agents, and works upwards synthetically. There must be no global controller responsible for the behavior of every agent. In contrast, each agent has its own autonomous behavior and the global behavior emerges from out of all the local interactions among individual behaviors [2].

In this paper, we study the relationships between real complex phenomena, their modeling and visual representations. We first present three different case studies: the modeling of colliding galaxies in the field of astronomical research, the modeling and simulation of human intelligence in computer science, and the creation of digital avatars in computer graphics. We then outline the relationships between reality, models and representations using a semiotic approach. We show that, in contrast with the classical idea in science, visual representation is as important as the modeling, and that and that there is no possible equivalence between the real phenomena and its model and representation.

2. Galaxies

2.1. *Reality*

From globular clusters to spiral galaxies, cosmological evolution shows a wide variety of patterns and complex behaviors. A typical example is the well-known "Mice" structure, also known as NGC 4676.

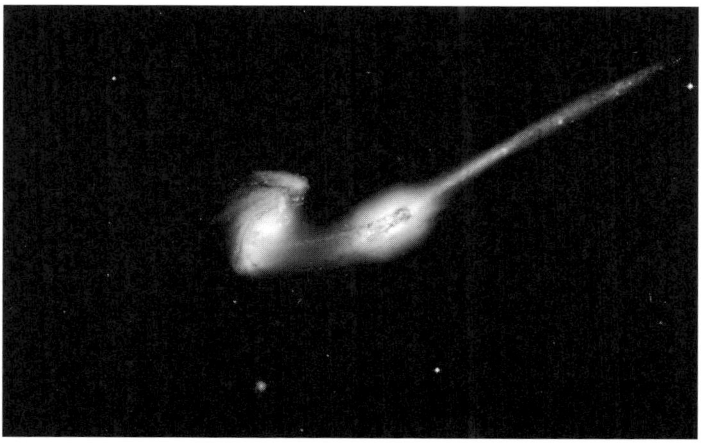

Fig. 1: The Advanced Camera for Surveys (ACS) on NASA/ESA Hubble Space Telescope has taken this image of NGC4676 in April 2002.

NGC 4676 is the result of two colliding spiral galaxies in the constellation Coma Berenices, located 300 million light-years away. The name "Mice" was given because of the two long tails of stars, gas and dust thrown out of the galaxies as a result of the collision. The nuclei of the two galaxies will eventually merge into a single giant galaxy. The Mice is a very complex system involving a huge number of physical interactions occurring at different levels from particles to stars. The study of this phenomenon is important because most galaxies have or will have at least one significant collision in their lifetimes.

2.2. *Models*

The Mice collision has been modeled by many research teams. Most of them, like Hibbard and Barnes [3], have used the classical point-mass particle approach. A point mass particle is an idealized object with a nonzero mass and no other properties or structure such as spatial extension.

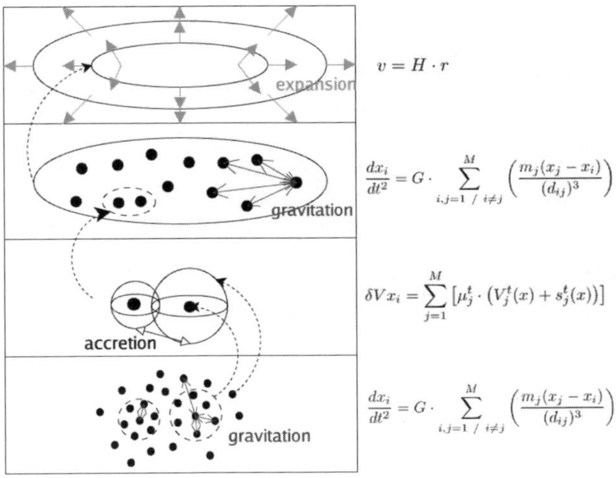

Fig. 2: Schematic representation of a hierarchical multi-agent model used for studying colliding galaxies phenomena such as "the Mice".

The simplicity of this approach makes it convenient for modeling galaxies as a large number of particles, applying the gravitational force between them. The drawback of this straightforward approach is that the system's dynamical properties highly depend on the number of point-mass particles used. A realistic number of particles should be around 1041 for a typical spiral galaxy which is not possible. The solution is to reduce the number of point-mass particles, but some dynamical behaviors do not appear until a very high number. Due to computation limitations, the average number of particles currently used in simulation is between 128^3 and 512^3.

We have proposed recently a hierarchical multi-agent model (see figure 3) which keeps a strong bond with physics [4]. The system is composed of four complexity levels, each one with its own class of agent and its related physical laws. Each agent of a given level is an emerging aggregate of agents at the lower level. We have shown evidences that with such hierarchical model, the emerging dynamical properties of the system are less sensitive to the number of agents.

2.3. *Representations*

As widely recognized, visualization has become a challenging aspect for analyzing results from cosmological studies [5]. Upcoming observations will produce huge amounts of data every night. Related simulations produce also terabytes of data and this number will certainly rise to petabytes within a decade. This growth in size and complexity of the available data stream from observations and simulations makes analyzing and interpreting this information a challenging problem.

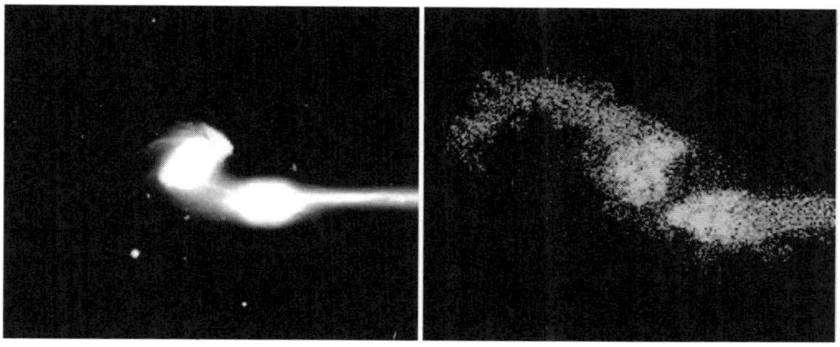

Fig. 3: An image of the Mice from the Hubble Telescope (left) and from a simulation using our hierarchical multi-agent model (right).

The first step in this analysis process is to actually look at these data using a visualization system. The human's eyes and brain still remains one of the most sensitive analysis tools, albeit optimized for certain types of qualitative analysis. Even a simple visual comparison between simulation images and those from the real physical system can give useful information about the accuracy of the model and its computer implementation. However, due to the increase in size and complexity, this first step becomes highly nontrivial. In order to face this problem, one solution is to provide a new generation of quantitative analysis tools driven by the visualization process, including the possibility of real-time simulation trials and on-the-fly analysis measures.

3. Intelligence

3.1. *Reality*

Intelligence has been most widely studied in humans, but also in animals and plants. There have been numerous definitions and hypotheses about intelligence, however with no consensus yet. Intelligence is a term describing various capacities of the mind and thus it can be defined in different ways, including the capacities for abstract thought, understanding, communication, reasoning, learning, planning, emotional intelligence and problem solving.

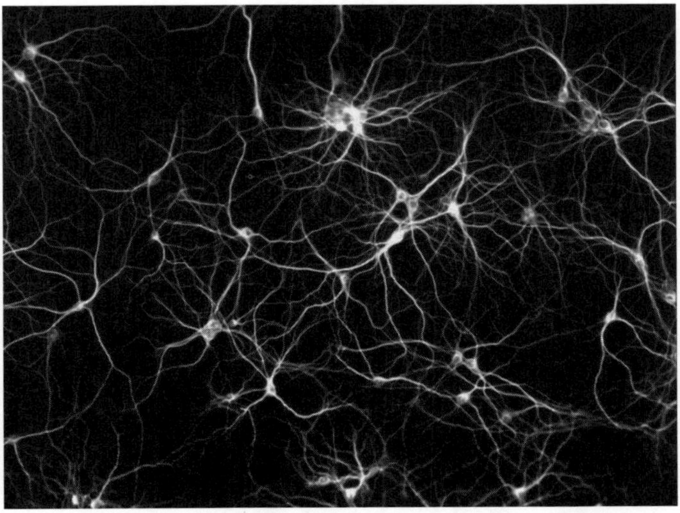

Fig. 4: The human brain has been estimated to contain 90 billion non-neuronal (glial) cells as well as 90 billion neurons, of which about 10 billion are cortical pyramidal cells. These cells pass signals to each other via as many as 1000 trillion synaptic connections (source Wikimedia Commons).

It is now admitted that human intelligence is a global property of the brain emerging from the huge number of nonlinear interactions that occur between billions of neurons cells. The complex architecture of the human brain is the result of millions years of co-evolution within the earth environment.

3.2. *Models*

There have been numerous models of intelligence. The field of artificial intelligence (AI) has been one of the most prolific since its birth in the middle 1950s. We can roughly class them in four categories:

Symbolic : In the early years of AI, most models were based on the theory that human intelligence could be reduced to symbol manipulation [6]. The main outcome of this approach was the development of expert systems based on rule inference and symbolic processing.

Neuron-based : The study of artificial neural networks began in the decade before the field AI research was founded. It was based on a formal model of the neuron cell in the form of a simple threshold automaton [7]. Researchers focused on elaborating different types of neural networks and their related learning algorithms. Pattern recognition is the main application field for this approach.

Reactive : In the 1990s, behavioral robotic architectures and Artificial Life [2] focused on "intelligence without representation" instead of the classical symbolic computational models [8]. The idea was to work on bio-inspired adaptive systems addressing their basic perceptual and sensorimotor tasks.

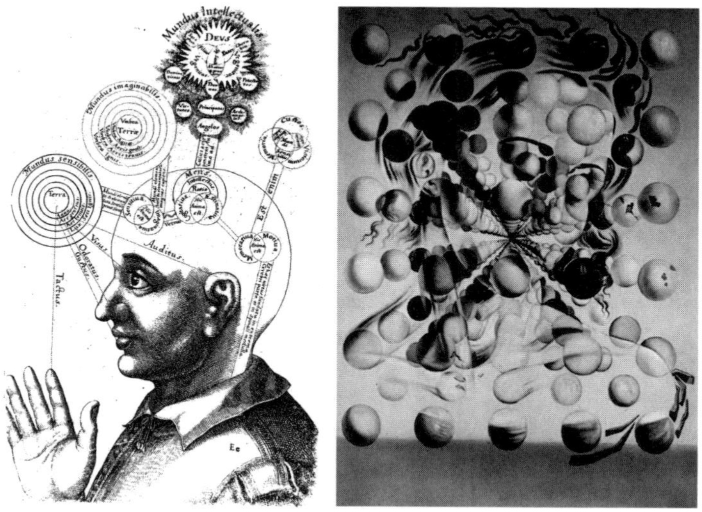

Fig. 5: On the left: an image from Robert Fludd in 1619 showing a model of human intelligence (source Wikimedia Commons). On the right, Salvator Dali's painting in 1952 "Galatea of the Spheres" is a good illustration of the Complex Artificial Intelligence modeling approach.

Distributed : During the same period of time, many projects were dedicated to the development of Distributed Artificial Intelligence and Multi-Agent Systems. The key concept is the abstraction called a software agent [9]: a virtual autonomous entity that has an understanding of its environment and acts upon it. An agent communicates with other agents to achieve a common goal, that one agent alone could not achieve.

More recently we have proposed a new approach to machine intelligence called Complex Artificial Intelligence. While taking advantages of the previous ones, this modeling approach considers intelligence as a global property of biological complex systems that emerges from the large amount of non-linear interactions between agents and the environment at the edge of order and chaos [10].

We have applied this approach to the "Ms House" project: an intelligent character composed of multiple distinct personalities [11]. Each individual personality is implemented as an autonomous bio-inspired nano-agent which is able to react to the user's input by computing an appropriate message. The character's identity emerges from the interactions between these nano-agents and the user.

3.3. *Representations*

The visual representation of AI has been historically drawn by science-fiction rather than by science. In the past, research labs have generally implemented basic user interfaces in their experiments. In the meantime, the image of AI in the public has been the one of a large supercomputer threatening humanity [12].

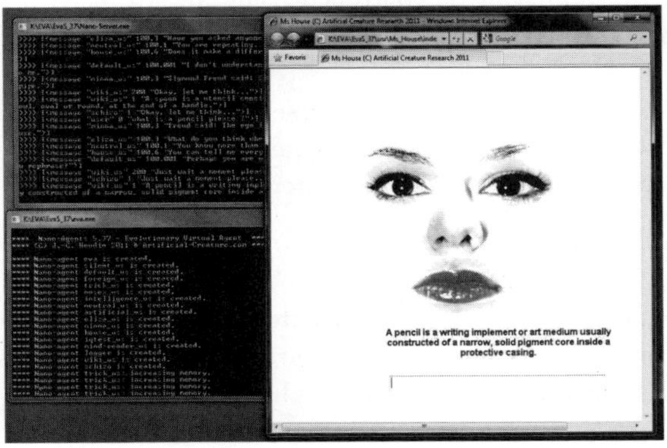

Fig. 6: A screenshot of the "Ms House" prototype showing the intelligent character answering the question "What is a pencil?" by searching information in real-time on the web.

We have conducted many experiments showing empirical evidences that the visual representation of the AI is as important as the behaviors, sometimes more. The current trend is to create the interface together with the intelligent behaviors within the rules of social interaction [13]. One solution is to implement human avatars as virtual embodiments of intelligent software agents.

4. Avatars

4.1. *Reality*

In Hinduism, an avatar (avatãra) is a Sanskrit word referring to a deliberate descent (incarnation or manifestation) of a deity from heaven to earth. In computer science, an avatar is a representation of a user in a computer program. The first use of avatars was certainly as small images on Internet forums, representing users in addition to their pseudonyms.

The actual notion of an avatar is a 2D or 3D, fixed or animated representation of a user in online virtual worlds and video games. These graphical representations

Fig. 7: The user's avatar can be controlled using an advanced perception system which and maps his movements on the character's 3D model (Image ©Microsoft).

facilitate the exploration of a virtual universe, or act as a focal point in conversations between users. Most avatars in video games are limited to the player's visual representation as a stereotyped hero character.

4.2. *Models*

Since a few years, the term avatar has been extensively used in order to describe very different kinds of models. We can define four main categories of avatars, inspired by the initial classification proposed by D. Thalmann [14]:

Visual avatar : a graphical representation of the user in the virtual world with a straightforward mapping between the user's movements and the avatar's behaviors.

Guided avatar : a graphical representation of a user in a virtual world with a high level control on its behaviors.

Semi-autonomous avatar : a guided avatar that may also act without the intervention of the user for a short period of time, for example when the user is not available.

Autonomous avatar : acts on its own without any intervention of a user. It is able to give to itself its proper laws, its conduct, opposite to the other avatars which are driven by the outside.

4.3. *Representations*

The visual representation of an avatar may take numerous forms and use many different technological solutions. It can be implemented as a simple 2D icon up to a 3D virtual human with realistic rendering. In some cases, the avatar is not a visual representation in a computer program, but rather a robotic system.

One of the main problems for the representation of avatars is the "uncanny valley" [15]. This theory states that as a robot is made more humanlike in its

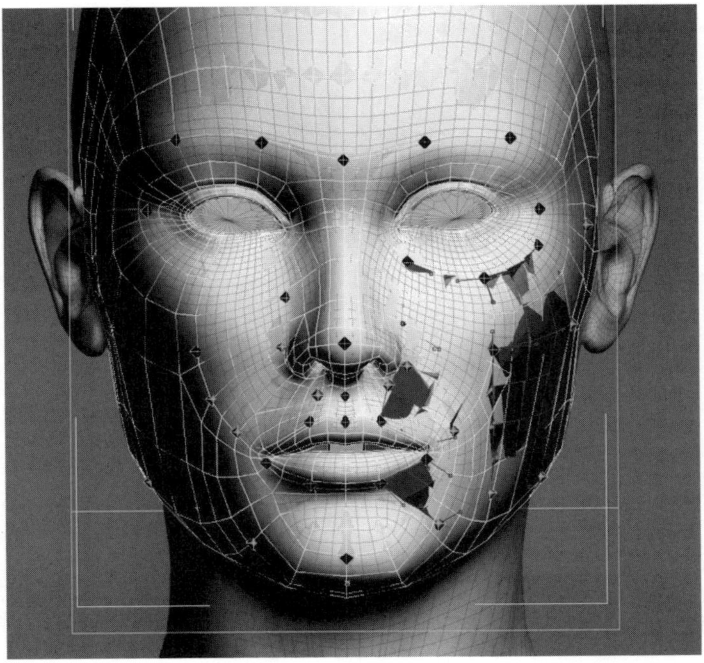

Fig. 8: The modeling of an avatar at the lower level is based on a set of vertex and polygons. This image shows the creation of a model for the Virtual Geisha project.

appearance and motion, the emotional response from a human being to the robot will become increasingly positive and empathic, until a point is reached beyond which the response quickly becomes that of strong revulsion. However, as the appearance and motion continue to become less distinguishable from a human being, the emotional response becomes positive once more and approaches human-to-human empathy levels.

5. Discussion

In order to understand the relationships between reality, models and representations, we need a theoretical framework that clarifies these concepts. In the nineteenth century, Charles Sanders Peirce defined "semiotic" as the "formal doctrine of signs", which abstracts "what must be the characters of all signs used by an intelligence capable of learning by experience" [16]. One classical paradigms used by semioticians is the triangle of reference, also known as the triangle of meaning [17]. This triangle shows the relations between three basic elements: the sign that is the form of a message; the referent, what the sign stands for; and the interpretant, the sense made of the sign. In the case of natural language, the sign is a word, the in-

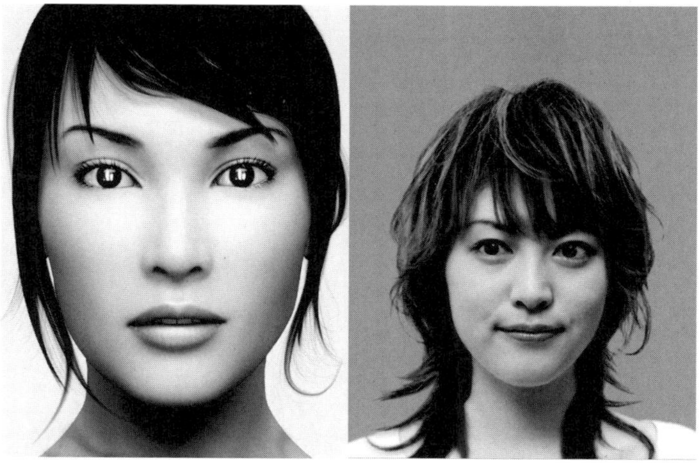

Fig. 9: These two images show realistic avatars that are perceived by some people as empathic and as weird by others. The left image is one of the early rendering of our Virtual Geisha project. The right image is an avatar produced by the motion portrait technology that automatically creates a 3D face model from a single picture (©Motion Portrait).

terpretant refers to a meaning of the word in the mind of a person, and the referent is the real world object.

We can transpose this triangular relation in the context of modeling complex phenomena. Thus, the referent is the real world physical phenomenon, the interpretant is the abstract model of this phenomenon, and the sign is a visual representation.

There could be a debate about the place of the computer program that implements the model and creates representations. One possible view is that it is an operational implementation of the model, so it can be considered as part of the model. Another view is to state that we have in facts two triangular relations, a first one with the real phenomenon and a second one with the computer implementation as referent. Both referents produce representations of the studied phenomenon: observations or synthetic images. Anyway, the important point is that these two views show the importance of visual representations. In most cases, not to say all, we have no direct access to the real phenomenon, so we can only analyze and compare visual representations and related measures.

There is a common idea that simulation in the near future will produce representations that will be so close to the one of the real phenomena, that those exposed to simulation will be hard put to tell the difference. Even if a simulation is

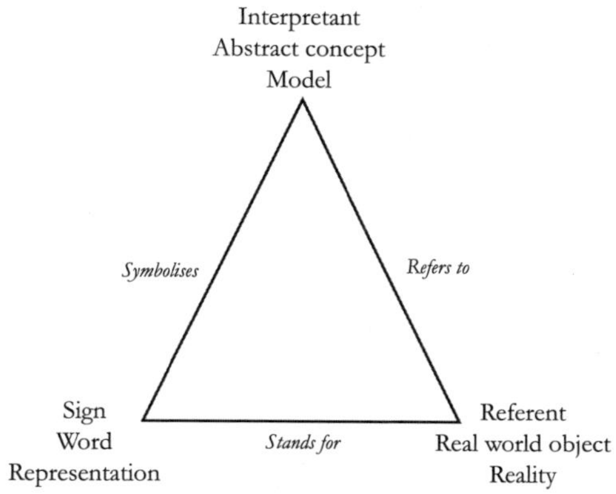

Fig. 10: The original semiotic triangle from Ogden and Richards, its variant applied to natural language, and our interpretation for complex systems.

a "degraded" version of reality, the degree of "failure" or degradation will become disappearingly small. The main arguments are the technological advances and that the human sensory systems have a limit to their resolution capacities.

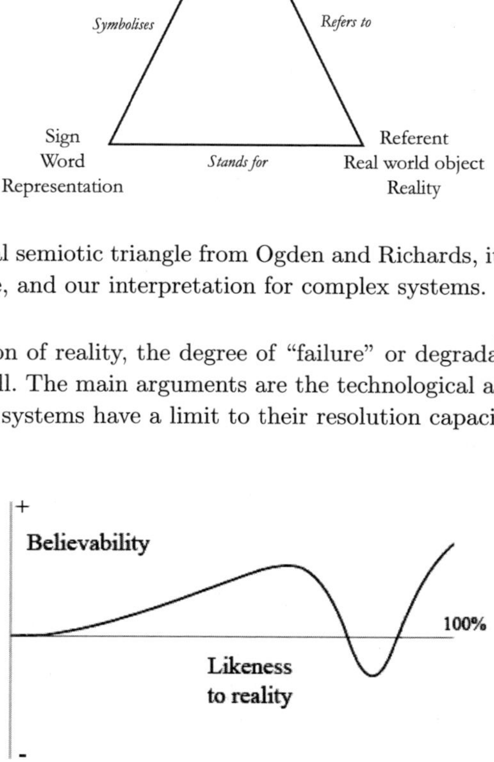

Fig. 11: A generalization of the "uncanny valley" in the framework of complex systems modeling and simulation of complex systems.

In the case of astronomical studies, no one will ever argue that a model and its implementation can be equivalent to (as real as) the real phenomena. However, for some other cases, the situation is not so clear. In particular, there have been a lot of philosophical debates about the equivalence of a computer program simulating the human intelligence and a (real) human mind since the proposition of an "imitation game" by A. Turing in 1950, also known as the Turing-test [6]. We argue that for all complex systems, there is no possible equivalence in practice between the real phenomenon and its modeling.

The main reason is a "complexity barrier" between the real phenomenon and its synthetic counterpart. This barrier is due to the huge difference in terms of quantitative and qualitative complexity. In addition, we think that the "uncanny valley" problem, originally formulated for humanoid robots, can be generalized to all modeling and simulation of complex systems. The importance of details in a visual representation increases exponentially when it becomes close to the one of a real phenomena. In other words, if a very small detail can transform an empathic avatar into a monster, it can also substantially decreases the benefits of a model and its representation.

6. Conclusion

In this paper, we have addressed the relationships between real complex phenomena, models and representations. We have first presented three experiments we have conducted in different scientific fields: the modeling of colliding galaxies in the field of astronomical research, the modeling and simulation of human intelligence in computer science, and the creation of digital avatars in computer graphics. A semiotic approach has then enabled us to clarify these relationships. In many research fields, representation was not in the past an important preoccupation for researchers whose main focus was rather on mathematical models. We have argued that visual representation is now a crucial need for all modeling and simulation in science.

References

[1] S. Wolfram, *A new kind of science* (Wolfram Media, Champaign, 2002).

[2] C. Langton, Artificial life, in *Proceedings of the First Artificial Life Conference*, (Addison-Wesley, Redwood City, 1989).

[3] J. Barnes and J. Hibbard, `http://www.ifa.hawaii.edu/barnes/research`.

[4] J.-C. Torrel, C. Lattaud and J.-C. Heudin, Complex systems in cosmology: The antennae case study *Lecture Notes of the Institute for Computer Sciences* **5** (Social Informatics and Telecommunications Engineering, 2009).

[5] J. Ahrens, K. Heitmann, S. Habib, L. Ankeny, P. McCormick, J. Inman, R. Armstrong and K. Ma, Quantitative and comparative visualization applied to cosmological simulations *Journal of Physics, Conference Series 46* 2006.

[6] A. Turing, Computing machinery and intelligence *Mind* **LIX**1950.

[7] W. McCulloch and W. Pitts, A logical calculus of ideas immanent in nervous activity *Bulletin of Mathematical Biophysics* **5**1943.

[8] R. Brooks, Intelligence without representation *Artificial Intelligence* **47**1991.

[9] M. Minsky, *The society of mind* (Simon and Schuster, New York, 1988).

[10] J. Heudin, Complex Artificial Intelligence, Invited talk at Robolift conference, Lyon (2011).

[11] J. Heudin, A schizophrenic approach for intelligent conversational agents *3rd International Conference on Agents and Artificial Intelligence* (Rome, 2011).

[12] J. Heudin, *Les créatures artificielles* (Odile Jacob, Paris, 2008).

[13] B. Reeves and C. Nass, *The Media Equation* (CLSI Publications, Stanford, 1996).

[14] D. Thalmann, T. Capin, N. Thalmann and I. Panszic, International conference on

artificial reality and tele-existence *Conference on Virtual Reality Software and Technology* (Rome, 1995).
[15] M. Mori, Bukimi no tani - the uncanny valley *Energy* **7**1970.
[16] C. Peirce *Collected Papers of C. S. Peirce, ed. by C. Hartshorne, P. Weiss, A. Burks, 8 vols., Harvard University Press Cambridge* 1998.
[17] C. Ogden and I. Richards, *The Meaning of Meaning: A Study of the Influence of Language Upon Thought and of the Science of Symbolism* (Routledge and Kegan, London, 1923).

ASTRONOMICAL IMAGES AND DATA MINING IN THE INTERNATIONAL VIRTUAL OBSERVATORY CONTEXT

F. PASIAN*
INAF–Osservatorio Astronomico di Trieste
Via G.B.Tiepolo 11, Trieste, I-34143, Italy
**E-mail: pasian@oats.inaf.it*

M. BRESCIA
INAF – Osservatorio Astronomico di Capodimonte
Salita Moiariello 16, Napoli, I-80131, Italy

G. LONGO
Universitá degli Studi "Federico II", Dipartimento di Scienze Fisiche
Via Cinthia, Napoli, I-80126, Italy

In the past ten years, the concept of Virtual Observatory (VObs) has increasingly gained importance in the domain of astrophysics, as a way of seamlessly accessing data in different wavelength domains stored in digital archives. There are many reasons why the VObs is useful for the development of science: to monitor time variability of phenomena, to compare phenomena in different bands, to increase return for investment (by fostering data re-use for scientific, educational and outreach purposes), to perform statistical analysis and mining on large quantities of data. The International Virtual Observatory Alliance (IVOA) has paved the way for the VObs to become a really useful tool for the scientific community, by promoting standards, by defining data interoperability methods, by fostering the needed coordination among data providers. But the VObs is more than just archives and standards: it is also infrastructure, basic software tools, advanced applications, evolution of methods and techniques, cross-fertilization with other communities. Discovering information in wide-field images and mining large archives are key items towards the use of the VObs as a tool for developing science. Data mining, or knowledge discovery in databases, while being the main methodology to extract the scientific information contained in Massive Data Sets (MDS), needs to tackle crucial problems since it has to orchestrate complex challenges posed by transparent access to different computing environments, scalability of algorithms, reusability of resources. To achieve a leap forward for the progress of astrophysics in the data avalanche era, the community needs to implement an infrastructure capable of performing data access, processing and mining in a distributed but integrated context.

1. Introduction: Astronomical Archives

Astronomy in recent years has expanded drastically its observing and modeling capabilities, increasing proportionally its demands for computational power and data access efficiency. To understand completely the implications of this statement, one should remember that Astronomy is mostly an observational science, where any "experiment" is a snapshot of the fraction of the Universe being observed, and cannot be repeated as such. Astronomy measures $I(\lambda)$, light intensity as a function of wavelength (or frequency, or energy). But, since most phenomena are in fact variable, the measured intensity is $I(\lambda, t)$ which is convolved with the transfer function of the complex system (telescope, instrument, sky) actually involved in the observation. It is therefore easy to understand why every single observation needs to be kept, obviously creating a data preservation issue. The increasing complexity of instrumentation and the need to optimize observing procedures (e.g. the meteorological conditions may be unacceptable for one type of observation while is suitable for others) lead to a change in concept on modern observatories. Classical observing (the scientist located at the telescope is given the whole night to perform his/her observations) is replaced by flexible scheduling and service observing, i.e. the observations and the basic data handling are performed by observatory staff. In this case, the scientist owning the observations extracts the data from disk storage, written in the FITS standard data format, which has been used by the community since 1977. Data might have been already partially processed (i.e. calibrated, with removal of the instrument signature).

The scenario described above has a pretty obvious extension, from data storages to full-fledged archives. Observatory policies have been defined so as to allow the data to become public typically after one year from the date of observation. Data taken for a specific scientific purpose can therefore be re-used for different purposes. This approach, progressively emerged in the past fifteen-twenty years, has allowed the possibility of developing new science. Examples are monitoring the time variability of phenomena, comparing phenomena in different energy bands (the so-called multi-wavelength astronomy), performing statistical analysis or data mining on large quantities of data. From the technical point of view, the possibility of re-processing the raw data when a better knowledge of instrumental effects is achieved. An important effect of archives, and of the consequent re-use of data, is increasing the return for the investment: since a second of observation on modern facilities is worth roughly 1 US$, and the over-subscription factor is on the average a factor of 5, the possibility of re-using top-class data to do science, or for educational and outreach purposes has also a positive social impact. But archiving is an activity that is essential in order to cope with the data avalanche that is hitting the astrophysics community. As a matter of fact, the cumulative photon collecting area of telescopes world-wide has increased by a factor of 8 in the past 25 years, and the detector technology has allowed the production of panoramic chips which have multiplied the data flow. To give an example, the archive of the European Southern Observatory (ESO), the leading European institution in the field of astronomy, has grown in size by a factor of 10 in three years. And of course, there are many dozens of ground- and space-based telescopes world-wide, covering the full electromagnetic spectrum, and equipped with instruments providing data which are telescope and band-dependent.

2. The Virtual Observatory

2.1. *Data and information services in Astronomy*

Besides data, many other sources of information for the Astrophysical community can be identified and can be summarized as follows:

(1) Telescopes (ground- and space-based, covering the full electromagnetic spectrum) are managed by Observatories;
(2) Instruments (which are telescope and band dependent) are managed by Observatories and/or Consortia;
(3) Data analysis software is usually instrument-dependent: it is managed by Observatories, Consortia and individual Researchers;
(4) Computing is supported by dedicated service Centers, or Organizations that can often be Observatories;
(5) Archives are managed by Observatories and/ or Agencies;
(6) Publications are managed by Journals; there is however a single point of search and access, the ADS system (which is the main source of bibliographic information for the whole of the Astronomical community);
(7) Data curation, i.e. management of metadata (i.e. the description of data, tables and catalogues) is performed by Data curators;
(8) Public Outreach is managed by Observatories and/or Agencies.

2.2. *The Virtual Observatory: definitions*

The Virtual Observatory (VObs) is an innovative, still evolving, system to allow all the above data and information systems to interoperate. It takes advantage of the astronomical data explosion (e.g., use statistical identification to perform multi-wavelength, multi-parameter analysis), to allow astronomers to interrogate multiple data centers in a seamless and transparent way and to utilize at best astronomical data, to permit remote computing and data analysis and to foster *new science*. The VObs aims at achieving the goal of having *all astronomical databases inside the scientist's PC* just as in the Web all documents are easily retrievable inside one's own PC.

The situation depicted in the above section is pretty complex, and the desire has been felt to unify the sources of information within a federated framework. Achieving the ambitious goal of allowing the different sources of information to interoperate is not trivial: as a matter of fact, the current situation shows both positive and negative aspects. On the positive side, one can consider that observational data are freely available *on-line* in general 1 year after observation; both data and catalogues are normally stored in *astronomical archives*. Results are published in academic journals, all *available on-line* with *pointers* to data archives; as mentioned above, there is one *single entry point for journals* (ADS). The processing and analysis software *maintained* Observatories/Archives is made available *on-line*. However, the different astronomical archives have *widely different access/search interfaces and standards/conventions*; and they mainly serve raw data. The widely specialized, *analysis software* for the various sub-branches is in general*complex*, thus yielding a *steep learning curve*, but this is needed, since multi-wavelength is now the norm to produce science. Publication-Archive links often point to *raw, unprocessed data* and are not complete. Furthermore, object *metadata are not homogeneously defined.*

2.3. *The International Virtual Observatory Alliance*

All of these considerations require that the various players speak the same language. It is therefore necessary that common standards and protocols are defined and adopted within the whole community. This issue is tackled within the International Virtual Observatory Alliance (IVOA), which coordinates 19 world-wide national or super-national VObs initiatives (shown in Figure 1).

Fig. 1: The members of the International Virtual Observatory Alliance are national or super-national VObs projects and initiatives.

The IVOA mission is *"to facilitate the international coordination and collaboration necessary for the development and deployment of the tools, systems and organizational structures necessary to enable the international utilization of astronomical archives as an integrated and interoperating virtual observatory"*. The work is carried out by means of teleconferences, "TWiki" pages, and bi-annual meetings (the last one was carried out in Nara, Japan, in October 2010, the next will be in Naples, Italy, in May 2011). The IVOA goals are the standardization of data, metadata and software, the definition of data interoperability methods, and the production of lists of available data and computing services (each provided by the individual VObs projects).

IVOA is an initiative that is based on the concept of collaborating and sharing. The absence of specific funding does not allow to force any decisions, which are basically taken unanimously. This practice is in principle good, since it builds consensus in the community: however, it brings as a consequence a slow convergence in the definition of standards. From the organizational point of view, the IVOA structure is composed of an Executive Board which includes the representatives from all national VObs projects and a dozen of Working and Interest Groups coordinated by a Technical Coordination Group. There is an estimated total of 300-400 individuals involved in IVOA development activities. Among the various groups, it is worth mentioning the recent creation (2009) of a Data Mining and KDD Interest Group.

2.4. *Status of the Virtual Observatory*

The VObs is progressively evolving to a truly operational phase [1]. Its paradigm to utilize multiple archives of astronomical data in an interoperating, integrated and logically centralized way, allows to "observe a virtual sky" by position, wavelength and time. Not only data actually observed are included in this concept: theoretical and diagnostic data are included as well. VObs represents a new type of a scientific organization for the era of information abundance:

(1) it is inherently distributed, and web-centric;
(2) it is fundamentally based on a rapidly developing technology;
(3) it transcends the traditional boundaries between different wavelength regimes, agency domains;
(4) it has an unusually broad range of constituents and interfaces and
(5) it is inherently multidisciplinary.

The international VObs has opened a new frontier to astronomy. In fact, by making available at the click of a mouse an unprecedented wealth of data and by implementing common standards and procedures, the VObs allows a new generation of scientists to tackle complex problems, which were almost unthinkable only a decade ago. Astronomers may now access a "virtual" parameter space of increasing complexity (hundreds or thousands features measured per object) and size (billions of objects).

2.5. *Extending the Virtual Observatory concept*

In the last couple of years, the VObs received support at an International level: positive statements on the importance of the initiative were made by the Organization for Economic Co-operation and Development (OECD), by the European Strategy Forum on Research Infrastructures (ESFRI), and by ASTRONET, the European initiative aiming at enhancing the coordination and cooperation between national funding and research management organizations in Europe who are responsible for astronomical research. In the US, the VObs initiatives are supported by NASA and NSF. This recognition and support is important since, to make sense, the Virtual Observatory needs to be an international effort. In this framework, an important comment was made in the 2006 ESFRI document, where the Virtual Observatory was indicated as the example to follow to create an international federation of libraries. As a matter of fact, the VObs concepts can be, and are being, re-used in different domains. Among the various interested communities, it is worth mentioning that the planetologists have their own International Planetary Data Alliance (IPDA), the Solar community has its own VObs and DataGrid, and the High-Energy Physics (HEP) community is interested in the VObs mechanisms.

3. Data Mining and the Virtual Observatory

As discussed in the previous section, up to now the VObs has mainly been a mechanism for data access, retrieval and manipulation. However, to obtain results, scientists need to deal with computational challenges related to the handling, processing and modeling of large quantities of data.

3.1. *Massive data processing in astrophysics*

Processing of huge quantities of data (large detectors, mosaics, images with high time resolution) is typical of the optical and solar communities. The amount of computations needed to process the data is impressive, but often "embarrassingly parallel" since based on local operators, with a coarse grained level of parallelism. In such cases, the "memory footprint" of the applications allows one to subdivide data in chunks, so as to fit the RAM available on the individual CPUs and to have each CPU to perform a single processing unit. There are many software resources (legacy applications, libraries, etc.) widely used in past and current projects and experiments that form a wide range of tools, services and facilities used by key astronomical projects and environments, and by the individual astronomer.

Which kind of resources is necessary to tackle the processing and analysis of such large quantities of data in astrophysics? In most cases "distributed supercomputers", i.e. a local cluster of PCs such as a Beowulf machine, or a set of PCs distributed over the network, can be an effective solution to the problem. In this case, the Grid paradigm can be considered to be an important step forward in the provision of the computing power needed to tackle the new challenges. As seen earlier, the concept of "distributed archives" is already familiar to the average astrophysicist, and a leap forward has been made by the VObs which was capable of organizing the data repositories to allow efficient, transparent and uniform access. In more than a sense, the VObs is an extension of the classical Computational Grid; it fits perfectly the Data Grid concept, being based on storage and processing systems, and metadata and communications management services.

3.2. *Grid-based data mining: the DAME system*

As underlined above, most of the implementation efforts for the VObs has concerned the storage, standardization and interoperability of the data together with the computational infrastructures. In particular it has focused on the realization of the low level tools and on the definition of standards. It is important to extend this fundamental target by integrating it in an infrastructure, joining service-oriented software and Grid resource-oriented hardware paradigm, including the implementation of advanced tools for MDS exploration, Soft Computing, Data Mining (DM) and Knowledge Discovery in Databases (KDD). Moreover, DM services often run synchronously. This means basically that they execute jobs during a single HTTP transaction. This might be considered useful and simple, but it does not scale well when it is applied to long-run tasks. Typical long-running activities are the following:

(1) any archive query traversing a massive DB table;
(2) a data-mining job running from a batch (sequential) queue and
(3) a pipeline workflow with several computing-intensive steps, applied sequentially for many (and massive) data sets.

In any of these cases, the system is stressed if the activity lasts longer than a few minutes and becomes unreasonably fragile if it lasts longer than a few hours. With synchronous operations, all the entities in the chain of command (client, workflow engine, broker, processing services) must remain up for the duration of the activity. If any component goes down or stops then the context of the activity is lost and must be restarted.

To overcome this limitation, a new-generation DM system called DAME [2] has been devised. One of its main design strategies is to permit asynchronous access to the infrastructure tools, allowing running of activity jobs and processes outside the scope of any particular web-service operation and without depending on the user connection status. The user, via client web applications, can asynchronously find out the state of the activity, has the possibility to keep track of his jobs by recovering related information (partial/complete results) without having the need to maintain open the communication socket. Moreover, the system is able to automatically perform a sort of garbage collection for cleaning up resources, swap areas and temporary system tools used during the activity run phase. Furthermore, as it will be discussed in what follows, the DAME design takes into account the fact that the average scientist cannot and/or does not want to become an expert also in Computer Science. In most cases he/she already possesses individual algorithms for data processing and analysis and has implemented private routines/pipelines to solve specific problems. These tools, however, often are not scalable to distributed computing environments. DAME aims at providing a user friendly web-based tool to encapsulate one's own algorithm/procedure into the package, automatically formatted to follow internal programming standards. The natural computing environment for MDS processing is a distributed infrastructure (Grid/Cloud), but again, we need to define standards in the development of higher level interfaces, in order to:

(1) isolate the end user (astronomer) from the technical details of VObs and Grid/Cloud use and configuration;
(2) make it easier to combine existing services and resources into experiments.

Data mining is usually conceived as an application (deterministic/stochastic algorithm) to extract unknown information from noisy data. This is basically true but in some way it is very reductive with respect to the wide range covered by mining concept domains. More precisely, in DAME, data mining is intended as techniques of exploration on data, based on the combination between parameter space filtering, machine learning and soft computing techniques associated with a functional domain. The functional domain term arises from the conceptual taxonomy of research modes applicable on data [3–5]. Dimensional reduction, classification, regression, prediction, clustering, filtering are examples of functionalities belonging to the data mining conceptual domain, in which the various methods (models and algorithms) can be applied to explore data under a particular aspect, related to the associated functionality scope. The analytical methods based partially on statistical random choices (crossover/mutation) and on knowledge experience acquired (supervised and/or unsupervised adaptive learning) could realistically achieve the discovery of hidden laws behind focused phenomena, often based on nature laws, therefore the simplest. During the R&D phase of the DAME project, aimed at defining and characterizing rules, targets, ontology, semantics and syntax standards, a functional breakdown structure was derived. It provides a taxonomy between possible data exploration modes, made available by the infrastructure as data mining experiment typology (use case).

4. One Step Further: Integration of the VObs, Data Processing and Data Mining Facilities

To offer scientists a useful service, all of the components of the informatics infrastructure need to be thought as integrated, or at least fully interoperable. In other words, the various infrastructure components (applications, computing, data) should interact seamlessly exchanging information, and be based on a strong underlying network component. The "big science" challenges in astrophysics call for an expansion of the computing infrastructures – and of network and data management infrastructures as well. The astronomical community is interested in using, and participating in defining, competitive computing infrastructures so to perform better research. But there is furthermore the need to integrate network, data and computing infrastructures, or at least to let them interoperate. To fulfill this requirement applications, computing power, data repositories and databases holding metadata or catalogues should be accessed as a single utility. It needs to have a Virtual Observatory interface, which is more familiar to the community [6].

4.1. *VObs-Grid interoperability*

The documents of the ASTRONET initiative [7, 8] have defined the Virtual Observatory as *"the e-infrastructure project in Europe"*. As a matter of fact, an e-infrastructure is a mechanism that, based on solid networking foundations, allows Computing, Data and Applications of various kinds (Data Reduction, Data Analysis, Theory, Numerical Simulations) to interoperate. Interoperability is also the key word of the Virtual Observatory. Integration of the VObs with computational infrastructures, and in particular with Grid facilities, is of key importance. Effort in this direction has been carried out, also within EU-funded projects. In particular, specific IVOA standards have been developed on top of Grid middleware: a FITS driver has been built on top of *gridftp* [9]; the IVOA Single Sign-On standard has been implemented by means of the Grid Authentication & Authorization tools, VOSpace on top of the Data Management mechanisms, VObs Workflows using Job Management, VObs registries using the Grid Information Systems [10]. Furthermore, a "native" way of accessing databases from the Grid through a Query Element (similar in structure to the CE) was developed [11].

4.2. *Implementation in a distributed context*

One major issue to tackle is the capability of such an integrated e-environment to operate within a truly distributed environment. The key problem is that data processing and mining on distributed MDSs implies the distribution of the code as well. As an example, let's consider the cross-check, or the merging, of science results obtained comparing two surveys (e.g. one based on ground-based and another one on space-borne instrumentation) covering a good fraction of the sky. In this case the VObs can help in identifying the location of the relevant data and in making the data themselves interoperable. But it would be unrealistic to move all of the data to the interested astronomer's home institution to perform the DM function. It is actually the data processing, DM or KDD codes that need to move, not the data.

The archive centers are therefore bound to become something more complex in the future: complete full-fledged *service centers*, offering the possibility to provide processing capabilities. This is already happening e.g. for on-the-fly (re-) calibration of data with up-to-date parameters in case of observations [12], or for production of simulations on-demand [13].

But there is an additional step to perform: such centers should be willing to host *user-driven* processing applications. This is not difficult to achieve technically, given the good control of Grid and virtualization concepts and techniques that is becoming common practice, and the continuously decreasing cost of computing and storage hardware. DAME has proven the feasibility of the approach for DM applications. It is actually a "political" issue, tied to cost of resources and accounting.

5. Conclusions

Modern scientific data mainly consist of huge data sets gathered by a very large number of techniques and stored in much diversified and often incompatible data repositories. More in general, in the e-science environment, it is considered as a critical and urgent requirement to integrate services across distributed, heterogeneous, dynamic "virtual organizations" formed by different resources within a single enterprise. In the last decade, Astronomy has become an immensely data-rich field due to the evolution of detectors (plates to digital to mosaics), telescopes and space instruments; new instrumentation is going to further enhance this growth. As a consequence, data archives have become an absolute must to cope with the data avalanche. Data Mining techniques have also increased immensely their capabilities, and are now capable of extracting information from this enormous quantity of data, provided that appropriate processing resources are available. The DAME system is a good example of what can be achieved. The effort to integrate and make the various data archives interoperable has been defined as the Virtual Observatory. The goal is to allow the users to feel all astronomical databases to be just "one click away". Achieving this goal would bring a degree of "democratization" in Astronomy, since first-class data are being made available to scientists in developing countries, to students of all levels, to amateurs. To make sense, the Virtual Observatory needs to be an international effort, which requires involvement at the *project* but also at the *data centre* level. This does not come for free, and work is required at the scientific and technical level to integrate one's own data archive in the VObs and make it interoperable with the others.

There is the need to expand network, data and computing infrastructures, and to integrate them, or at least let them interoperate thoroughly. This effort once again needs to be truly international, thus needing appropriate agreed-upon common policies. In this framework, the archive centers should become *service centers*, offering the possibility of hosting user-driven processing applications, in particular DM applications. In such a way astronomers will have the capability of using new powerful tools and techniques to improve their knowledge and understanding of astrophysical phenomena.

6. Acknowledgments

This work is partially supported by INAF (National Institute of Astrophysics) through the VObs.it initiative and by the European Commission through the EuroVO-AIDA and EuroVO-ICE projects, and is endorsed by the International Virtual Observatory Alliance (IVOA). The Italian participation in EuroVO and IVOA is partially funded by INAF through VObs.it. The DAME project, run jointly by the Department of Physics of the University "Federico II", INAF Astronomical Observatory of Napoli, and the California Institute of Technology, is financed through grants from the Italian Ministry of Foreign Affairs, the European FP7 EuroVO projects and by the USA-National Science Foundation.

References

[1] R. Hanisch, The virtual observatory: Retrospective and prospectus, in *Astronomical Data Analysis Software and Systems XIX*, eds. Y. Mizumoto, K.-I. Morita and M. Ohishi (San Francisco: Astronomical Society of the Pacific, 2010).

[2] M. Brescia, G. Longo, G. S. Djorgovski, S. Cavuoti, R. D'Abrusco, C. Donalek,

A. Di Guido, M. Fiore, M. Garofalo, O. Laurino, A. Mahabal, F. Manna, A. Nocella, G. d'Angelo and M. Paolillo, *ArXiv e-prints* (2010).

[3] R. D'Abrusco, A. Staiano, G. Longo, M. Brescia, M. Paolillo, E. D. Filippis and R. Tagliaferri, *The Astrophysical Journal* **663**, p. 752 (2007).

[4] R. D'Abrusco, G. Longo and N. A. Walton, *Monthly Notices of the Royal Astronomical Society* **396**, 223 (2009).

[5] M. Brescia, S. Cavuoti, G. d'Angelo, R. D'Abrusco, C. Donalek, N. Deniskina, O.Laurino and G. Longo, *Journal of the Italian Astronomical Society* **13** (2009).

[6] F. Pasian and G. Longo, e-infrastructures for astronomy: An integrated view, in *Astronomical Data Analysis Software and Systems XIX*, eds. Y. Mizumoto, K.-I. Morita and M. Ohishi (San Francisco: Astronomical Society of the Pacific, 2010).

[7] P. de Zeeuw and F.J. Molster (eds.), *A Science Vision for European Astronomy* (ASTRONET, 2007).

[8] M.F. Bode, M.J. Cruz and F.J. Molster (eds.), *The ASTRONET Infrastruc-ture Roadmap: A Strategic Plan for European Astronomy* (ASTRONET, 2008).

[9] G. Taffoni, A. Barisani, C. Vuerli, W. Pence and F. Pasian, Grid data source engine: Grid gateway to the virtual observatory, in *Astronomical Data Analysis Software and Systems XV*, eds. C. Gabriel, C. Arviset, D. Ponz and E. Solano (San Francisco: Astronomical Society of the Pacific, 2006).

[10] G. Taffoni, C. Vuerli, F. Pasian and G. Rixon, Bridging VO and computational grid applications within euro-VO, in *Astronomical Data Analysis Software and Systems XVII*, eds. R. W. Argyle, P. S. Bunclark and J. R. Lewis (San Francisco: Astronomical Society of the Pacific, 2008).

[11] Leopoldo Benacchio and Fabio Pasian (eds.), *Grid-enabled astrophysics* (Polimetrica International Scientific Publisher, 2007).

[12] F. Stoehr, D. Durand, J. Haase and A. Micol, The HST cache, in *Astronomical Data Analysis Software and Systems XVIII*, eds. D. A. Bohlender, D. Durand and P. Dowler (San Francisco: Astronomical Society of the Pacific, 2009).

[13] F. Pasian, G. Taffoni, C. Vuerli, P. Manzato, F. Gasparo, S. Cassisi, A. Pie-trinferni and M. Salaris, BaSTI: the virtual observatory meets the computational grid, in *Astronomical Data Analysis Software and Systems XVII*, eds. R. W. Argyle, P. S. Bunclark and J. R. Lewis (San Francisco: Astronomical Society of the Pacific, 2008).

DAME: A WEB ORIENTED INFRASTRUCTURE FOR SCIENTIFIC DATA MINING AND EXPLORATION

S. CAVUOTI*

Department of Physics, Universita' degli Studi Federico II,
Via Cintia 26, 80125 Napoli, Italy
** E-mail: s.cavuoti@gmail.com*
http://dame.dsf.unina.it

M. BRESCIA

INAF - Osservatorio Astronomico di Capodimonte,
Via Moiariello 16, 80131 Napoli, Italy
E-mail: brescia@na.astro.it

G. LONGO, M. GAROFALO and A. NOCELLA

Department of Physics, Universita' degli Studi Federico II,
Via Cintia 26, 80125 Napoli, Italy

Nowadays, many scientific areas share the same broad requirements of being able to deal with massive and distributed datasets while, when possible, being integrated with services and applications. In order to solve the growing gap between the incremental generation of data and our understanding of it, it is required to know how to access, retrieve, analyze, mine and integrate data from disparate sources. One of the fundamental aspects of any new generation of data mining software tool or package which really wants to become a service for the community is the possibility to use it within complex workflows which each user can fine tune in order to match the specific demands of his scientific goal. These workflows need often to access different resources (data, providers, computing facilities and packages) and require a strict interoperability on (at least) the client side. The project DAME (DAta Mining & Exploration) arises from these requirements by providing a distributed WEB-based data mining infrastructure specialized on Massive Data Sets exploration with Soft Computing and machine learning methods. It results as a multi-disciplinary platform-independent tool perfectly compliant with modern KDD (Knowledge Discovery in Databases) requirements and Information & Communication Technology trends.

Keywords: Astroinformatics; data mining, knowledge discovery in database; machine learning; virtual observatory.

1. Introduction

In almost all fields, modern technology allows to capture and store huge quantities of heterogeneous and complex data often consisting of hundreds of features for each record and requiring complex metadata structures to be understood. A need has therefore emerged for a new generation of software tools, largely automatic, scalable and highly reliable. Strictly speaking, Knowledge Discovery in Databases (KDD) is about algorithms for inferring knowledge from data and ways of validating the obtained results, as well as about running them on infrastructures capable to match the computational demands. In practice, whenever there is too much data or, more generally, a representation in more than 5 dimensions [1], DAME copes with this problem by implementing the KDD models under the S.Co.P.E. (Scientific Cooperation for Pluri-disciplinary Experiments) grid [9] and by taking into account the fact that background knowledge can make it possible to reduce the amount of data that needs to be processed by adopting a learning rule based on the fact that in many cases most of the attributes turn out to be irrelevant when background knowledge is taken into account [2].

Data Mining requires a lengthy fine tuning phase which is often not easily justifiable to the eyes of a non experienced user.

Such complexity is one of the main explanations for the gap still existing between the new methodologies and the huge community of potential users which fail to adopt them. In order to be effective, in fact, Knowledge Discovery in Databases requires a good understanding of the mathematics underlying the methods, of the computing infrastructures and of the complex workflows which need to be implemented, and most users (even in the scientific community) are usually not willing to make the effort to understand the process and prefer to recur to traditional approaches which are far less powerful but much more user friendly [3].

DAME, by making use of the Web application paradigm, of extensive and user friendly documentation and of a sample of documented and realistic use cases, represents a first attempt to bring the KDD models to the user hiding most of their complexity behind a well designed infrastructure. In this paper we describe the prototype (beta release) version of the DAME (Data Mining and Exploration) web application, nowadays available under final commissioning which addresses many of the above issues and aims at providing the scientific community with a user friendly and powerful data mining platform.

2. General Description

DAME was initially conceived to work on astrophysical Massive Data Sets data as a tool offered to the community and aimed at solving some of the above quoted problems by offering a completely transparent architecture, a user friendly interface and the possibility to access a distributed computing infrastructure.

DAME starts from a taxonomy of data mining methods (hereinafter called functionalities) and collects a set of machine learning algorithms (hereinafter called models) that can be associated to one or more functionalities depending of the specific problem domain. This association "functionality-model" represents what in the following we shall refer to as "experiment domain".

At a lower level, any experiment launched on the DAME framework, externally configurable through dynamical interactive web pages, is treated in a standard way, making completely transparent to the user the specific computing infrastructure used and the specific data format given as input. Dimensional reduction, classification, regression, prediction, clustering, filtering, are examples of functionalities belonging to the data mining conceptual domain, in which the various methods (models and algorithms) can be applied to explore data under a particular aspect, connected to the associated functionality scope. In its first implementation the infrastructure prototype has been focused on the classification, regression and clustering functionalities.

A special care was devoted in each single phase of the design and development to produce a complete and exhaustive documentation both technical and user oriented. As The concept of "distributed archives" is familiar to most scientists. In the astronomical domain, the leap forward was the possibility to organize through the Virtual Observatory (hereafter VObs) [4] the data repositories to allow efficient, transparent and uniform access. In other words, the VObs was intended to be a paradigm to use multiple archives of astronomical data in an interoperating, integrated and logically centralized way, so to be able to "observe" and analyze a virtual sky by position, wavelength and time [14]. The link between data mining applications and the VObs data repositories is currently still under discussion since it requires (among the other things) the harmonization of many recent achievements in the fields of VObs, grid, cloud, HPC (High Performance Computing), and Knowledge Discovery in Databases. DAME was conceived to provide the VObs with an extensible, integrated environment for data mining and exploration. In order to do so, DAME had to support the VObs standards and formats, especially for data interoperability and to abstract the application deployment and execution, so to provide the VObs with a general purpose computing platform taking advantage of the modern technologies.

In order to gradually accomplish such requirements, DAME intended to give the possibility to remotely interact with data archives and data mining applications via a simple web browser [5]. Thus, with web applications, a remote user does not require to install program clients on his desktop, having the possibility to collect, retrieve, visualize and organize data, configure and execute mining applications through the web browser and in an asynchronous way. An added value of such approach being the fact that the user does not need to directly access large computing and storage power facilities. Connected with the dichotomy between supervised and unsupervised learning, in the DAME data mining infrastructure, the choice of any machine learning model is always accompanied by the functionality domain.

In what follows we shall therefore adopt the following terminology:

- Data mining model: any of the data mining models integrated in the DAME suite. It can be either a supervised machine learning algorithm or an unsupervised one, depending on the available Base of Knowledge (BoK, i.e. the set of training or template cases available) and the scientific target of the user experiment;

- Functionality: one of the functional domains in which the user wants to explore the available data (for example, regression, classification or clustering). The choice of the functionality target can limit the choice of the data mining model;
- Experiment: it is the scientific pipeline (including optional pre-processing or preparation of data) and includes the choice of a combination of data mining model and a functionality;
- Use Case: for each data mining model, different running cases are exposed to the user. These can be executed singularly or in a prefixed sequence. Being the models derived from the machine learning paradigm [6], each has training, test, validation and run use cases, in order to, respectively, perform learning, verification, validation and execution phases. In most models there is also the "full" use case, that automatically executes all listed cases as a sequence.

The functionalities and models, already available and under deployment in the DAME web application, are the following:

- Classification and Regression (Supervised): three different versions of the classical MultiLayer Perceptron (MLP) trained, respectively by Back Propagation, Genetic Algorithms and Quasi Newton Algorithms; Support Vector Machines (SVM);
- Dimensional Reduction (unsupervised): Principal Probabilistic Surfaces (PPS);
- Clustering and Image Segmentation (unsupervised): Self Organizing Maps (SOM).

3. The Architecture

The DAME design architecture is implemented following the standard LAR (Layered Application Architecture) strategy, which leads to a software system based on a layered logical structure, where different layers communicate with each other via simple and well-defined rules:

- Data Access Layer (DAL): the persistent data management layer, responsible of the data archiving system, including consistency and reliability maintenance.
- Business Logic Layer (BLL): the core of the system, responsible of the management of all services and applications implemented in the infrastructure, including information flow control and supervision.
- User Interface (UI): responsible of the interaction mechanisms between the BLL and the users, including data and command I/O and views rendering.

A direct implication of the LAR strategy adopted in DAME is the Rich Internet Application (RIA) [7], consisting in applications having traditional interaction and interface features of computer programs but usable via simple web browsers, i.e.

not needing any installation on user local desktop. RIAs are particularly efficient in terms of interaction and execution speed. By keeping this in mind, the main concepts behind the distributed data mining applications implemented in the DAME Suite are based on three issues:

- Virtual organization of data: extension of the already remarked basic feature of the VObs;
- Hardware resource-oriented: this is obtained by using computing infrastructures, like grid, which enable parallel processing of tasks, using idle capacity. The paradigm in this case is to obtain large numbers of instances running for short periods of time;
- Software service-oriented: this is the base of typical cloud computing paradigm [8]. The data mining applications implemented runs on top of virtual machines seen at the user level as services (specifically web services), standardized in terms of data management and working flow.

The complete Hardware infrastructure of the DAME Program, where the grid sub-architecture is provided by the S.Co.P.E. supercomputing facility [9, 13] is incorporated into the more general cloud scheme, including a network of multi-processor PCs and workstations, each of them internally dedicated to a specific function. The integrity of the system, including the grid public access, is guaranteed by a registration procedure, giving the possibility to access all facilities from just one account. In particular, a robot certificate is automatically handled by the DAME system to provide transparent access to the S.Co.P.E. grid resources [10]. Depending on the computing and storage power, requested by the job and by the processing load currently running on the network, an internal mechanism redirects the jobs to a job-queue in a pre-emptive scheduling scheme. The interaction with the infrastructure is completely asynchronous and a specialized software component (DR, DRiver) has the responsibility to store off-line job results in the user storage workspaces, that can be retrieved and downloaded in subsequent accesses. This hybrid architecture, renders it possible to execute simultaneous experiments that gathered all together, bring the best results. Even if the individual job is not parallelized, we obtain a running time improvement by reaching the limit value of the Amdahl's law (N):

$$\frac{1}{1 - P + \frac{P}{N}} \tag{1}$$

where P is the fraction of a program that can be made parallel (i.e. which can benefit from parallelization), and (1 - P) is the fraction that cannot be parallelized (remains serial), then the resulting maximum speed-up that can be achieved by using N processors is obtained by the law expressed above.

For instance, in the case of the AGN (Active Galactic Nucleus) classification experiment [11, 12], each of the 110 jobs runs for about a week on a single processor. By exploiting the grid, the experiment running time can be reduced to about one week instead of more than 2 years.

The DAME software architecture is based on five main components: Front End (FE), Framework (FW), Registry and Data Base (REDB), Driver (DR) and Data Mining Models (DMM).

The Front End (FE) is the component directly interfacing the end user with the infrastructure (through the web browser). and it can be considered structured in two main parts (also addressable as a "client" and a "server", even though both of them are resident on the remote side with respect of the user): the main GUI (Graphical User Interface) of the Suite and the internal interface (hereinafter Front End Server) with the inner infrastructure.

4. Conclusions and Future Developments

DAME is an evolving platform and new modules and specific workflows as well as additional features are continuously added. The modular architecture of DAME can also be exploited to build applications, finely tuned to specific needs. Examples available so far and accessible through the DAME website, being VOGCLUSTERS (Virtual Observatory Globular Clusters), a VObs web application aimed at collecting and make available all existing data on galactic globular clusters for data and text mining purposes, and NExt-II (Neural Extractor) for the segmentation of wide field astronomical images. The main product is a Data Mining Web Application suite available at address: http://dame.dsf.unina.it/beta_info.html

Moreover, it is foreseen the introduction of MPI (Message Passing interface) technology, by investigating its deployment on a multi-core platform, based on GPU+CUDA computing technique [15]. This could improve computing efficiency in data mining models, such as Genetic Algorithms, naturally implementable in a parallel way. In conclusion, we are confident that DAME may represent what is generally considered an important element of e-science: a stronger multi-disciplinary approach based on the mutual interaction and interoperability between different scientific and technological fields (nowadays defined as X-Informatics, such as Astro-Informatics).

Acknowledgments

All partners acknowledge the financial support of the Italian Ministry of Foreign Affairs for the Italy-USA bi-lateral grant "Building an e-science Data Mining Infrastructure", the European Union through the projects VO-Tech and VO-AIDA and the Ministry of University and Research through the PON S.Co.P.E. GRID project. SGD, AAM, and CD acknowledge a crucial support from the Fishbein Family Foundation, and a partial support from the NASA grant 08-AISR08-0085

References

[1] S. Odenwald, Cosmology in More Than 4 Dimensions, *Astrophysics Workshop*, N.R.L. (1987).

[2] G. Paliouras, Scalability of Machine Learning Algorithms, M. Sc. Thesis, University of Manchester (1993).
[3] T. Hey, S. Tansley, K. Tolle, *The Fourth Paradigm: Data-Intensive Scientific Discovery* (Microsoft Research, 2009).
[4] F. Genova, G. Rixon, F. Ochsenbein, C.G. Page, Interoperability of archives in the VO, in *Proc. of SPIE Conference Virtual Observatories*, eds. Alexander S. Szalay (2002), pp.20-26.
[5] M. Brescia, G. Longo, F. Pasian, *Nuclear Instruments and Methods in Physics Research* **623**, 2 (2010).
[6] C.M. Bishop, *Pattern Recognition and Machine Learning"* (Springer Verlag, Heidelberg, 2006).
[7] A. Bozzon et al., *ACM Transaction on the Web (TWEB)* **4** (2010).
[8] J. Shende, Service-Oriented Architecture and the Cloud, Cloud Expo, SYS-CON Media, http://cloudcomputing.sys-con.com/node/1524663 (2010).
[9] L. Merola, The SCOPE Project, in *Proc. of the Final Workshop of GRID Projects PON RICERCA 2000-2006* (Catania, Italy, 2008).
[10] N. Deniskina, et al., GRID-Launcher v.1.0, Memorie della Societa Astronomica Italiana, **80**, (Munich, 2009) p.571.
[11] S. Cavuoti, et al., Photometric AGN classification in the SDSS, in *Proc. of Italian E-science 2008 Conference* (Naples, Italy, 2008).
[12] M. Brescia, et al., Astrophysics in S.Co.P.E, Memorie della Societa Astronomica Italiana **13**, 56 (2009).
[13] M. Brescia, et al., DAME: A Distributed Web Based Framework for Knowledge Discovery in Databases, in *Proc. of 54th SAIT Conference Next Decade Prospective for the Italian Astronomy* (Napoli, 2009).
[14] G. Fabbiano, D. Calzetti, C. Carilli, S. G. Djorgovski, et al., Recommendations of the VAO-Science Council, arXiv:1006.2168v1 [astro-ph.IM] (2010).
[15] B. Maier, High Performance Computing with CUDA, American Physical Society, Joint Fall 2009 Meeting of the Texas Sections of the APS, AAPT, and SPS Post deadline, abstract #C5.006 (2009).

GALACTIC PHASE SPACES

D. CHAKRABARTY*

Department of Statistics, University of Warwick,
Coventry CV4 7AL, U.K.
**E-mail: d.chakrabarty@warwick.ac.uk*

Galaxies are argued to manifest complexity, thereby contradicting models of smooth parametric galactic phase space densities. An estimation of chaos in models of our galaxy is forwarded to suggest strength and possible causes of non-linearities in phase space. A Bayesian nonparametric methodology that is designed to acknowledge uncertainties in measured data, when applied to an observed external galaxy, indicates a non-linear galactic phase space density that could result from a bistable system potential. The effect of such a phase space structure on the inverse modelling of phase space data is discussed.

Keywords: Non-linear dynamics; Hypothesis testing; Bayesian non-parametrics.

1. Introduction

A dynamical system is given by its phase space W that is marked by an evolution function $g(\mathbf{w}, t)$ that determines system evolution with time t ($t \in T$). Thus, in general $g : T \times W \longrightarrow W$. Thus, characterisation of the dynamical system would entail estimation of the probability density function $f(\mathbf{w}, t)$ and of $g(\mathbf{w}, t)$, given the available information. In the treatment of astronomical systems - such as galaxies - as non-dissipative dynamical systems, the approximation of the stellar fluid as collisionless is invoked at the very outset. We assume our astronomical dynamical system as Hamiltonian, and write $g : \mathbb{R}^3 \longrightarrow \mathbb{R}^3$. The spatial coordinates $\mathbf{x} \in \mathbb{R}^3$ and velocities $\mathbf{v}$ ($\mathbf{v} = \dot{\mathbf{x}}$) of all system particles are in W. We also assume that the system has reached stationarity, allowing us to write the *pdf* of W as $f(\mathbf{w})$. The information that is available for the estimation of the phase space density structure and the evolution function can be data of certain phase space coordinates of individual system particles, sampled from a sub-manifold U in W ($\{\mathbf{u}_k\}_{k=1}^{N}$, where $U \subseteq W$), and/or the observable function $\Psi(\mathbf{u})$. It is the nature of galactic observations that constrain $\mathbf{u}_k$ to $x_1^{(k)}, x_2^{(k)}, v_3^{(k)}$, where $x_1^{(k)}, x_2^{(2)}$ are the coordinates of the kth particle in the image of the system and $v_3^{(k)}$ is the component of $\mathbf{v}_k$ along the line that connects the observer to the kth particle.

The other idiosyncrasy of galaxies is that for such a gravitational system, $g(\mathbf{x})$ is the gravitational field due to all the gravitating matter in the system and is an expression of the gravitational mass density $\rho(\mathbf{x})$: $\nabla \cdot g(\mathbf{x}) = -4\pi G\rho(\mathbf{x})$, where G is the (known) Universal Gravitational constant. The main contributor towards $\rho(\mathbf{x})$

is the dark matter constituent of the system; such dark matter density cannot be observed photometrically since dark matter neither emits nor reflects light, Thus, in place of the more readily available, photometic astronomical observations, we need to inversely model the effects of the total gravitating mass of a galaxy, in order to estimate $\rho(\mathbf{x})$. It may be mentioned that once $\rho(\mathbf{x})$ - and thereby $g(\mathbf{x})$ - is estimated, the evolution of a sample of phase space coordinates can be computed, by plugging in $g(\mathbf{x})$ in Newton's or Einstein's equations of motion. The estimation of such evolution is uncertain insofar as the probabilistic nature of the estimate of $\rho(\mathbf{x})$, i.e. $\hat{\rho}(\mathbf{x})$, is concerned.

In this paper, we explore the estimation of the total gravitational mass density of galaxies, using inverse modelling of the aforementioned "effect"s of galactic gravitational mass. Importantly, the focus of this paper is the effect of non-linearities in $\rho(\mathbf{x})$ and $f(\mathbf{w})$, on the estimation of macroscopic system parameters. In particular, we will examine the effect of chaoticity - brought about by non-linearities in $\rho(\mathbf{x})$ - on galactic relaxation, using the Milky Way as an example. Once the possibility of a multimodal, non-linear galactic phase space density is demonstrated, we will indicate the fundamental risk involved in the estimation of $\rho(\mathbf{x})$ using kinematic data, when the galactic phase space is such.

2. Phase Space of the Milky Way

Traditionally, galaxies are treated as built of regular orbits, evolving towards an integrable potential. If this were true, it would indeed be very difficult to explain how - as is typically considered in statistical mechanics - chaotic mixing could have brought the system to the current equilibrium state. Thus, it was considered interesting to test for the chaos-inducing efficiency in a galactic model that is initially based on regular orbits, exhibiting a uniformly distributed phase space structure, and then quantify the relative strength of chaoticity. If such chaoticity is identified, its source is sought - stochasticity in rapidly changing potential as distinguished from resonance-coupling effects - in our own galaxy (Chakrabarty 2007). A useful corollary of such a pursuit is estimation of some Milky Way parameters.

Here we discuss the estimation of relevant Milky Way parameters $S \in \mathbb{R}^n$ in general, (with n=2 in Chakrabarty, 2007), given the velocity data of stars in our galaxy, from within an ϵ neighbourhood of the Sun, $\mathcal{D} = \{v_i\}_{i=1}^{N}$. It is of importance to appreciate that the estimated spatial location of us - the observer of this data set, is very well approximated by the location of Sun on the Milky Way disc on astronomical scales. In fact, given the near isotropic sampling around the Sun, if $\epsilon \ll R_\odot$, where $R_\odot$ is a fiduciary value for the radial location of the Sun w.r.t. the Galactic centre (viz. the IAU standard value that we want to improve upon), *the location of the Sun can be approximated as a summary of the distribution of the locations of the stars in this observed sample.* This inequality does hold for the data used in Chakrabarty (2007) in which $\epsilon/R_\odot \approx 0.0125$.

The estimation of $\hat{\mathbf{s}}$ as $\mathbf{s}_i$ follows from the identification of the simulated phase space density $\nu(\mathbf{w}|\mathbf{s}_i)$ that offers "maximal" support to the hypothesis that $\mathcal{D} \sim \nu(\mathbf{w}|\mathbf{s}_i)$, $i = 1, \ldots, N_{sim}$, where there are N_{sim} number of simulated phase space densities generated (discussed in details in Section 2.2).

2.1. *Simulating phase space data*

In Chakrabarty (2007), the Milky Way is modelled as an axisymmetric disc galaxy. Thus, the modelling is done on a 2-D disk, i.e. $\mathbf{x} = (x_1, x_2)^T$ where $x_1 = R\cos\phi$ and $x_2 = R\sin\phi$, where R and ϕ are the radial and azimuthal coordinates. In this example about our galaxy, given that the particle velocities are measured w.r.t. the Sun, i.e. we work with the heliocentric particle velocity $\mathbf{v_r} = (U, V)^T$. The solar velocity is considered known within uncertainties offered in astronomical literature.

A sample of phase space coordinates $\{\mathbf{w_0}\}$, simulated from a realistic phase space density $f(\mathbf{w})$ at $t = 0$, is evolved in a parametric gravitational field that models the background axisymmetric Galactic gravitational field $g_0(R)$, perturbed by the non-axisymmetric, perturbing fields $(g_b(R, \phi - \Omega_b t)$ and $g_s(R, \phi - \Omega_s t))$ of two perturbers that model two rotating structural features of the Milky Way disk, namely the bar and the spiral pattern. Here the bar (and spiral) is rotating with frequency Ω_b (and Ω_s), so that the rotation induces an extra azimuth of $\Omega_b t$ in time t. The net gravitational field that the galactic particles evolve in is then $g(R, \phi, t|\Psi)$ where Ψ represents the fixed model parameters including perturbation strengths, spiral details, etc.

2.2. *Generating the simulated phase space densities*

At the end of the simulation time, orbits are recorded stroboscopically in a frame that rotates with the bar, at times when $(\Omega_b - \Omega_s)t$=0. By "orbits" is implied the stellar phase space coordinates R, ϕ, U, V values at a given t. If the estimation of the system parameter $\mathbf{s}$ is sought, then the recorded orbits are put on a regular $S_1 - S_2 - \ldots - S_m$ grid, where $m =$ dims. Chakrabarty (2007) sought estimates for 2 system parameters, the radial (s_1) and azimuthal (s_2) locations of the observer, (i.e. the Sun), with $s_1 = 0$ at the Galactic centre (GC) and $s_2 = 0$ along the bar major axis. Then, the recorded orbits are placed in a regular polar $s_1 - s_2$ grid, with bin widths δR, $\delta\phi$. Afterall, the sample of stars with the spatial coordinate values $\{R_k, \phi_l\}_{k=i-1,l=j-1}^{k=i,l=j}$ will have a summary spatial location within the ij^{th} $s_1 - s_2$ bin i.e. if this sample is the observed sample, then the solar location from which this sample is observed, must be within this bin (see Section 2); here $i = 1, \ldots, i_{max}$, $j = 1, \ldots, j_{max}$, $i_{max} = (R_{max} - R_{min})/\delta R$, $j_{max} = (\phi_{max} - \phi_{min})/\delta\phi$.

Discrete $U - V$ data recorded in each $s_i - s_j$ cell are noted, $(\forall i = 1, \ldots, i_{max}, j = 1, \ldots, j_{max})$. The *pdf* of $U - V$ space, for the ith value of s_1 and j^{th} value of s_2 is $\nu(U, V|s_1^{(i)}, s_2^{(j)})$, abbreviated as $\nu_{ij}(U, V)$. Then $\nu_{ij}(U, V)$ is estimated using these simulated $U - V$ data and a kernel density estimation technique that employs an

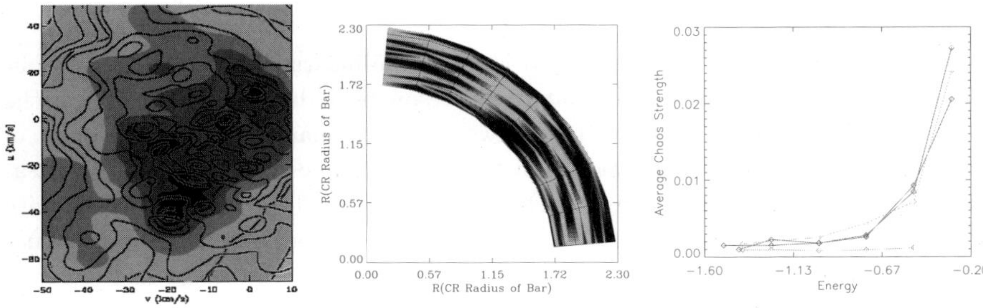

Fig. 1: *Left:* The velocities recorded in the ij^{th} $s_1 - s_2$ cell are used to estimate the ij^{th} simulated *pdf* of $U - V$ space that is overlaid in unfilled contours over $\nu_0(U, V)|\mathcal{D}$ (in coloured filled contours). The non-linear distribution of probability mass is evident in both density structures. *Middle:* Distribution of the support in data $\mathcal{D}$, to the null that the observed data are drawn from the ij^{th} simulated phase space density, as *p*-value of the test statistic - shown in gray-scale over ranges of s_1 and s_2 used in Chakrabarty (2007). *Right:* Strength of chaos in resonance overlap model (in red), in models in which the resonances due to the 2 perturbers occur outside respective resonance widths (in cyan and blue), as distinguished from Milky Way model missing contribution from the perturbing potential marked by a high spatial gradient (in green) (from Chakrabarty & Sideris, 2008).

adaptive, bivariate Gaussian kernel. The density estimation is motivated by the bandwidths used for estimating the density $\nu_0(U, V)$ from the observed data $\mathcal{D}$.

2.3. *Testing*

Thus, we have $R_{max} \times \phi_{max}$ number of $s_1 - s_2$ bins, and a simulated density at each bin. If $\{U_0^{(i)}, V_0^{(i)}\}_{i=1}^{N} \equiv \mathcal{D}$ is sampled from $\nu_0(U, V)$ then Chakrabarty (2007) tested for the null $H0 : \nu_0(U, V) = \nu_{ij}(U, V)$. To test for this null, a test statistic $\mathcal{S}_{ij}$ is defined in the ij^{th} $s_1 - s_2$ bin, with the aim that the null should be rejected for low *p*-values. We say $\mathcal{S}_{ij} := \mathcal{L}_{ij}^{-1}$ where $\mathcal{L}_{ij}$ is the likelihood, i.e $\mathcal{L}_{ij} = \prod_{q=1}^{N_{ij}} \nu_{ij}(U_q, V_q)$, where heliocentric velocities of N_{ij} stars are recorded in the ij^{th} $s_1 - s_2$ bin, $i = 1, \ldots, i_{max}$, $j = 1, \ldots, j_{max}$. This definition of the likelihood assumes that the data are *i.i.d* inside each $s_1 - s_2$ bin. The *p*-value for the ij^{th} bin is $\alpha_{ij} = \Pr(\mathcal{S}_{ij} \geq S_0 | H0)$ where $S_0^{-1} := \prod_{i=1}^{N} \nu_0(U_0^{(i)}, V_0^{(i)})$. In the context of this nonparametric inference, α_{ij} is calculated empirically, by drawing N_{tot} number of random (U, V) samples (of size N) from ν_{ij}; if for N_{great} of these N_{tot} samples, the sample test statistic exceeds S_0, then $\alpha_{ij} = N_{great}/N_{tot}$.

The i, j for which α_{ij}=1, are recorded: $\{s_1^{(m)}, s_2^{(n)}\}_{m=1,n=1}^{m_{max},n_{max}}$, $m_{max} \leq R_{max}$, $n_{max} \leq \phi_{max}$. s_1 and s_2 in these ranges correspond to the observer locations at which the best support for the null is recovered - these ranges are then advanced as the interval estimates for the solar location.

3. How Much Chaos?

The frequencies Ω_s and Ω_b affect the degree of non-linearity in the solar neighbouhood by controlling the locations at which resonances due to the spiral and the bar respectively occur. Now, overlap of two major resonances implies the onset of global chaos. As the relative spatial distance between resonances of the two features is controlled by the ratio $\Omega = \Omega_s/\Omega_b$, this ratio crucially controls chaoticity in the Galactic disk. We set up 4 distinct Milky Way models at 4 different values of Ω. One value of Ω is chosen to cause the resonances due to the bar and spiral to physically overlap while 2 other Galactic models involve Ω values for which the two resonances lie outside each other's respective resonance widths. For the 4th model, neither resonance overlap, nor rapid gradients of the perturber's gravitational field was included. The latter is ensured by excluding the spiral pattern of the Galaxy as a perturber. With these models, a quantification of chaos in the Milky Way disk was undertaken (Chakrabarty & Sideris, 2008). The results are shown in Figure 1; we find that without the spiral, the disk is not rendered chaotic. Also, the strength of chaos is nearly the same in all models that include the spiral, whether resonance overlap happens or not. This indicates the possibility that galactic systems are not neccessarily built of regular orbits and that chaos is induced as response to large gradients in potential.

4. Non-linear Probability Mass Distribution in Phase Space in External Galaxy

The Milky Way example above indicated the possibility that the idea that galactic phase spaces are smooth, monolithic structures is a misplaced notion. The complexity manifest in the Galaxy is treated as general, to inspire a non-linear phase portrait in external galaxies too. Here, the effect of such a phase space structure on the estimation of $\rho(\mathbf{x})$ is discussed in the context of the example distant galaxy NGC 3379 (Chakrabarty 2009).

If in a general dynamical system, the data are sampled from the sub-space U of W, the estimate $f(\hat{\mathbf{w}}, t)$ will be prior-driven unless the dimensionality of the function space is reduced to be $\leq$ that of the data space, as in any inverse problem. However, if the system admits only vague priors, dimensionality reduction will be demanded. Additionally, the simultaneous estimation of $g(\mathbf{x}, \mathbf{v}, t)$ could be achieved if $g(\cdot)$ could be embedded into the structure of $f(\cdot)$; this could be done by invoking the dependence of $f(\cdot)$ on a functional ψ of $\rho(\cdot)$, i.e. $f = f(\psi[\rho], \mathbf{w}, t)$.

The above situation is very much the essence of the application to galaxies as dynamical systems. To enhance sparsity in the models, first we slap stationarity on the unknowns, thus reducing the problem to a stationary inverse problem though the general Bayesian inferrential framework presented in Chakrabarty (2009) supports the learning of $f(\mathbf{w}, t)$ and $g(\mathbf{x}, \mathbf{v}, t)$. We also suggest a velocity independent evolution, which is acceptable for the collisionless system that a galaxy is. Next, we approximate the triaxial geometry of the system (astronomical literature suggests an ellipsoidal geometry) with a spherical one, i.e. assume the gravitational mass density is $\rho(r)$, where $r^2 = \sum_{i=1}^{3} x_i^2$. Next, using the governance of the phase space density by the Vlasov equation (in the gravitational case), we express $f(\mathbf{w})$ as a function of the integrals of motion K_i, $i = 1, 2, \ldots$. In fact, we choose to describe f by only 2 such integrals of motion, keeping the constraint on dimensionality of function space in mind. These integrals are chosen to expand the scope of modelling non-linearlities in phase space, such as anisotropy, which is acknowledged by the dependence of f on the particle angular momentum L. For the kth particle, angular momentum is $L_k = \mathbf{x}_k \wedge \mathbf{v}_k$; the asymmetric dependence on the components of $\mathbf{x}$ and $\mathbf{v}$ acknowledges anisotropy. The other integral of motion is the total energy E: for the kth particle $E_k = \Phi[\rho(r_k)] + \sum_{i=1}^{3}(v_i^{(k)})^2$, where $\Phi(r)$ is the gravitational potential energy, a functional of $\rho(r)$ as dictated by Gauss' Law.

Writing $f(E, L)$ is thus conducive to the embedding of one unknown - $\rho(r)$ - into the structure of the other unknown, $f(\cdot)$. Thus, we are now ready to perform estimation of both the uknown functions. This is attempted by first projecting $f(E[\rho(r_k), \mathbf{v}_k], L(\mathbf{x}_k, \mathbf{v}_k))$ into the manifold $U \subset W$ from which the data is sampled. Thus, if x_1, x_2, $, v_3$ are measured for the kth particle, then

$$\nu(x_1^{(k)}, x_2^{(k)}, v_3^{(k)}) = \int f(E[\rho(r_k), \mathbf{v}_k], L(\mathbf{x}_k, \mathbf{v}_k)) dx_3^{(k)} dv_1^{(k)} dv_2^{(k)} \qquad (1)$$

For this integral to be computed, we need to establish the mapping $\chi : E, L \longrightarrow W/U$. Such can be achieved using Stoke's theorem.

The likelihood is written as $\mathcal{L} = \mathcal{L}' \circ \eta(v_3|\sigma_\delta)$, where $\eta(v_3|\sigma_\delta)$ is the distribution of the measurement uncertainties in v_3. We can neglect the same for x_1 and x_2. The error distribution is given by astronomers and assumed normal with 0 mean and a dispersion σ_δ that is again provided by astronomers for the given observational apparataus. Here "$\circ$" denotes convolution of $\mathcal{L}'$ and $\eta(\cdot)$. This is how measurement errors are incorporated into the estimation. Also, assuming that the data are *i.i.d.*, $\mathcal{L}' = \prod_{k=1}^{N} \nu_k(x_1^{(k)}, x_2^{(k)}, v_3^{(k)})$.

Once the likelihood $\mathcal{L}$ is computed, it is used to write the joint posterior probability of the unknowns, given the data, i.e. $\Pr(f(\cdot), \rho(\cdot)|\{x_1^{(k)}, x_2^{(k)}, v_3^{(k)}\}_{k=1}^{N})$, using truncated normal priors. We sample from this high dimensional posterior using an adaptive Metropolis-Hastings (Haario et. al, 2005).

An application of this methodology to samples $\mathcal{D}_1$ and $\mathcal{D}_2$, of 2 different types of galactic particles in the galaxy NGC 3379 was performed by Chakrabarty (2009). The findings suggest that estimates for $\rho(r)$ from the two data sets are significantly different, over an extended range in r. This implies 2 distinct gravitational mass density structures for the same galaxy, which is of course impossible - this dichotomous result is thus incompatible with the information about our system and stems from an erroneous model assumption. Assuming stationarity in $f(\cdot)$ and $\rho(\cdot)$ and having tested for sphericity, the erroneous assumption is identified as both data sets being drawn from the same stochastic region of W that is defined by a given function of particle E and L. Such an assumed unique function is impossible if the galactic phase space comprises regions that do not communicate with each other and $\mathcal{D}_1$, $\mathcal{D}_2$ have been sampled from such disjoint stochastic volumes of W. Such isolated regions of W can exist for different reasons, but the easist way of ensuring such a phase space structure is if the system gravitational potential $\Phi(r)$ is bistable.

The ramification of such a phase space structure is that $\hat{\rho}(r)$ estimates based on data sets drawn from the isolated parts of W will in general not concur; also no such $\hat{\rho}(r)$ will be the galactic density. This underlies the fundamental risk involved in trying to perform estimation of $\rho(r)$ in a galaxy, without heeding to the inherent non-linearities in the galactic phase space. While having kinematic data from distinct particle populations is indicative of this, merging such kinematic data with independent measurement on a functional - say, $M[\rho(r)]$ as M_0 - and its uncertainties (δM) can be useful; such measurement, along with the uncertainties, could be incorporated into the modelling in different ways, for example by supplementing the aforementioned $\mathcal{L}'$ with the term $(M[\tilde{\rho}(r)] - M_0)/\delta M$, where the functional is computed using the density $\tilde{\rho}(r)$ that is the currently accepted density in the current step within Metropolis (Chakrabarty et. al, *under preparation*).

5. Summary

Galactic W is advanced as non-linear in general and the reason for this is suggested to be steep gradients in the potential, in addition to overlap of resonances of distinct dynamical components of the system. The effect of non-linearities in the galactic $f(\mathbf{w})$ might be to provide fallacious estimates of the system potential, when input data sets are sampled from distinct, isolated basins of attraction within W.

References

[1] D. Chakrabarty, *Different Traces give Different Gravitational Mass Distributions* (Warwick Statistics Tech report, i.e. CRiSM Working Papers Series, 2009).
[2] D. Chakrabarty, *Astronomy & Astrophysics* **467** (2007), p. 145.
[3] D. Chakrabarty & Sideris, I., *Astronomy & Astrophysics*, **488** (2008) p. 161.
[4] H. Haario, E. Saksman & J. Tamminen, *Computational Statistics*, **20** (2005) p. 265.

FROM DATA TO IMAGES: A SHAPE BASED APPROACH FOR FLUORESCENCE TOMOGRAPHY

O. DORN* and K.E. PRIETO

School of Mathematics, The University of Manchester,
Oxford Road, Manchester, M13 9PL
** E-mail: oliver.dorn@manchester.ac.uk*
http://www.maths.manchester.ac.uk/ odorn/

Fluorescence tomography is treated as a shape reconstruction problem for a coupled system of two linear transport equations in 2D. The shape evolution is designed in order to minimize the least squares data misfit cost functional either in the excitation frequency or in the emission frequency. Furthermore, a level set technique is employed for numerically modelling the evolving shapes. Numerical results are presented which demonstrate the performance of this novel technique in the situation of noisy simulated data in 2D.

Keywords: Fluorescence tomography, level set technique, shape reconstruction, linear transport equation, radiative transfer equation.

1. Fluorescence Tomography

Fluorescence tomography is a new modality for biomedical imaging which has received increased attention recently [1, 7, 10]. The domain of investigation is irradiated by laser light of a specific frequency, called the excitation light. This light at excitation frequency travels through the tissue in a random walk fashion, being scattered typically severel hundreds to thousands of times. During this random walk, the laser light can be absorbed by fluorophore molecules (markers) which are present in the tissue. These markers either are present for natural reasons, or they have been introduced into the body in order to observe anatomical features or physiological processes. When such an absorption process occurs, the fluorescence molecules remain for a very short time in an excited stage, until they release part of the captured energy in form of emitted radiation of a frequency which is typically lower than the excitation frequency. This so-called emission light travels now through the body, again in a random walk fashion, and can be captured by detectors when it leaves the body through the outer boundaries. The goal is to reconstruct from these measurements the internal fluorophore concentration and characteristics.

The correct model for light propagation in tissue at fluorescence excitation and emission frequencies is the linear transport equation, often also called the radiative transfer equation or the linear Boltzmann equation [3]. Its specific form is:

$$\frac{1}{c}\frac{\partial u}{\partial t} + \theta \cdot \nabla u + (\mu_a(x) + \mu_s(x))\, u \tag{1}$$

$$= \mu_s(x) \int_{S^{n-1}} \eta(\theta \cdot \theta')\, u(x,\theta',t)\, d\theta' \, + \, q,$$

with the initial condition

$$u(x,\theta,0) = 0 \qquad \text{in} \quad \Omega \times S^{n-1} \tag{2}$$

and the boundary condition

$$u(x,\theta,t) = 0 \qquad \text{on} \quad \Gamma_- . \tag{3}$$

Here,

$$\Gamma_\pm \,:=\, \big\{(x,\theta,t) \in \partial\Omega \times S^{n-1} \times [0,T], \quad \pm\nu(x)\cdot\theta \,>\, 0\big\}\,.$$

Ω is a convex, compact domain in $\mathbf{R}^n$, $n = 2, 3$, with smooth boundary $\partial\Omega$. In our numerical experiments, we will only consider the case $n = 2$, but the algorithm extends in a straightforward way to $n = 3$. $\nu(x)$ denotes the outward unit normal to $\partial\Omega$ at the point $x \in \partial\Omega$, and $u(x,\theta,t)$ describes the density of particles (photons) which travel in Ω at time t through the point x in the direction θ. The velocity c of the particles is assumed to be normalized to $c = 1\text{cms}^{-1}$. The quantity $\mu_a(x)$ is the space dependent absorption coefficient, $\mu_s(x)$ is the space dependent scattering coefficient.

Often not the radiative transfer equation is used for modelling light propagation in tissue, but its diffusion approximation, due to the high complexity of this integro-differential equation. However, it is well-known that in many applications the diffusion approximation is not satisfactory since it is not able to correctly model the regions close to a directional laser source, close to the boundary of the domain, and regions of low scattering coefficients. Therefore, we will do the theoretical and numerical modelling here using the full time-dependent linear transport equation.

The basic linear transport equation (1)-(3) models the propagation of laser light through tissue without any fluorophore present, as it occurs for example in the application of Diffuse Optical Tomography (DOT) [2, 4–6]. For an illustration, see figure 1. In this application, the outgoing radiation is measured and the goal is to reconstruct from its temporal and spatial distribution along the outer boundary the spatial distribution of $\mu_a(x)$ and $\mu_s(x)$ inside the domain.

However, in the application of fluoresence tomography, this equation is coupled to a second linear transport equation which models the creation and propagation of emission photons in the same domain. See Fig. 2 for an illustration of this phenomenon. The coupled system reads now as follows.

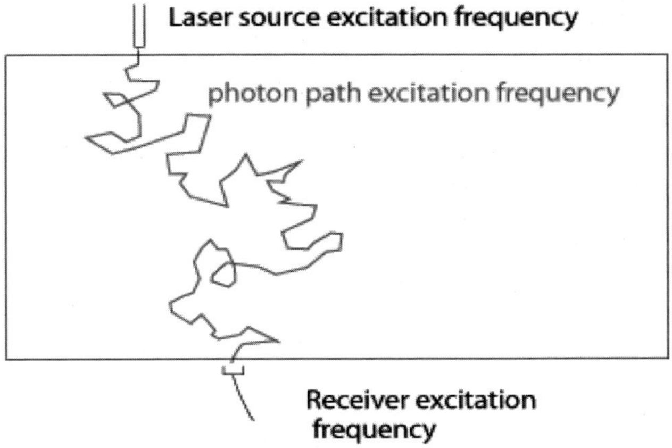

Fig. 1: Random walk of laser light through tissue. The figure illustrates the possible propagation of one excitation photon through the domain with no fluorophore present.

$$\frac{\partial u^{(e)}}{\partial t} + \theta \cdot \nabla u^{(e)} + \left[\mu_a^{(e)} + \mu_s^{(e)} + \mu_{fl}\right] u^{(e)} \qquad (4)$$

$$- \mu_s^{(e)} \int_{S^{n-1}} \eta(\theta \cdot \theta') u^{(e)}(x, \theta', t) d\theta' = q^{(e)}$$

$$\frac{\partial u^{(f)}}{\partial t} + \theta \cdot \nabla u^{(f)} + \left[\mu_a^{(f)} + \mu_s^{(f)}\right] u^{(f)} \qquad (5)$$

$$- \mu_s^{(f)} \int_{S^{n-1}} \eta(\theta \cdot \theta') u^{(f)}(x, \theta', t) d\theta' = q^{(f)}$$

where

$$q^{(f)} = \frac{\gamma}{2\pi} \mu_{fl} \int_{S^{n-1}} u^{(e)}(x, \theta', t) d\theta' \qquad (6)$$

with appropriate initial and boundary conditions similar to (1)-(3) for both $u^{(e)}$ and $u^{(f)}$.

Let us consider a synthetic example which is shown in Fig. 3.

The rectangular domain Ω is filled with a background medium in which some fluorophore concentration is embedded. The absorption and scattering coefficients of the background are $\mu_a(x) = \mu_a^{(0)} + \delta\mu_a(x)$ and $\mu_s(x) = \mu_s^{(0)} + \delta\mu_s(x)$, where $\mu_a^{(0)}$ and $\mu_s^{(0)}$ are known constants and $\delta\mu_a(x)$ and $\delta\mu_s(x)$ are unknown small fluctuations which can be seen in the upper two images of Fig. 3. We assume here for simplicity that $\mu_a(x)$ and $\mu_s(x)$ are the same for excitation and emission frequencies, such that we omit the upper indices $^{(e)}$ and $^{(f)}$ for these two quantities in the following. Embedded in this background are the three regions filled with fluorescent material

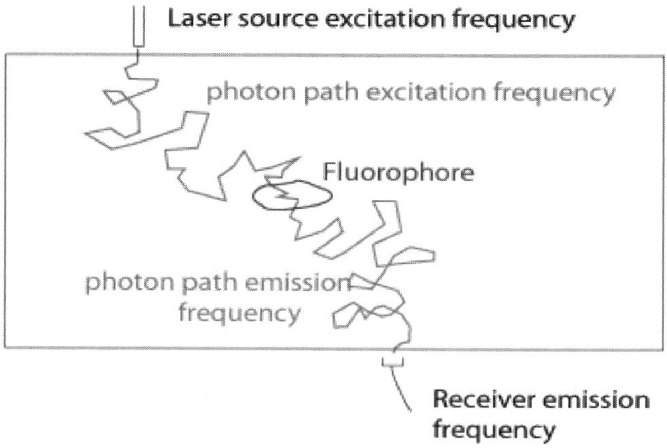

Fig. 2: A typical fluorescence tomography scenario. The figure illustrates the possible propagation of one excitation photon through the domain which travels through a region where fluorophore is present. The photon is captured and gives rise to a new photon of different (emission) frequency.

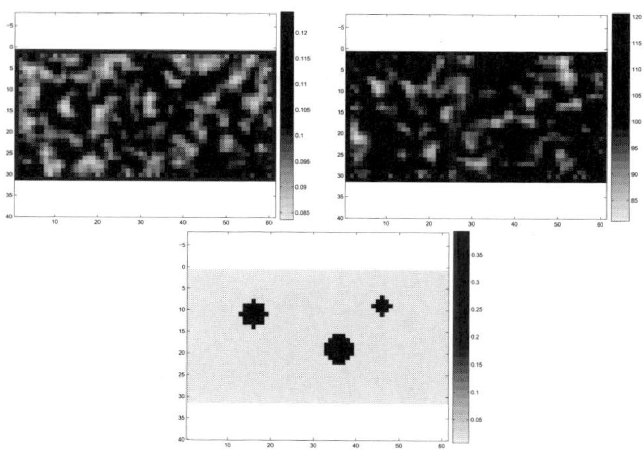

Fig. 3: Top left: true absorption profile (excitation and emission wavelength). Top right: true scattering profile (excitation and emission wavelength). Bottom: true fluorescence μ_{fl} profile.

whose distribution is $\mu_{fl}(x)$. We assume here for simplicity that this distribution has the form $\mu_{fl}(x) = \mu_{FL}\chi_D(x)$ where μ_{FL} is a known scalar constant and $\chi_D(x)$ is the unknown characteristic function for the domain D which is filled with the fluorophore. Since the contrast μ_{FL} is here assumed to be known, only the shape of the domain D needs to be reconstructed from the data.

When applying an excitation source at the boundary of the domain, it gives rise to a time-dependent evolution of excitation and emission photons inside of Ω. Fig. 4 shows in the left column different time snapshots of this distribution of the excitation photons for one given source position, and on the right hand side the corresponding emission photon distribution, as modelled by a finite differences time domain implementation of the system (4)-(9) in 2D.

Fig. 5 displays the corresponding data as a function of detection time and detector position.

Fig. 6 shows a plot of the time-dependence of the data at excitation and emission frequency measured at one particular detector during the detection time.

The goal is to reconstruct the fluorescence distribution $\mu_{fl}(x)$ from this data set.

2. Shape Reconstruction Problem

We need a slightly more formal description of our problem: Find D in the equation

$$A(m) = g \tag{7}$$

where

$$m(x) = \begin{cases} m_{int} & \text{for} \quad x \in D \\ m_{ext} & \text{for} \quad x \notin D \end{cases} \tag{8}$$

The nonlinear operator A maps the shape (boundary) to the measured data. The shape appears as a spatially varying coefficient in the underlying transport equations (4)-(9).

The goal is to find the correct shape by minimizing the least squares cost

$$J(m) = \|A(m) - g\|^2 + \text{`regularization'} \tag{9}$$

Our strategy is to find a velocity function $F(x,t)$ (where t is an artificial evolution time) normal to the current shape boundary such that the cost $J(m)$ decreases when deforming the boundary according to this flow in the outward normal direction.

We see that there are two basic problems to solve in the shape evolution approach:

(1) Finding a (normal) velocity function $F(x,t)$ which reduces the least squares data misfit in excitation and/or emission wavelength (we will use an adjoint technique solving adjoint radiative transfer equations).

(2) Moving the shape *computationally* according to the velocity function (we will use a level set technique).

First we need to calculate a descent direction to the above given cost functional. We can verify easily that, given an arbitrary flow field $F(x,t)$, the infinitesimal change of the cost due to the corresponding flow at a given time step is

$$\frac{\partial J}{\partial t} = \int_{\partial D(t)} \left[(A'(m))^*(A(m) - g) \right] (m_{int} - m_{ext}) F(x,t)\, ds(x) \tag{10}$$

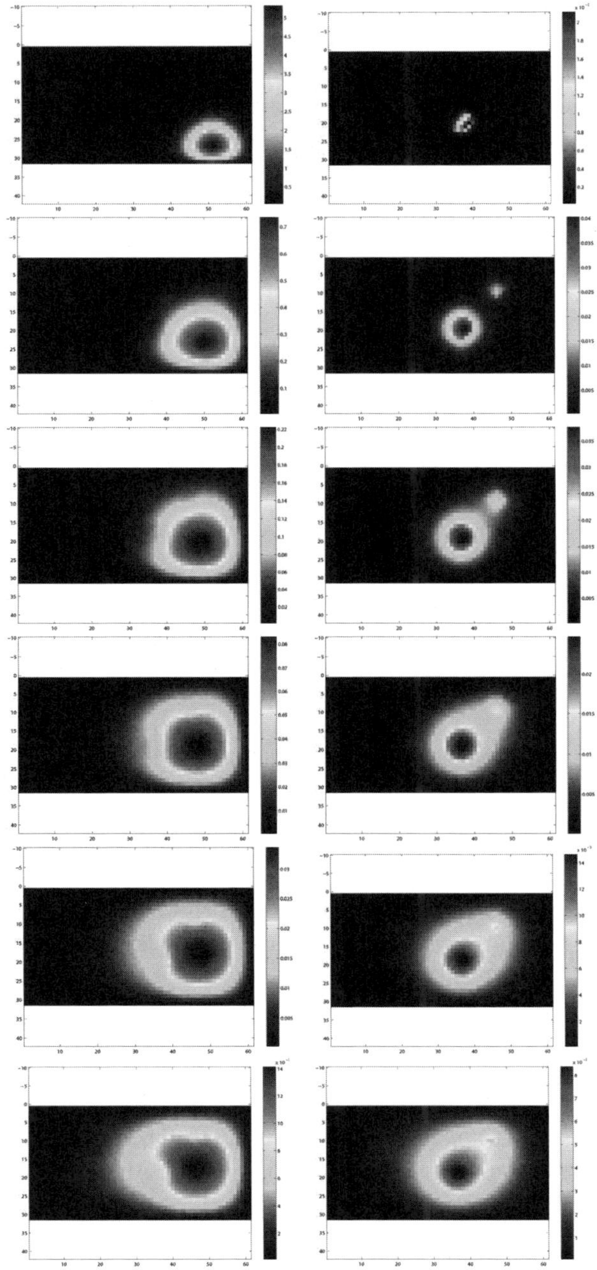

Fig. 4: Evolution of photon density for a source located at one specific boundary point. Left column: excitation wavelength, right column: emission wavelength. In each column from top to bottom: time steps 8, 24, 40, 56, 72, 88.

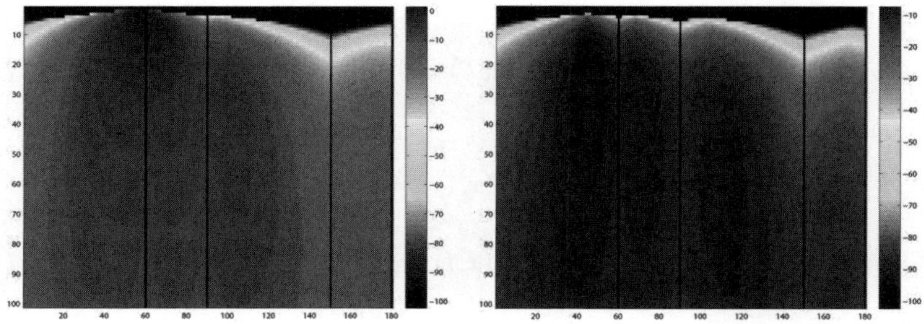

Fig. 5: Data in logarithmic scale (i.e. $\log(data)$) due to a source at the same boundary point as in Fig. 4. Left: excitation wavelength, right: emission wavelength.

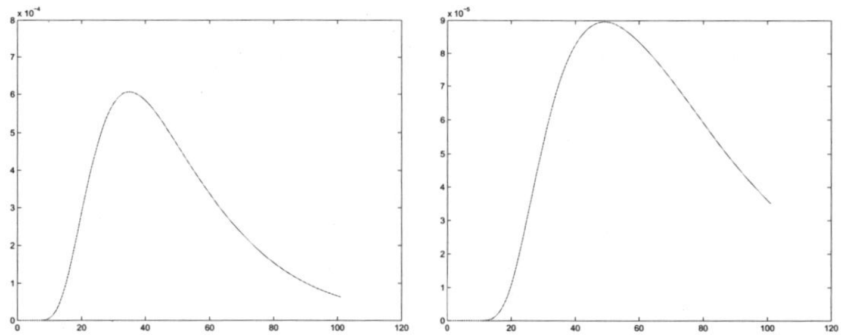

Fig. 6: Plot of time-dependent data recorded at one particular detector position due to the source at the above boundary point. Left: excitation wavelength, right: emission wavelength.

In order to obtain a descent direction, we choose

$$F(x,t)|_{\partial D(t)} = -\left[(A^{'}(m))^{*}(A(m)-g)\right](m_{int}-m_{ext})|_{\partial D(t)} \tag{11}$$

and extend this to the whole domain.

The following result (not proven here) gives an explicit recipe how to calculate this descent direction efficiently by an adjoint scheme.

Let $z^{(e)}$ and $z^{(f)}$ be solutions of the following coupled system of adjoint radiative transfer equations

$$-\frac{\partial z^{(e)}}{\partial t} - \theta\cdot\nabla z^{(e)} + \left[\mu_a^{(e)} + \mu_s^{(e)} + \mu_{fl}\right] z^{(e)} - \mu_s^{(e)} \int_{S^{n-1}} \eta z^{(e)} d\theta' \tag{12}$$

$$= \frac{\gamma}{2\pi}\mu_{fl}\int_{S^{n-1}} z^{(f)} d\theta',$$

$$-\frac{\partial z^{(f)}}{\partial t} - \theta\cdot\nabla z^{(f)} + \left[\mu_a^{(f)} + \mu_s^{(f)}\right] z^{(f)} - \mu_s^{(f)} \int_{S^{n-1}} \eta z^{(f)} d\theta' = 0, \tag{13}$$

with 'initial' condition

$$z^{(e)}(x,\theta,T) = z^{(f)}(x,\theta,T) = 0 \qquad \text{on} \quad \Omega \times S^{n-1} \tag{14}$$

and boundary conditions

$$z^{(e)}|_{\Gamma_+} = \tilde{z}^{(e)}(x,t) := \zeta^{(e)}(x,t), \tag{15}$$

$$z^{(f)}|_{\Gamma_+} = \tilde{z}^{(f)}(x,t) := \zeta^{(f)}(x,t). \tag{16}$$

Define

$$I^{(e)}(x) := -\int_{[0,T]}\int_{S^{n-1}} u^{(e)} z^{(e)}\, d\theta dt, \tag{17}$$

and

$$I^{(f)}(x) := \frac{\gamma}{2\pi}\int_{[0,T]} \Big[\int_{S^{n-1}} u^{(e)} d\theta\Big]\Big[\int_{S^{n-1}} z^{(f)}\, d\theta\Big] dt, \tag{18}$$

Then a descent direction for the least squares data misfit is given by

$$F(x) = -\Big[I^{(e)}(x) + I^{(f)}(x)\Big]\,\mu_{FL} \tag{19}$$

For numerically tracking the evolving shapes during the minimization of the cost

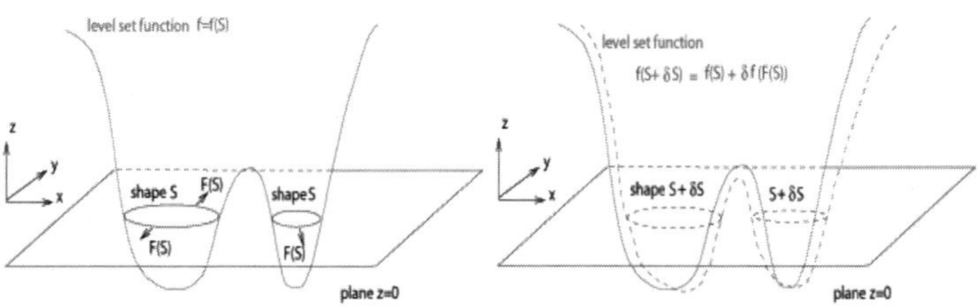

Fig. 7: Level set representation of evolving shapes. Left: shape S and level set function f at a given time step t subject to normal velocity field $F(S)$; Right: correction of the level set function by $\delta f(F(S))$ yields the shape $S+\delta S$ and level set function $f+\delta f$ at time step $t+\delta t$.

functional, we use a level set technique [5, 8, 9, 11, 12]. This is an implicit characterization of shapes as illustrated in Fig. 7. In the following we address some practical aspects of this technique.

- In order to evolve the level set function f such that the zero level set follows the flow F, we need to numerically solve the following Hamilton-Jacobi type equation (Osher and Sethian, 1988):

$$\frac{\partial f}{\partial t} + F|\nabla f| = 0. \tag{20}$$

- In each time-step $t^{(n)}$ of this artificial evolution we need to solve one forward (4)-(9) and one adjoint coupled transport problem (12)-(16) in order to determine $F(t^{(n)}) = F[u^{(n)}, z^{(n)}]$.
- Practically, we will calculate discrete time-steps of the Hamilton-Jacobi type evolution equation:

$$\frac{f^{(n+1)} - f^{(n)}}{\Delta t^{(n)}} + F^{(n)}|\nabla f^{(n)}| = 0. \tag{21}$$

$$f^{(n+1)} = f^{(n)} + \delta f^{(n)} \tag{22}$$

$$\text{with} \quad \delta f^{(n)} = -(\Delta t^{(n)}\,|\nabla f^{(n)}|)\,F^{(n)} \tag{23}$$

- $\Delta t^{(n)}$ might be chosen quite large: 'line search strategy'.

3. Numerical Results

A numerical experiment is shown for demonstrating the performance of our shape based reconstruction scheme in a 2D setup. The domain of interest has the size 3cm×6cm. Laser sources are arranged at the bottom of the domain, and detectors are positioned along the top of the domain. The background scattering coefficient is $\mu_s^{(0)} = 100\text{cm}^{-1}$ with about ±20% random fluctuations. The background absorption coefficient is $\mu_a^{(0)} = 0.1\text{cm}^{-1}$, also with about ±20% random fluctuations. The values $\mu_s^{(0)}$ and $\mu_a^{(0)}$ are assumed known, but the small fluctuations are unknown.

Three disc-shaped regions filled with fluorophore of $\mu_{FL} = 0.4\text{cm}^{-1}$ are embedded into this background which need to be reconstructed from the data. As initial guess we use a small disc located at the center of the domain. We treat the inverse problem in two different ways. First, we assume that we only have the data available which correspond to the excitation frequency. This situation is similar to the situation encountered in the application of DOT [2, 4, 5]. Fig. 8 shows the evolution of the shapes during this reconstruction.

Then, we assume that we only have the data available which correspond to the emission frequency. This is the typical situation in fluorescence tomography [1, 7, 10]. The corresponding shape evolution is shown in Fig. 9.

4. Discussion and Future Work

We observe that the shape based algorithm using level sets is capable to reconstruct good approximations to the fluorophore distribution in tissue from both, the excitation and the emission frequency data, in our simulations. Therefore, it is a promising approach for this type of imaging problems. In order to assess the real potential of this technique, a more realistic 3D setup needs to be implemented and, as far as real data are available, it needs to be tested against real data. A critical point will play here the available signal to noise ratio at excitation and emission frequency. Furthermore, an interesting problem arises in Fluorescence Lifetime Imaging where the emission photons are released by the fluorophore only when a small characteristic

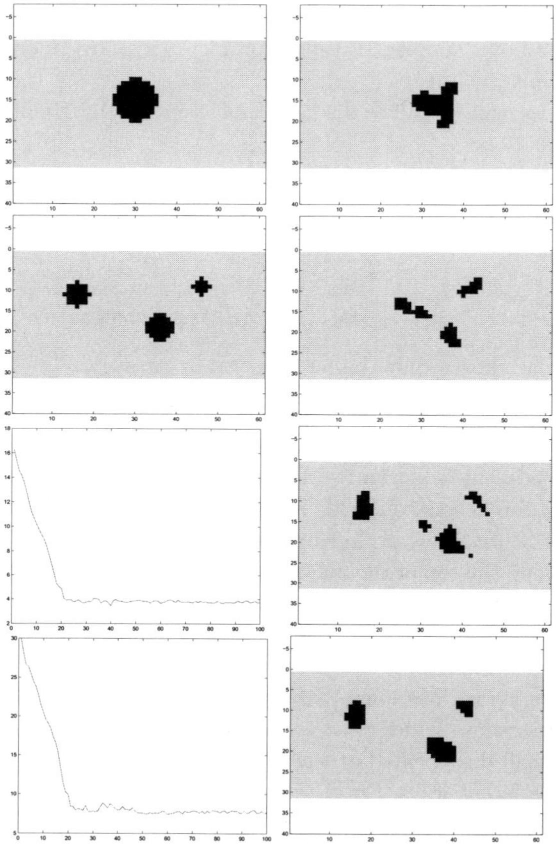

Fig. 8: Reconstruction of μ_{fl} *using only the data at excitation wavelength.* Left column, top to bottom: initial guess, correct distribution, cost evolution at excitation wavelength and at emission wavelength. Right column from top to bottom: shape evolution after 4, 10, 40 and 90 steps of the level set algorithm.

time has passed after capturing the excitation photon. This characteristic life time carries important information and therefore needs to be reconstructed jointly with the fluorophore concentration from the gathered data. These interesting problems will be addressed in our future research.

Acknowledgment

The work of O.D. was supported by the European Union through the Sixth Framework Programme as part of the project on 'Integrated technologies for in-vivo molecular imaging', contract No. LSHG-CT-2003-503259. The work of K.E.P. was funded by the Mexican Consejo Nacional de Ciencia y Tecnología (CONACYT).

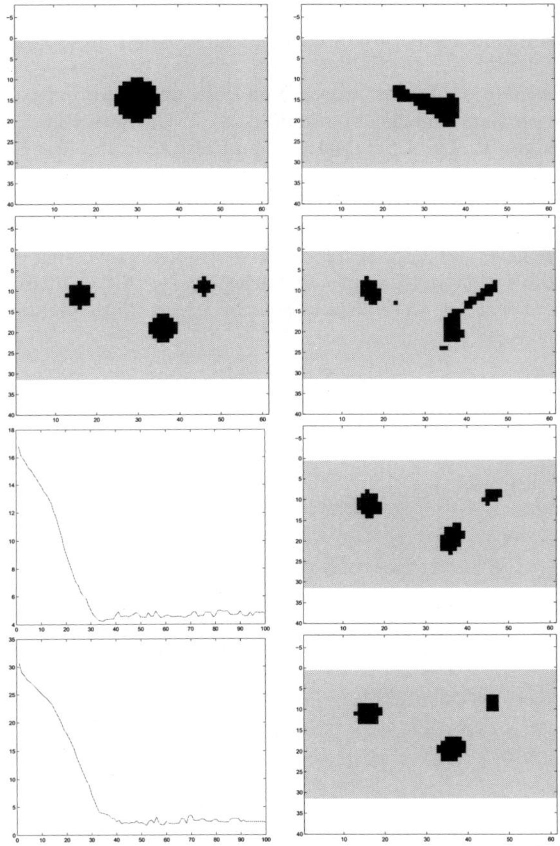

Fig. 9: Reconstruction of μ_{fl} *using only the data at emission wavelength.* Left column, top to bottom: initial guess, correct distribution, cost evolution at excitation wavelength and at emission wavelength. Right column from top to bottom: shape evolution after 10, 30, 50 and 90 steps of the level set algorithm.

References

[1] Álvarez D, Medina P and Moscoso M, Fluorescence lifetime imaging from time resolved measurements using a shape-based approach, Optics Express 17 (11), pp.8843-8855 (2009).

[2] Arridge, S.R.: Optical tomography in medical imaging. Inverse Problems, **15** (2), R41–R93 (1999).

[3] Case, K.M., Zweifel, P.F.: Linear Transport Theory. Plenum Press, New York (1967)

[4] Dorn, O.: A transport-backtransport method for optical tomography. Inverse Problems, **14**, 1107–1130 (1998)

[5] Dorn, O.: Shape reconstruction in scattering media with voids using a transport model and level sets. Canad. Appl. Math. Quart., **10** (2), 239–275 (2002).

[6] Natterer, F., Wübbeling, F.: Mathematical Methods in Image Reconstruction. Mono-

graphs on Mathematical Modeling and Computation 5, SIAM 2001.
[7] Ntziachristos V, Fluorescence molecular imaging, Annu. Rev. Biomed. Eng. 8, 1-33 (2006).
[8] Osher, S., Sethian, J.: Fronts propagation with curvature dependent speed: Algorithms based on Hamilton-Jacobi formulations. J. Comput. Phys., **56** 12–49 (1988)
[9] Osher S., Fedkiw R: Level Set Methods and Dynamic Implicit Surfaces. Springer, New York (2003)
[10] Rudge TJ, Soloviev VY, Arridge SR, Fast image reconstruction in fluorescence optical tomography using data compression, Optics Letters **35**, 753-765 (2010).
[11] Santosa, F.: A Level-Set Approach for Inverse Problems Involving Obstacles. ESAIM: Control, Optimization and Calculus of Variations, **1**, 17–33 (1996)
[12] Sethian, J.A.: Level Set Methods and Fast Marching Methods (2nd ed). Cambridge University Press (1999)

THE INFLUENCE OF TEXTURE SYMMETRY IN MARKER POINTING: EXPERIMENTING WITH HUMANS AND ALGORITHMS

M. CARDACI

Dipartimento di Psicologia e CITC, Universitá degli Studi di Palermo
Ed. 15 viale delle Scienze, 90100 Palermo, Italy
**E-mail: cardaci@unipa.it*

M.E. TABACCHI

Dipartimento di Matematica ed Informatica, Università degli Studi di Palermo
via Archirafi 34, 90123 Palermo, Italy
and Istituto Nazionale di Ricerche Demopolis
via col Romey 7, 91100 Trapani, Italy
E-mail: metabacchi@demopolis.it—unipa.it

Symmetry plays a fundamental role in aiding the visual system, to organize its environmental stimuli and to detect visual patterns of natural and artificial objects. Various kinds of symmetry exist, and we will discuss how internal symmetry due to textures influences the choice of direction in visual tasks. Two experiments are presented: the first, with human subjects, deals with the effect of textures on preferences for a pointing direction. The second emulates the performances obtained in the first through the use of an algorithm based on a physic metaphor. Results from both experiments are shown and comment.

Keywords: Symmetry; Perception.

1. Introduction

In Human Perception, symmetry plays a fundamental role in aiding the visual system, to organize its environmental stimuli and to detect visual patterns of natural and artificial objects. As Gestalt points out, we tend to perceive objects as symmetrical structures around a center, even if perfect symmetry is absent in nature. More specifically symmetry facilitates perceptual grouping [1]. Indeed, when we perceive similar but disconnected elements, approximately symmetrical to each other, we tend to organize them as a coherent whole. Moreover is well known that in figure/ground segregation process, symmetrical images are generally considered figure, rather than ground. Symmetry detection also heavily influences shape recognition. E.g., an observer may give a different description of a shape if it is embedded in a context with prevalent horizontal or vertical symmetry [2].

It must also be noted that the concept of symmetry by itself is not unitary and various kinds of symmetry exist (rotational symmetry, bilateral symmetry, translational symmetry, skew symmetry, see [3]). Furthermore, human judgement on symmetry is quite far from its mathematical counterpart, for a number of reasons: a critical one is gravity, which is always present in human activities, and makes humans peculiarly sensible to the axial orientation of a figure [4]; another one is subsumed by the good Gestalt principles - the human visual system tends to highlight symmetries (and even to imagine inexistent regularities based on real symmetries), but natural occurring symmetries are just an approximation of formal, mathematical ones [5–7].

As there is a difference in the visual experience of symmetries in humans and the underlying mathematical concepts, we will discuss a method to emulate the former using the latter. and specifically approximate reasoning, and local/global algorithms. We will specifically focus on the rotational symmetry. A figure has rotational symmetry when it can be rotated less than 360° around its central point, or axis, and still match the original figure.

In this paper we compared the results obtained from an experiment on human subjects with the output of an algorithms based on a charge field computational model.

2. Human Judgment of Textured Rotational Symmetry

From a human point of view, the vertical axis is predominant in rotational symmetry. Since early Mach's intuitions on the predominance of vertical symmetry axis in humans, the axial symmetry effects have been fully confirmed by the literature: even in infants symmetry detection is easier and faster when the symmetry axis is vertical [8]. Researches showed how detecting symmetry in human observers is influenced also by scanning strategies and attentional cueing effects [9]. Concerning this, several studies stressed the facilitating effects produced by external frames or cue lines aligned at the symmetry axis [10–12]. Other studies showed that in symmetry detection scanning strategies are also affected by qualitative features of the figure, i.e. presence of curvature minima along boundaries, concavity of the figure and number of vertices mismatches [13, 14]. Moreover, symmetry judgment is faster and more accurate if the figure contains multiple symmetries [15, 16]. Other less studied variables could be relevant. In particular, the internal structure (texture) of stimulus and its interaction with the vertical bias has often been neglected.

2.1. *Protocol*

We present the results obtained from a pilot experiment with human subjects on equilateral triangles.

2.1.1. *Stimuli*

We have used four different equilateral triangles, as shown in Figure 1. The triangles were generated by manipulating internal texture: TR1 and TR2 (respectively b and c) globally maintain the three main symmetries, and are called tri-symmetric, whereas TR3 and TR4 (d and e) maintain only one symmetry and are called mono-symmetric. All the test triangles have the same number of black areas (4) and the same ratio between gray and black areas (7:9). We call A, B and C their vertices starting from top in clockwise order. We have already used these stimuli in a similar task: detecting preferences for vertical symmetry [17].

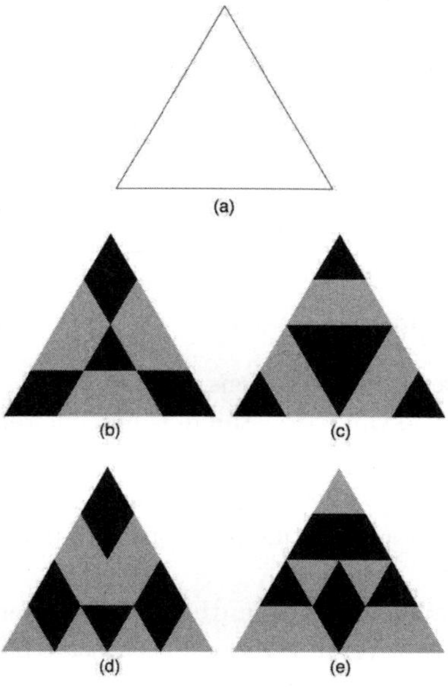

Fig. 1: The stimuli used in our experiment. (a) is the shape of the triangle, (b) is TR1, (c) is TR2, (d) is TR3, (e) is TR4.

2.1.2. *Subjects*

50 naive university students, representative of the university population for age and gender, unaware of the experiment scopes and aims. Privacy dealt in accordance with Italian privacy laws.

2.1.3. *Procedure*

The triangles were presented on screen in randomized order (width: 500 pixels). Each triangle was set on a black background, centered on the screen and rotated at a random angle. Outside the triangle a circular white marker (widht: 20 pixels) was also positioned on a imaginary circle, whose center coincided with the triangle's center. Each stimulus was presented two times, for a total of 8 trials. The procedure was very simple: the subject was requested to point the triangle towards the external marker, using the mouse wheel to rotate it. All data were automatically registered by the software.

2.2. *Results*

For each stimulus we measured the angle between the original position and the final position. Results are reported in Table 1.

Table 1: Mean values for the traveled angle

	Mean ($\partial°$)
TR1-TR2	32.3
TR3-TR4	84.7

The following comments can be done:

- Results follow theoretical expectations.
- The mean value for tri-symmetric stimuli is much lower than the one for mono-symmetric. As such, human choice focused on vertical symmetry when preponderant over texture, and on rotational symmetry elsewhere.
- The compactness factor (travel the shorter road to point the marker) is trumped by vertical symmetry.
- There are no discernible effects from the field and even when the field-vertical is nearer the marker, the texture-vertical is chosen (this is confirmed by further analysis on marginal cases, not reported here). Environmental fluctuations may have a negative effect on the number of species. This effect is due to physiological stress, which may even cause the extinction of some species.

3. Reproducing Human Performance through Algorithms

In order to reproduce the human performance of the experiment detailed in the previous section we have chosen to use a modified version of FiST algorithm. We have already employed FiST (Field Symmetry Transform), i.e. in [18], and given a detailed, formal description of its innards in [19]. FiST is a global evaluator of bilateral symmetry preferences for greyscale images, based on a physical model, easily implementable and parallelizable; the peculiarity of this model is that, since it is based on a field metaphor, it lends itself naturally to situations in which there exists a preferential symmetry or perceived direction, such as human biases or preferences for vertical symmetries. Informally FiST operates by treating the image as a 2d body on a flat surface, composed by punctiform charges, each corresponding to a pixel, whose values are proportional to the greyscale values of the pixels. An equal-spaced orthogonal grid, which can have a different resolution from the one of the image, is superimposed, and the vectors corresponding to the energy field created by the charges is computed for each of the grid's intersections. Two distributions, one for direction and the other for intensities, are then obtained, and then further data analysis, depending on the task for which the transform is employed, is carried out; the field can be perturbed prior to creation of the statistics in order to account for biases and preferences for a particular direction of symmetry. Formally, FiST is defined as follows:

$$FiST_a(G) = h(|\mathbf{e}_k|, f(e_k)) : \mathbf{e_k} \in E/a\frac{2\pi}{n} \leq \theta_{\mathbf{e_k}} < (a+1)\frac{2\pi}{n}$$

where n is the number of equal intervals of size $2/n$ in which vectors of the field are clustered after their computation: the choice of n depends on the quantity of information needed, and should be a balanced compromise: too small and it could lead to grouping together symmetries stemmed from nearby axes, too big and values that should belong together but are at a small distance because of rounding errors or imprecisions inherent in discrete computation are not registered as such.

$\mathbf{e}$ is a vector in E, the field grid. The grid is superimposed to the image, but its resolution (the distance between each intersection) is independent from the number of pixels of the image itself, and should be chosen according to image complexity and details; e is its module, θe its direction.

f and h are balancing functions, chosen according to the weight that modules and number of vectors in a cluster should have in the distribution. We have chosen the most elementary functions, a mean for f and a multiplicative factor for h. The total vector e in each point of intersection of the field grid of coordinates (s, t) is obtained due to the superimposition principle, while each independent contribution by the pixel $g(i, j)$, to the field vector point e of coordinates (s, t) is computed using Coulomb's law.

3.1. *Protocol*

In order to emulate more closely the experiment from the previous section, we have modified FiST in an iterative fashion. Instead of the search for a maximum symmetry point by analysis of the FiST graph, we have incorporated the marker from the human experiment in the image (at a fixed position), chosen a starting position for the triangle using a random rotation angle, and then iterated in both direction until a minimum value for the difference between the triangle and the triangle+marker version is found. The number of iterations becomes the algorithmic counterpart to the traveled angle in the human experiment.

3.1.1. *Results*

Mean results form 50 runs of FiST on the TR1-TR4 stimuli presented in Figure 1, each repeated two times, using a 10px-equivalent grid, and a 0.1° iteration angle, are presented in Table 2. A significative correlation with results from human experiment in Table 1 can be observed.

Table 2: Results from the application of an iterative version of FiST to TR1-TR4 stimuli.

	Mean Iterations
TR1-TR2	287
TR3-TR4	910

4. Conclusions

Symmetry is a genuine dynamic visual feature [20]. As such, the different aspects and subtleties that influence human choices has to be integrated in a comprehensive and well defined framework if we wish to push toward a coherent automatic detection. The experiments presented hereby show how symmetry detection, as many natural human tasks (see [21] for an example of a more cognitive type), is subject to a strong influence from both endogenous and exogenous factors, and in order to emulate the performance of human subjects we may resource to less precise but more flexible approaches [22]. This is a pilot study, due both to the size of the human sample and to the elementary characteristics of the stimuli. Further research will be devoted to implement the same family of algorithms in a more Gibsonian environment, and to expand the selection of tasks presented in the human experiment.

References

[1] M. Wertheimer, *Psychological Research* **4**, 301 (1923).
[2] S. E. Palmer, *Acta Psychologica* **59**, 67 (1985).
[3] P. Wenderoth, *Perception* **23**, 221 (1994).
[4] R. Ferguson, Modeling orientation effects in symmetry detection: The role of visual structure, in *Proceedings of the 22nd Conference of the Cognitive Science Society*, (Erlbaum, Hillsdale, New Jersey, 2000).
[5] G. Kanizsa, The role of regularity in perceptual organization, in *Studies in perception: Festschrift for Fabio Metelli*, (Martello-Gianti, Milano, 1975).
[6] R. van Lier and J. Wagemans, *Journal of experimental psychology. Human perception and performance* **25** (1999).
[7] R. J. van Lier, P. A. van der Helm and E. L. J. Leeuwenberg, *Journal of Experimental Psychology: Human Perception and Performance* **21**, 571 (1995).
[8] M. H. Pornstein and S. J. Krinsky, *Journal of Experimental Child Psychology* **39**, 1 (1985).

[9] D. P. Carmody, C. F. Nodine and P. J. Locher, *Percept Mot Skills* **45**, 1267 (1977).
[10] A. Herbert, G. Humphrey and P. Jolicoeur, *Canadian journal of experimental psychology* **48** (1994).
[11] H. Pashler, *Journal of Experimental Psychology: Human Perception and Performance* **16**, 150 (1990).
[12] A. B. Sekuler and M. B. Swimmer, *Canadian Journal of Experimental Psychology* **54**, 42 (2000).
[13] R. Ferguson, *Proceedings of the Eighteenth Annual Conference of the Cognitive Science Society*, p. 12 (1996).
[14] G. C. Baylis and J. Driver, *Visual Cognition* **1**, 377 (1994).
[15] F. L. Royer, *Journal of Experimental Psychology: Human Perception and Performance* **7**, 1186 (1981).
[16] J. Wagemans, L. Van Gool and G. d'Ydewalle, *Attention Perception & Psychophysics* **50**, 413 (1991).
[17] M. Cardaci and M. E. Tabacchi, *Sistemi Intelligenti* **3**, 441 (2009).
[18] B. Pinna and M. Tabacchi, A fuzzy approach to the role of symmetry in shape formation: The illusion of the scalene triangle, in *Fuzzy Logic and Applications*, eds. V. Di Gesù, S. Pal and A. Petrosino, Lecture Notes in Computer Science, Vol. 5571 (Springer Berlin-Heidelberg, 2009) pp. 197–204.
[19] M. E. Tabacchi, Evaluation of planar symmetry preferences through a physics-based operator, in *Practices of Cognition*, ed. M. Cruciani (Università degli Studi di Trento, 2010) pp. 217–226.
[20] B. Zavidovique, V. DiGesu and M. E. Tabacchi, *Symmetry* **2**, 554 (2010).
[21] M. Petrou, M. E. Tabacchi and R. Piroddi, *The Computer Journal* **53**, 1738 (2010).
[22] S. Termini and M. E. Tabacchi, *Fuzzy Sets Syst.* (2011).

A MULTISCALE AUTOCORRELATION FUNCTION FOR ANISOTROPY STUDIES

M. SCUDERI*

Dipartimento di Fisica e Astronomia, Universitá degli Studi di Catania & Istituto Nazionale di Fisica Nucleare, Sez. di Catania, Via S. Sofia 64
Catania, 95123, Italy
**E-mail: mario.scuderi@ct.infn.it*
http://www.dfa.unict.it/

M. DE DOMENICO

Laboratorio sui Sistemi Complessi, Scuola Superiore di Catania, Via Valdisavoia 9, Catania, 95123, Italy

Istituto Nazionale di Fisica Nucleare - Sez. di Catania, Via S. Sofia 64
Catania, 95123, Italy
E-mail: manlio.dedomenico@ct.infn.it

A. INSOLIA

Dipartimento di Fisica e Astronomia, Universitá degli Studi di Catania & Istituto Nazionale di Fisica Nucleare - Sez. di Catania, Via S. Sofia 64
Catania, 95123, Italy
E-mail: antonio.insolia@ct.infn.it

H. LYBERIS

IPN Orsay CNRS/IN2P3 and Université Paris Sud, Rue Georges Clemenceau 15
Orsay, 91406, France

Université Paris VII Denis Diderot, Paris 7, Paris Cedex 13
Paris, 75205, France

Dipartimento di Fisica Generale, Universitá di Torino, Via Pietro Giuria 1
Torino, 10125, Italy
E-mail: lyberis@ipno.in2p3.fr

In recent years many procedures have been proposed to check the anisotropy of a dataset. We present a new simple procedure, based on a scale dependent approach, to detect anisotropy signatures in a given distribution with particular attention to small dataset. The method provides a good discrimination power for both large and small datasets, even in presence of strong contaminating isotropic background. We present some applications to simulated datasets of events to investigate statistical features of the method and present and inspect its behavior under both the null or the alternative hypothesis.

Keywords: Anisotropy, Multiscale analysis, Cosmic Rays

1. Introduction

In many field involving data analysis, the search for anisotropy has played a crucial role. Many experimental data such as cosmological, astrophysical, atmospherical or geophysical ones require the search for clustering of objects and the measure of deviation from isotropy of a given angular distribution. During the last decades, many estimators, namely correlation function, have been proposed [1–3]. These methods apply to angular coordinates of objects as well to distributions of arrival directions of events.

Different estimators have been defined in the common anisotropy analysis, by Davis-Peebles [1] Landy-Szalay [2] and Hamilton [3].

Recently, new estimators have been introduced to study the anisotropy signature of a given direction distributions: the modified two-point Rayleigh [4], and shape-strength method derived from a principal component analysis of triplets of events [5].

Within the present work, we present a new fast and simple method for anisotropy analysis, which makes use of a multiscale approach and depends on one parameter only. The main advantage of our estimator is the possibility to analytically treat the results: the analytical approach drastically reduces computation time and makes available the possibility of applications to very large datasets. We test the method on several simulated isotropic and anisotropic arrival direction distributions and perform an extensive analysis of its statistical features under both the null and the alternative hypotheses.

2. MAF: Multiscale Autocorrelation Function

Let $\mathcal{S}$ be a region of a spherical surface defining a sky and let $P_i(\phi,\theta)$ $(i = 1, 2, ..., n)$ be a set of n points on $\mathcal{S}$. The sky $\mathcal{S}$ is partitioned within a grid of N equal-area (and almost-equal shape) disjoint boxes $\mathcal{B}_k$ $(k = 1, 2, ..., N)$ as described in Ref. 6. Each box $\mathcal{B}_k$ covers the solid angle

$$\Omega_k = \frac{1}{N}\int_{\theta_{\min}}^{\theta_{\max}}\int_{\phi_{\min}}^{\phi_{\max}} d\cos\theta d\phi = 2\pi(1-\cos\Theta)$$

where 2Θ is the apex angle of a cone covering the same solid angle. Let Ω be the solid angle covered by $\mathcal{S}$: N, Θ and Ω are deeply related quantities that define a scale.

The function $A(\Theta)$ that quantifies the deviation of data from an isotropic distribution at the scale Θ, is chosen to be the Kullback-Leibler divergence [7, 8].

$$A(\Theta) = \mathcal{D}_{\mathrm{KL}}(\psi(\Theta)||\overline{\psi}(\Theta)) = \sum_{k=1}^{N}\psi_k(\Theta)\log\frac{\psi_k(\Theta)}{\overline{\psi}_k(\Theta)} \tag{1}$$

where $\psi_k(\Theta)$ is the fraction of points in the dataset that fall into the box $\mathcal{B}_k$ and $\overline{\psi}_k(\Theta)$, generally a function of the domain meshing, is the expected number of points on $\mathcal{B}_k$ from an isotropic distribution of n points on $\mathcal{S}$. The Kullback-Leibler divergence is an information theoretic measure that quantifies the error in selecting the density $\overline{\psi}(\Theta)$ to approximate the density $\psi(\Theta)$ and it is strictly connected to maximum likelihood estimation (see Appendix A in Ref. 9).

It is straightforward to show that $A(\Theta)$ is minimum for an isotropic distribution of points or, in general, when $\psi(\Theta) \sim \overline{\psi}(\Theta)$. If $A_{\text{data}}(\Theta)$ and $A_{\text{iso}}(\Theta)$ refer to the data and to an isotropic realization with the same number of points respectively, we define *multiscale autocorrelation function* (MAF) the estimator

$$s(\Theta) = \frac{|A_{\text{data}}(\Theta) - \langle A_{\text{iso}}(\Theta)\rangle|}{\sigma_{A_{\text{iso}}}(\Theta)} \tag{2}$$

where $\langle A_{\text{iso}}(\Theta)\rangle$ and $\sigma_{A_{\text{iso}}}(\Theta)$ are the sample mean and the sample standard deviation, estimated from several isotropic realizations of the data. If $\mathcal{H}_0$ denotes the null hypothesis of an underlying isotropic distribution for the data, the chance probability at the angular scale Θ, properly penalized because of the scan on Θ, is the probability

$$p(\Theta) = \Pr(s_{\text{iso}}(\Theta') \geq s_{\text{data}}(\Theta)|\mathcal{H}_0, \forall\Theta' \in \mathcal{P}) \tag{3}$$

obtained from the fraction of null models giving a multiscale autocorrelation, at any angular scale Θ' in the parameter space $\mathcal{P}$, greater or equal than that of data at the scale Θ[a]. Under the null hypothesis $\mathcal{H}_0$, the estimator $s(\Theta)$ follows an half-gaussian distribution, independently on the value of the angular scale Θ and on the number of events on $\mathcal{S}$, as it will be successively shown in the text.

3. Dynamical Counting

The simplest definition of the counting algorithm involves the fixed grid introduced in Ref. 6. In some cases, such a *static counting* approach could not reveal an existing cluster. For instance, Figure 1(a) shows a typical scenario where some points of a given triplet fall into different cells. Indeed, the fixed grid may cut a cluster of points within one or more edges.

The smoothing, adopted in our study, deals with a new counting procedure for the estimation of the density $\psi_k(\Theta)$.

From the starting distribution of points, we define a new distribution defined as follows: for each point P_i, we consider a set of 8 points distributed around P_i as described in Figure 1(b), being the distance between each of this points and P_i equal to $\frac{\Theta}{2}$. Within this framework, from each point P_i we define a set of 9 points (P_i and the new 8 points), and assign to each point the weight $\frac{1}{9}$. Finally, we follow

[a]The null hypothesis is eventually rejected in favor of the alternative $\mathcal{H}_1 = \neg\mathcal{H}_0$ – being $\neg$ the negation operator – at the angular scale Θ, with probability $1 - p(\Theta)$.

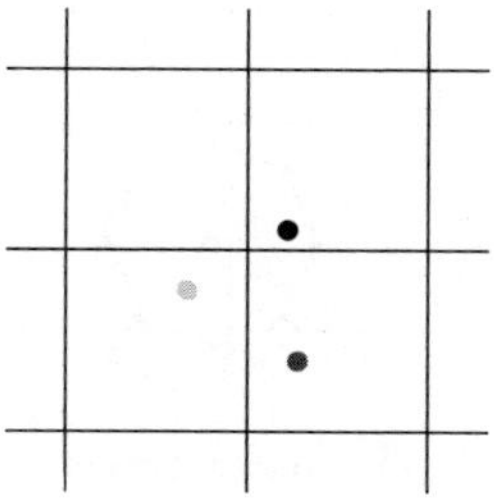

(a) Static counting

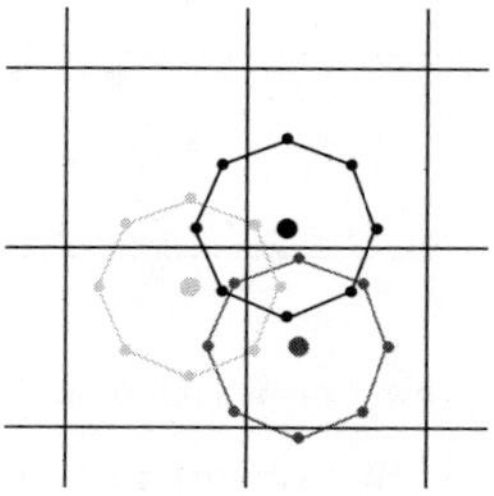

(b) Dynamic counting

Fig. 1: A cluster of three points falling into different boxes.

the standard procedure described in Section 2 by using the weighted distribution of points[b].

Our numerical studies show that such a *dynamical counting* approach gives more information on the amount of clustering in the data.

The main difference between the static and the dynamical counting lies in the value of the estimator when the procedure is applied to Monte Carlo realizations of the sky. For instance, let us consider the Figure 2, where we show a clustered (Figure 2(a)) and an unclustered (Figure 2(b)) set of points. The static counting is not able to recover the differences between the two configurations. Conversely, if the dynamical counting is applied, the *extended* points in Figure 2(a) are concentrated mainly in two adjacent boxes while in Figure 2(b) they are distributed on the neighbor cells, this lead to two different $\psi(\Theta)$. Monte Carlo skies producing the same clustered configuration shown in Figure 2(a), and of consequence the same weight distribution, are not frequently expected: in this case, the value of $s(\Theta)$ should be greater than that one estimated from the static method. The direct consequence of a greater value of the estimator $s(\Theta)$ is a lower chance probability and the main advantage of using the dynamical counting, instead of the static one, should be the lowest penalization of $s(\Theta)$ only if an anisotropy signal is really present.

In order to illustrate the importance of dynamical counting in the anisotropy

[b] In general, in the weighting procedure, we have to take into account many factors, as for example the exposure of the experiment. See Ref. 9 for more details.

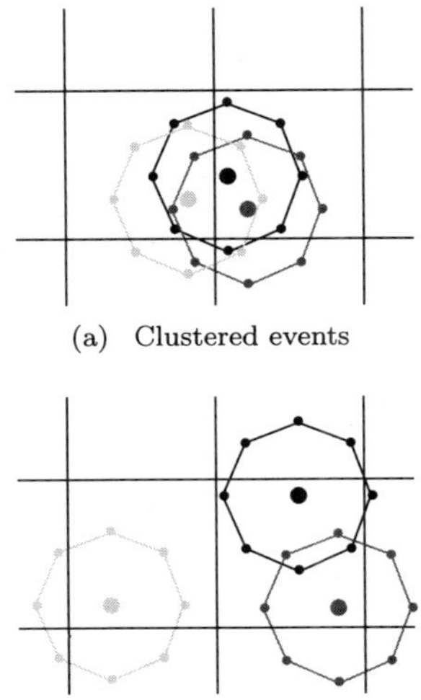

(a) Clustered events

(b) Unclustered events

Fig. 2: a) Three clustered points: the weighted points are concentrated in two adjacent boxes. b) Three unclustered points: the weighted points are distributed on the neighbor cells.

signal detection, we have generated 5000 isotropic and anisotropic skies of 100 events each. In each anisotropic sky, 60% of events are normally distributed, with dispersion ρ, around 10 random sources and 40% of events are isotropically distributed. For each angular scale Θ, we have estimated the average value of $s(\Theta)$ with the static and the dynamical counting, separately. Results are shown in Figure 3 for $\rho = 5°$ (a), $\rho = 10°$ (b), $\rho = 20°$ (c) and for the isotropic map (d). In the case of the anisotropic skies, the dynamical counting provides a greater estimation of $s(\Theta)$ than the static counting, leading to a smaller estimation of the corresponding chance probability and improving the signal-to-noise ratio.

4. Interpretation of MAF

In the specific case of MAF, the angular scale $\Theta^\star$, where the significance is minimum, turns to be the significative *clustering scale*: it is the scale at which occurs a greater accumulation of points respect to that one occuring by chance, with no regard for a particular configuration of points, e.g. doublets or triplets. To illustrate better the clustering scale detection feature of MAF, we have generated 5000 isotropic and anisotropic skies of 100 events each, as previously described at the end of Section 3. Figure 4 shows the average chance probability, with 68% region around the mean value, versus the angular scale for three values of the dispersion, namely $\rho = 5°$ (a), $\rho = 10°$ (b), $\rho = 20°$ (c), and for the isotropic map (d). As expected, chance probability is close to one and nearly flat in the case of the isotropic map, because all

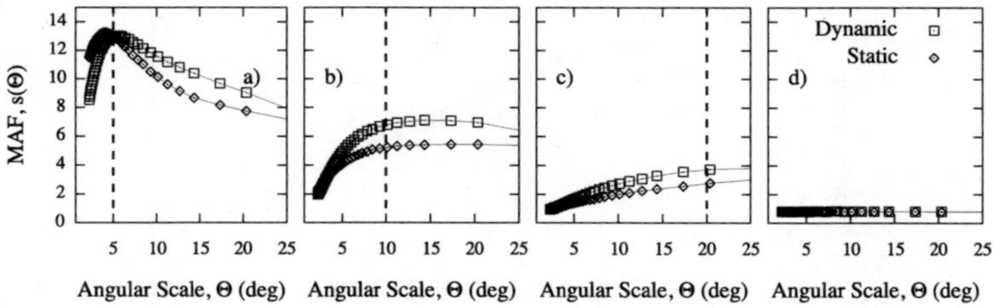

Fig. 3: MAF: average $s(\Theta)$ (solid line) estimated from 5000 isotropic and anisotropic skies of 100 events each. In each anisotropic sky, 60% of events are normally distributed, with dispersion ρ, around 10 random sources and 40% of events are isotropically distributed. The dashed line indicates the value of the dispersion adopted to generate the corresponding mock map: a) 5, b) 10 and c) 20 degrees; d) isotropic map.

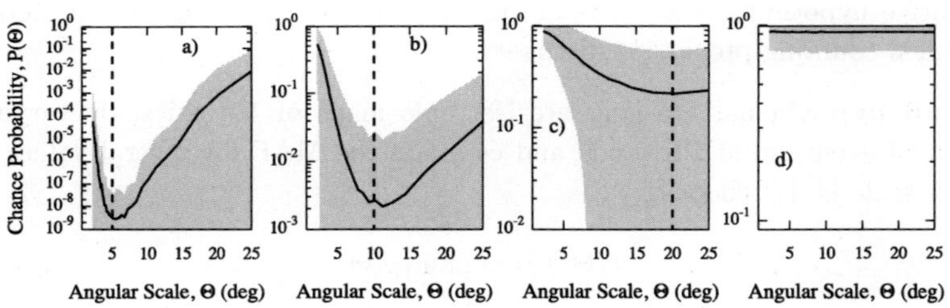

Fig. 4: MAF: average chance probability (solid line), with 68% region around the mean value, estimated from isotropic and anisotropic skies generated as explained in Figure 3. Dynamical counting is used. The dashed line indicates the value of the dispersion adopted to generate the corresponding mock map: a) 5, b) 10 and c) 20 degrees; d) isotropic map.

clustering scales are equally likely. Conversely, for all anisotropic maps, the average chance probability gets a minimum around the corresponding value of ρ. Thus, our estimator is able to recover the most significant clustering scale. It should be remarked that when the 20° dispersion is used, the angular scale of the minimum is less obvious because of the large fluctuations due to the isotropic contamination. Finally, it is worth noticing that we have observed that the curve around the value of ρ gets narrower by increasing the number of events.

5. Statistical Analysis of MAF

We have investigated the statistical features of MAF by inspecting its behavior under both the null or the alternative hypothesis (see Ref. 9 for further information). In particular, we have estimated the significance α (or Type I error), i.e. the probability to wrongly reject the null hypothesis when it is actually true, and the power $1-\beta$ (where β is known as Type II error), i.e. the probability to accept the

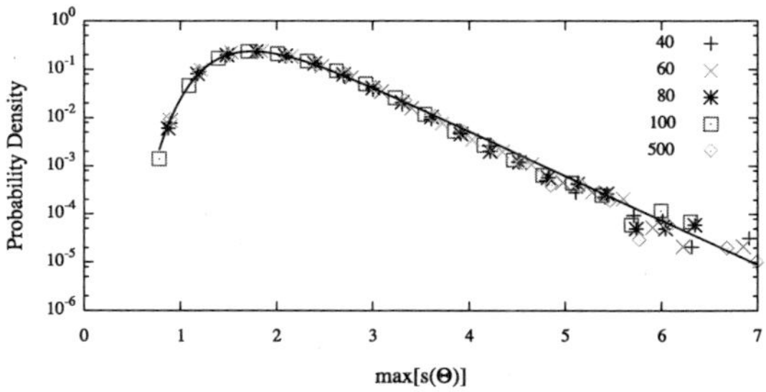

Fig. 5: MAF: distributions of $\max\{s(\Theta)\}$ for $n = 40, 60, 80, 100$ and 500 events. Solid line correspond to the least-square fit of the Gumbel density with parameters $\mu = 1.743 \pm 0.002$ and $\sigma = 0.470 \pm 0.002$ ($\chi^2/\text{ndf} = 1.1 \times 10^{-5}$).

alternative hypothesis when it is in fact true. In the following we will adopt the dynamical counting previously discussed.

Null hypothesis. We generate isotropic maps of 10^5 skies, by varying the number of events from 20 to 500 and estimate the MAF for several values of the angular scale Θ. By choosing:

$$\tilde{p}(\Theta^\star) = \arg\min_{\Theta} p(\Theta)$$

We find an excellent flat distribution of probabilities $\tilde{p}(\Theta^\star)$, as expected for analyses under the null hypothesis $\mathcal{H}_0$. In other words, MAF is not biased against $\mathcal{H}_0$, as required for good statistical estimators.

Because of the definition in Eq. (1) and of the central limit theorem, a Gaussian distribution is expected for the function $A(\Theta)$, and of consequence, the half-normal distribution

$$\mathcal{G}_{1/2}[s(\Theta)] = \frac{2}{\sqrt{2\pi}\sigma(\Theta)} e^{-\frac{s^2(\Theta)}{2\sigma^2(\Theta)}} \tag{4}$$

for $\sigma(\Theta) = 1$, is expected for the estimator $s(\Theta)$ defined as in Eq. (2), being normalized to zero mean and unitary variance. It follows that the (unpenalized) probability to obtain by chance a value of the MAF, greater or equal than a given value s_0, is just $1 - \text{erf}(\frac{s_0}{\sqrt{2}})$, being erf the standard error function, independently of the angular scale Θ.

Although this nice feature of the MAF estimator it is more important to identify the distribution of the penalized probability $p(\Theta)$, because of the required penalization due to the scan over the parameter Θ. For this reason, the distribution of $s_{\max} = \max\{s(\Theta)\}$ is of interest for applications.

Our numerical studies show that such a distribution exists and it corresponds to one of the limiting densities in the extreme value theory (see Appendix B in Ref. 9).

In Figure 5 are shown the probability densities of $s_{\max}$ for $n = 40, 60, 80, 100$ and 500 events: independently on n, each density is in excellent agreement with the Gumbel distribution of extreme values, for the parameters $\mu = 1.743 \pm 0.002$ and $\sigma = 0.470 \pm 0.002$. Such values correspond to the mean and to the standard deviation of the distribution, $\tilde{\mu} \approx 2.00$ and $\tilde{\sigma} \approx 0.59$, respectively. It follow that the probability to obtain a maximum value of $s(\Theta)$, at any angular scale Θ, greater or equal than a given value $\max\{s(\Theta)\}$ is

$$p(\max\{s(\Theta)\}) = 1 - \exp[-\exp(\frac{\max\{s(\Theta)\} - \mu}{\sigma})],$$

Alternative hypothesis. In order to investigate the behavior of MAF under the alternative hypothesis, we have generated anisotropic maps of 10^4 skies, by varying the number of events from 20 to 100.

In order to estimate the power of MAF, we build reasonable anisotropic maps of CR events reflecting in part the real-world scenario, keeping in mind that our purpose is to build an anisotropic set of events for statistical analysis and not to generate events mimicking real datasets with the best available approximation (see Ref. 9 for further information).

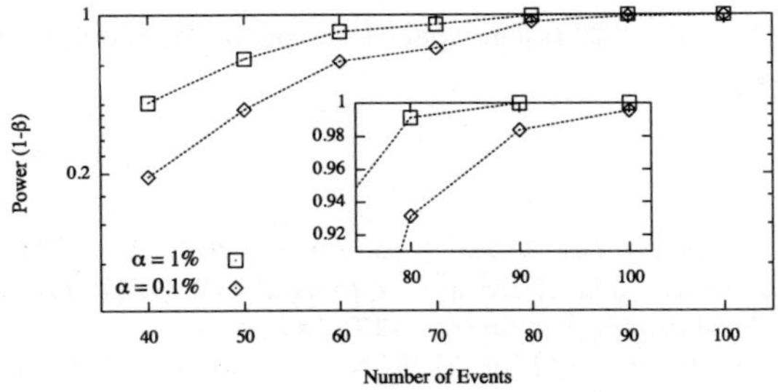

Fig. 6: MAF power vs. the number of events sampled from anisotropic mock maps generated as described in the text, for values of the significance corresponding to $\alpha = 0.1\%$ and $\alpha = 1\%$.

A sky is labelled as *anisotropic* if, for a fixed value of the significance α, the penalized chance probability as defined in Eq. (3) is lesser or equal than α, i.e. if the condition

$$\tilde{p}(\Theta^{\star}) = \arg\min_{\Theta} p(\Theta) \leq \alpha$$

holds for some angular scale $\Theta^{\star}$. In Figure 6 is shown the power for two values of the significance threshold, namely $\alpha = 0.1\%$ and $\alpha = 1\%$, estimated through the analytical approach. For applications, a power of 90% is generally required: under this threshold the method could miss to detect an existing anisotropy signal. In the case of the MAF, and for the considered anisotropic mock map, the power increases with the number of events n and it is able to detect the anisotropic signal for $n \geq 60$, with significance $\alpha = 1\%$. However, by decreasing the significance for the statistical test, the power requires a greater number of events to reach the 90% threshold, as expected.

6. Conclusion

We introduced a new statistical test, based on a multiscale approach, for detecting an anisotropy signal in the arrival direction distribution of UHECR. We showed that our procedure is suitable for the analysis of both small and large datasets of events.

The advantages of this new approach over other methods are multiples. First, the method allows an analytical description of quantities involved in the estimation of the amount of anisotropy signal in the data, avoiding thousands of Monte Carlo realizations needed for the penalizing procedure of results. Second, the method allows the detection of a physical observable, namely the clustering scale. Third, the method is unbiased against the null hypothesis and it provides a high discrimination power even in presence of strong contaminating isotropic background, for both small and large datasets. This method is suitable for the detection of the anisotropy signal in each data set involving a distribution of angular coordinates on the sphere, and it can be adapted to non-spherical spaces by properly redefining the dynamical counting algorithm.

References

[1] M. Davis and P. Peebles, *The Astrophysics Journal* **267**, 465 (1983).
[2] S. Landy and A. Szalay, *The Astrophysics Journal* **412**, 64 (1993).
[3] A. Hamilton, *The Astrophysics Journal* **417**, 19 (1993).
[4] M. Ave, L. Cazon, J. Cronin, J. de Mello Neto, A. Olinto, V. Pavlidou, P. Privitera, B. Siffert, F. Schmidt and T. Venters, *Journal of Cosmology and Astroparticle Physics*, p. 023 (2009).
[5] J. Hague, B. Becker, M. Gold and J. Matthews, *Journal of Physics G: Nuclear and Particle Physics* **36**, p. 115203 (2009).
[6] B. Stokes, C. Jui and J. Matthews, *Astroparticle Physics* **21**, 95 (2004).
[7] S. Kullback and R. Leibler, *The Annals of Mathematical Statistics* **22**, 79 (1951).
[8] S. Kullback, *The American Statistician* **41**, 340 (1987).
[9] M. De Domenico *et al*, *JCAP* (2011).

A MULTISCALE, LACUNARITY AND NEURAL NETWORK METHOD FOR γ/h DISCRIMINATION IN EXTENSIVE AIR SHOWERS

A. PAGLIARO*,[1] and F. D'ANNA

IASF/INAF - via Ugo La Malfa 153 - 90146 Palermo - Italy
[1]*INFN - Sezione di Catania - Viale A. Doria 6 - 95125 Catania - Italy*
**E-mail: pagliaro@ifc.inaf.it*

G. D'ALÍ STAITI

Università degli Studi di Palermo - Dipartimento di Fisica e Tecnologie Relative - Viale delle Scienze - 90128 Palermo - Italy and
INFN - Sezione di Catania - Viale A. Doria 6 - 95125 Catania - Italy

This paper presents a new method for the identification of extensive air showers initiated by different primaries. The method uses the multiscale concept and is based on the analysis of multifractal behaviour and lacunarity of secondary particle distributions together with a properly designed and trained artificial neural network. The separation technique is particularly suited for being applied when the topology of the particle distribution in the shower front is as largely detailed as possible. In the present work the method is discussed and applied in the experimental framework of ARGO-YBJ, to obtain hadron to gamma primary separation. We show that the presented approach gives very good results, leading, in the $1-10$ Tev energy range, to a clear improvement of the discrimination power with respect to the existing figures for extended shower detectors.

Keywords: Cosmic Rays; Extensive Air Showers; Multiscale Analysis; Wavelet Methods; Neural Networks

1. Introduction

The aim of this paper is to introduce a separation technique, on an event by event basis, based on the different topology of the showers produced by gamma and hadron-initiated showers, both as far as the space and time distribution in the shower front are concerned. A detailed picture of the shower is therefore mandatory for the method to be applied. The main idea is to exploit at best the differences in the multifractal behaviour of the space distribution and the differences in the arrival time distribution for showers initiated by different progenitors.

Our technique uses a multifractal-wavelet and lacunarity based analysis similar to the one used by Rastegaarzaeh and Samimi [1]. We improve Rastegaarzaeh and Samimi method by introducing a lacunarity parameter on time arrival of the showers and an artificial neural network that allows better separation. The concept of lacunarity quantifies the geometric arrangement of gaps in solid objects (lacunae). It can be extended to the description of distribution of data sets including, but not restricted to, those with fractal and multifractal distributions. We use lacunarity to describe gaps in time arrivals of showers.

The method applies at best in the experimental framework where a detailed picture of the shower front is given. We have been testing it using the ARGO-YBJ detector [5]. ARGO-YBJ is a compact array of RPC with a sampling area of about 100x110 m^2, consisting of a central carpet of about 5600 m^2 with 92% coverage, surrounded by a guard ring with coarse sampling. Thus, in the central carpet surface a fully detailed map of the shower front can be worked out, with a space resolution for multiple hit counting given by 7x56 cm^2 and a time resolution of the order of 1 ns in a slightly larger unit.

2. Multiscale Wavelet and Fractal Analysis

The wavelet transform is, by definition, the decomposition of a function on a basis obtained by translation and dilation of a particular function localized in both physical and frequency space. A wavelet analysis of a density field associates each point with a real number which represents the smoothed local density contrast at a given scale (see for example Pagliaro and Becciani [6], Gambera et al [7]). Fractal analysis, on the other hand, is concerned with the measurement of the local smoothness of the signals [2]. A deterministic fractal is defined using the concept of self-similarity. The fractal dimension D plays a central role. It is a measure of how the members are distributed in space. Intuitively, the larger the fractal dimension, the rougher the texture [3]. Self-similar multifractals are geometrical objects invariant by dilation. Multifractality is not characterized by a single fractal dimension, but by a function.

The large numbers of interactions in showers from the same progenitor are self-similar. For this reason showers may be characterized by their fractal dimension. Moreover, the distribution of the shower particles near the core has a self-similar character different from those far away. Different physics is involved in production of different secondary particles. For these reasons, extensive air showers have a multifractal behaviour and more than one fractal dimension.

In our method, for a distribution of particles on a plane, the number of particles inside a radius R is computed. If a scaling law of the form $N(R) \propto R^D$ holds and D is a single non-integer value, the distribution has a fractal distribution with dimension D. Multifractal behaviour can be revealed by studying the scaling laws for secondary particles at different core distances. Holschneider [8] has shown that if a function f has scaling law with exponent D around x_0: $f(R(x_0, \lambda\epsilon)) \sim \lambda^D f(R(x_0, \epsilon))$ then the wavelet transform $W(s,t) = \int g(x; s,t) f(x) dx$, where s is the single dilation parameter (scale) and t is the translational parameter, has the same scaling exponent, if the wavelet g has a zero average and decays fast at infinity. Thus the local scaling behaviour is represented by $W(s,t) \sim s^{D(t)}$. Therefore for any distribution function f the slope of the plot of $\log W(s,t)$ versus $\log s$ will give the fractal dimension of the distribution around point t for the range of the scale s.

3. Lacunarity

Lacunarity is the deviation of a fractal from translational invariance and can be extended to the description of spatial distribution of real data sets, including those with multifractal distributions. Lacunarity is defined as $\Lambda = E(\frac{M}{E(M)} - 1)^2$ where M is the mass of the fractal set (defined as the total number of points in the image) and $E(M)$ is the expected value of the mass computed for the fractal. This measures the discrepancy between the actual mass and the expected value. Lacunarity is small when texture is fine and large when texture is coarse. The mass of the fractal set is related to the length by $M(L) = kL^D$.

Among the algorithms proposed to measure lacunarity, we adopt the gliding box method of Allain and Cloitre [9]. We decided to use lacunarity to quantify heterogeneity in time arrivals of showers, so adding a third dimension to our spatial analysis. We assess time pattern lacunarity by gridding each time arrival array into squares as in McIntyre and Wiens [4]. In this protocol, a box is superimposed on the map. The number of full square contained inside the box (referred to as the box mass) is tallied. Then the box moves one unit up. The size of the box is initially four squares and then is enlarged adding squares until the box size equals the length of the array. A frequency distribution is then created of box masses. This frequency distribution is then converted to a probability distribution. Lacunarity is then computed from these probability values by determing the variance and mean of the number of full squares per gliding box: $\Lambda = [(variance/mean^2) + 1]$ and may assume any value between 1 and ∞. A pattern lacunarity 1 indicates that full squares are uniformly dispersed at a given scale. Values $\neq 1$ indicate non-uniformity in the distribution.

4. Simulations of Extensive Air Showers

To test our separation technique we analysed simulated showers initiated both by gamma-rays and protons. The showers were simulated by means of the CORSIKA code with QGSJET model [10], and the ARGO-G code [11] to simulate the pattern of the shower front hit as detected in the ARGO-YBJ detector. A detailed description of the ARGO-YBJ detector performances is beyond our scope. A total number of 1000 events were chosen for the analysis. The showers have zenith angles $0° < \theta < 10°$ and primary energies: $8 < E < 10 \cdot 10^{12} eV$ for the proton generated ones and $4 < E < 5 \cdot 10^{12} eV$ for the gamma generated ones. Differences in energies ensure that the numbers of hits on the analysis carpet are comparable. The energy spectrum was generated according to power laws with spectral index -2.7 for protons and -2.5 for gammas. The spatial distribution of the secondary particles at ARGO altitude ($4300m$ a.s.l.) is used for the analysis. The time analysis is performed on the shower from time zero (first detection on the carpet of a secondary particle) to time 2000 ns. The output of the simulated showers includes the effect of the detector response as far as space and time distribution on the carpet is concerned.

5. Spatial Separation Technique

Our spatial separation technique uses the differences in wavelet based multifractal behaviours of showers of different progenitors. These are: the mean and the standard deviation of individual Gaussian fits to the distributions of multifractal dimensions.

For each shower, the dependence with respect to the distance from the shower core of the fractal dimension is almost linear. This has a consequence: we need to specify the mean and the standard deviation just for two different mean distances from the core. So we choose to analyze two regions: a circle of fixed radius r_0 centred at the shower core (inner ring) and a ring with fixed inner and outer radii r_i and r_o (outer ring). In our analysis we choose the values: $r_0 = 7$ m, $r_i = 9$ m, $r_o = 12$ m. The radii have been chosen such that on average the two rings contain about an equal number of hits. Our shower core is computed on the wavelet transform of the spatial map on the maximum scale ($s = 2^5$). On that scale, in this analysis, the shower core is found as the coordinates $xcore$ and $ycore$ of the maximum value of the wavelet coefficients. The multifractal behaviour of each individual shower is then specified by four quantities: the mean of the Gaussian fit in the inner and the outer ring (μ_I, μ_O), the standard deviation in the inner and the outer ring (σ_I, σ_O). The set of scales are powers of two: $s = 2^r$ and the first scale always corresponds to the size of 1 pixel. The scale s may be considered as the resolution. We choose to investigate on the scales 2 to 32 that on the ARGO-YBJ carpet correspond to physical sizes ≈ 1.2 m and ≈ 20 m.

As seen, if a function f has scaling law with exponent D around x_0: $f(R(x_0, \lambda\epsilon)) \sim \lambda^D f(R(x_0, \epsilon))$ then the wavelet transform $W(s,t) = \int g(x; s,t) f(x) dx$ has the same scaling exponent. Thus: the local scaling behaviour is represented by $W(s,t) \sim s^{D(t)}$; for any distribution function f, the slope of the plot of $\log W(s,t)$ versus $\log s$ will give the fractal dimension of the distribution around point t for the range of the scale s. We compute a $log(W)/log(s)$ matrix for each scale s on the two regions selected (inner and outer ring). $D = log(W)/log(s)$ is the fractal dimension and has a Gaussian distribution. If μ_I, μ_O are the average values of D in the region and σ_I, σ_O its standard deviations we compute average values of $\mu_I, \mu_O, \sigma_I, \sigma_O$ on all the scales and obtain our parameters: $\overline{\mu}_I, \overline{\mu}_O, \overline{\sigma}_I, \overline{\sigma}_O$.

All these quantities have a nearly monotonic dependence on the mass number of the progenitor. However, they show large fluctuations, prohibiting de facto mass separation. So, to get a high-resolution separation technique, we need at least one more parameter and then define and train an artificial neural network.

6. Separation Technique on Time Arrivals

Our separation technique on time arrival of the shower use the lacunarity technique in the same two regions as the spatial separation: the inner and outer ring. First, we need to compute our time array as $T = T_{max} - T_{min}$ where T_{min} is the time arrival of the first secondary particle on the carpet and is set to 0 and T_{max} is the time arrival of the last secondary particle we include in our analysis. Maximum value of T is 2000 ns. Then we need to define the time scale on which we compute lacunarity. This is a crucial parameter. We call it t_{lac} and to compute $T = T_{max} - T_{min}$

A box of length t_{lac} is placed at the origin of the sets. The number of occupied sites within the box (box mass k) is then determined. The box is moved one space along the set and the mass is computed again. This process is repeated over the entire set, producing a frequency distribution of the box masses $n(k, t_{lac})$. This frequency distribution is converted into a probability distribution $Q(k, t_{lac})$ by dividing by the total number of boxes $N(t_{lac})$ of size t_{lac}: $Q(k, t_{lac}) = n(k, t_{lac})/N(t_{lac})$

The first and second moments of the distribution are computed, Z_1 and Z_2: $Z_1(t_{lac}) = \sum_k k \cdot Q(k, t_{lac})$, $Z_2(t_{lac}) = \sum_k k^2 \cdot Q(k, t_{lac})$. The lacunarity is now defined as: $\Lambda(t_{lac}) = Z_2/Z_1^2$. Lacunarity is computed both in the inner and the outer ring (Λ_I, Λ_O). We find that 5 ns is a good choice for the t_{lac} parameter.

7. A Standard Three Layer Neural Network

We assume therefore that the mass of the progenitor can be estimated with the use of an artificial neural network of the six variables. The neural network is a standard three layer perceptron with only one output neuron (1=hadron, 0=gamma). The neural network we choose is of the feed forward type and it is made of three perceptron layers. The input is made of six neurons: (1,2) Average of the fractal dimensions on the spatial scales in the inner and outer region: $\overline{\mu}_I, \overline{\mu}_O$; (3,4) Average of the standard deviation of the fractal dimensions on the spatial scales in the inner and outer region: $\overline{\sigma}_I, \overline{\sigma}_O$; (5,6) Lacunarity of the time arrivals arrays in the inner and outer region: Λ_I, Λ_O. The hidden layer is made of four neurons, while the output vector is defined in a one dimensional space and it is trained to be 0 for gamma initiated events and 1 for hadronic ones. Network was implemented and optimized by using the Stuttgart Neural Network Simulator Tool (SNNS) [12]. SNNS is a simulator for neural networks developed at the Institute for Parallel and Distributed High Performance Systems at the University of Stuttgart. The network was trained by using 1000 events from indipendent samples.

8. Test Run

In Fig. 1 we present the results of applying our method to 1000 simulated showers of two different primaries (gamma and hadron). As is seen in the histogram, the identification is achieved with a very high resolution and almost no overlap is present between showers from different primaries. More than 90% of the hadron showers have neural network output neuron > 0.5 and more than 90% of the gamma showers have neural network output neuron < 0.5.

The most important parameter in gamma/hadron discrimination is the Q (quality) factor. The Q factor is defined as $Q = \frac{\varepsilon_\gamma}{\sqrt{1-\varepsilon_h}}$ where ε_γ is the fraction of showers

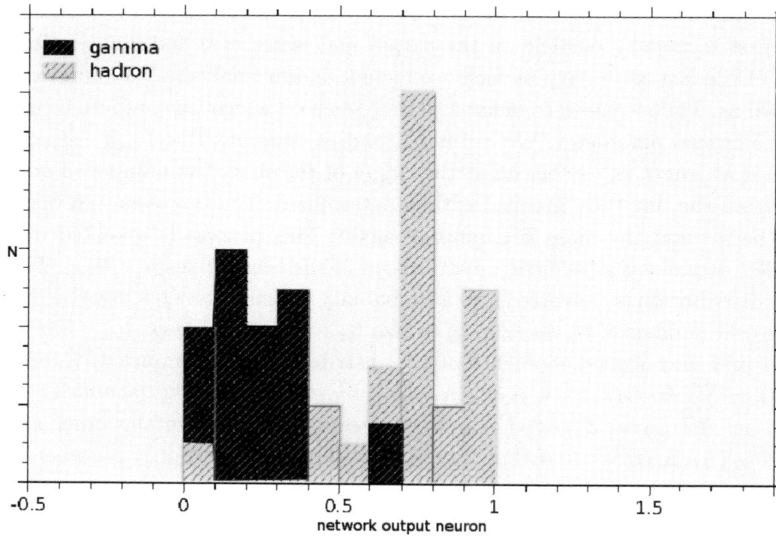

Fig. 1: Gamma-hadron separation for our test run with 1000 events (500 gammas and 500 hadrons)

induced by photons correctly identified by the discrimination criterion and ε_h is the fraction of showers induced by protons correctly identified by the discrimination criterion so that $1 - \varepsilon_h$ is the background contamination. In our test run a value of $Q \approx 2.47$ has been reached, which is among the largest obtained in the field.

9. Conclusions

A new technique for separating extensive air showers initiated by different progenitors, based on multifractal and lacunarity analysis, has been developed. The multifractal behaviour and lacunarity time structure of each shower has been represented by six variables. Due to large shower to shower fluctuations, the differences in any single one of these variables have a very poor separation power [13]. So, on these six quantities, a neural network analysis has been performed. Network were implemented and optimized by using the Stuttgart Neural Network Simulator Tool and trained by using events from indipendent samples.

It is well known that the most important parameter in gamma/hadron discrimination is the Q (quality) factor, defined as $Q = \frac{\varepsilon_\gamma}{\sqrt{1-\varepsilon_h}}$ where ε_γ is the fraction of showers induced by photons correctly identified by the discrimination criterion and ε_h is the fraction of showers induced by protons correctly identified by the discrimination criterion, so that $1 - \varepsilon_h$ is the background contamination (fraction of events induced by protons and erroneously identified as gamma induced ones).

Our approach gives very good results, leading to a clear improvement of the discrimination power with respect to the existing figures for extended shower detectors: the value $Q \approx 2.47$ obtained in our test run is among the largest obtained in the field. The technique shows up to be very promising and its application may have important astrophysical prospects in different experimental environment of extended air shower study.

Acknowledgments

The authors would like to thank the whole ARGO-YBJ collaboration for having provided sample of simulated events produced in a real experimental framework.

References

[1] G. Rastegaarzaeh and J. Samimi, *Journal of Physics G: Nuclear and Particle Physics* **27** (2001).
[2] B.B. Mandelbrot, *The Fractal Geometry of Nature*, (W.H. Freeman, New York, 1983).
[3] M. Tuceryan and A.K. Jain *Texture Analysis, Handbook of Pattern Recognition & Computer Vision*, (World Scientific, 1992).
[4] N.E. McIntyre and J.A. Wiens, *Landscape Ecology* **15**, 313 (2000).
[5] C. Bacci et al (ARGO-YBJ Coll.), *Astroparticle Physics* **17**, 151 (2002).
[6] A. Pagliaro and U. Becciani, *Parallel Algorithms and Applications* **14** (2000).
[7] M. Gambera, A. Pagliaro, V. Antonuccio, U. Becciani, *Astrophysical Journal* **488**, 136 (1997).
[8] M. Holschneider, *Journal of Statistical Physics* **50**, 963, (1988).
[9] C. Allain and M. Cloitre, *Physical Review A* **44**, 3552, (1991).
[10] D. Heck et al, CORSIKA (cosmic ray simulation for KASKADE) Report FZKAS 6019 (Forschungszentrum Karlsruhe 1998).
[11] http://argo.le.infn.it/analysis/argog/
[12] http://www.ra.cs.uni-tuebingen.de/SNNS/
[13] H. Rebel, *Nuclear Physics A* **663**, 847 (2000).

BAYESIAN SEMI-PARAMETRIC CURVE-FITTING AND CLUSTERING IN SDSS DATA

S. MUKKHOPADHYAY, S. ROY and S. BHATTACHARYA*

Indian Statistical Institute
203 Barrackpore Trunk Road, Kolkata, 700108, India
**E-mail: sourabh@isical.ac.in*

The Bayesian semi-parametric curve-fitting procedure, based on a new, flexible, fast, and efficient mixture analysis idea assuming unknown number of components is applied to analyze SDSS data to study the relationship between apparent magnitude and redshift for quasars and the possibility of clusterings. The cosmological data analysis provides strong evidence against linear relationship, and clearly indicates the possibility of *clustering* of quasars specially at high redshift. This sheds new light not only on the issue of evolution, existence of acceleration or deceleration and environment around quasars (say, radio loud and radio quiet) at high redshift but also help us to estimate the cosmological parameters related to acceleration or decceleration.

Keywords: Cluster analysis; Cosmology; Dirichlet process; Model validation; Markov chain Monte Carlo; Non-linear regression; Reversible jump Markov chain Monte Carlo

1. Introduction

Among many different ways of testing models of cosmological sources, especially quasars, one is through investigations of the distributions, ranges and more importantly the correlations among the relevant physical characteristics, such as luminosity, spectra, redshifts or cosmological distances. The impossibility of direct measurements to quasi stellar objects prevents one from validating any direct relationship between distance and redshift where the measurable quantities are the apparent magnitude (m), redshift (z) which, as related to luminosity function or, even the probability distribution of absolute magnitude ($\mathcal{M}$), as predicted by a given cosmology and the angular diameter of the object. The advancement of the computing power has enabled theorists to simulate the realistic physical situations, beyond scope of experimental techniques. Various statistical models have been considered by statisticians (for example, [1] and [2]), [3], [4].

In this paper, adopting a new and efficient Bayesian semi-parametric curve-fitting procedure based on mixtures (discussed in [4]) we study the SDSS (2007) data set, which, unlike SDSS (2005) (used in [3]), is not truncated. The fitted curve for quasar redshift data exhibits a linear trend for small redshift and non-linearity for high redshifts. This has important implications in the context of Hubble relation for quasar redshifts. In fact, we obtained strong evidence against the validity of linear law like Hubble's law, given this data set. We also carried out a change point analysis to understand the possible physical origin of these non-linearities at various redshift range. Finally, analysis of the posterior distribution of clusterings associated with our model clearly indicates the existence of two distinct clusters at high redshift ($z > 2.8$) which are dependent on luminosity.

2. The Data and Overview of Mixture Models

Our massive cosmological data set, consisting of 96307 observations on logarithm of redshift (z) and apparent magnitude (m) for quasars (quasi-stellar objects) collected from SDSS data, did not reveal any clear-cut parametric relationship between the two variables of interest. Exploratory analyses clearly ruled out the (bivariate) normality assumption for the data. Indeed, our quantile-quantile plots of each of the two variables showed that the marginal distributions of both the variables are far from univariate normal.

To resolve this problem we use idea of mixture models (discussed in [5] and [6]).

However, a technical problem associated with classical analysis of mixture models is associated with the number of mixture components included in the model. In the classical statistical literature there does not seem to exist any rigorous procedure of selecting an adequate number of mixture components. On the other hand, the Bayesian paradigm offers elegant solutions to this problem. Among the contributions of Bayesians in this topic, notable are those of Escobar [7] (henceforth, EW) and Richardson [8] (henceforth, RG). The former use Dirichlet process to indirectly induce (random) variability in the number of components, while the approach of RG directly acknowledges uncertainty about the number of components and puts a prior distribution on the same, thus rendering the problem variable-dimensional. The methodology of RG relies on reversible jump Markov chain Monte Carlo [9] (RJMCMC) for drawing inference.

But the RJMCMC turns complicated for dimension more than one and approach by Escobar [7] is not computationally fast for massive data sets. We attempt to avoid most of the difficulties noted above by adopting a model in which the complexity RJMCMC is not present and the computation is much faster. This has been elaborated in a recent paper of the present authors (see [4]).

3. Fitted Bayesian Cosmological Curve and Change Point Analysis

The fitted curve and the associated pointwise 95% credible intervals are shown in Figure 1a; the solid line in the middle represents the estimated Bayesian cosmological curve, and the pointwise 95% credible intervals are shown in solid lines above and below it. The difference in the nature of the lines in the same curve occurs due to variaton in the nature of red shift of the quasars of different ages. The number of different such quasars is reflected in the number of distinct components of the mixture model. The different distinct components of the mixture correspond to distributions of absolute magnitude for different ages of the clusters. The above discussion points towards a need for detailed cluster analysis of the data set, where not only the number of clusters, but the entire clustering is of interest to astro-physicists.

The obtained non-linear curve is linear for the first half ($log(z) \leq -2.0$) with intercept 18.7840 and slope 0.5136 (1.182608 with respect to logarithm with base 10, which is of interest to astro-physicists). After that, however, non-linearity is exhibited. But we also note that the form of non-linearity can be approximated by linear line segments indicating presence of change points. A close look will reveal presence of four change points, but to get a solid basis of belief, we performed a detailed change point analysis, assuming four change points.

Figure 1a also shows that the curve obtained by the change point analysis, which is shown in dotted line, nicely approximates our fitted semiparametric Bayesian curve (the solid line in the middle) at all places except at the extreme lower end of the x-space, where there are hardly any information about the curve. Moreover, the entire change point curve falls within the (pointwise) 95% credible intervals associated with the semiparametric Bayesian curve. This is not surprising, since the changepoint curve may be looked upon as a special case of our approach, with one of the linear functions having weight unity and all others having zero weight.

Apart from the semiparametric Bayesian curve and the change point curve, we also fitted the least squares regression line, obtained by assuming a simple linear regression of $\log(z)$ on M. This linear regression is related to Hubble's law, and the implications of the slope of this straight line will discussed later in detail. For now we note that the least squares regression line falls well within the pointwise 95% credible intervals of our Dirichlet process-based semiparameteric Bayesian curve. This shows that the linear regression, although not optimal (in the sense that normality assumption does not hold for this data set, for example), is not ruled out by our semiparametric method.

4. Application of the Clustering Idea to the Cosmology Data Set

By clustering of the data we mean a suiatble partition of the data set into different distinct groups. Our new clustering technique is all about choosing a good repres-ntative clustering among set of all possible clusterings of the data depending up on a pre-defined thresold value of variation within each clustering, namely $\epsilon > 0$. On application of the central clustering ideas, we observe that for different values of $\epsilon > 0$ we have different central clusterings, clearly indicating multimodality of the posterior distribution of clusterings. Following the technique described in [4] applied for obtaining the global central clustering, we obtain the global central clustering. The radius of 95% credible region of the global mode is 0.35, which is reasonably low. The modal value of the distribution of distance of the clusterings from the central clustering has mode at 0.2.

We note that the central clustering in our case consists of 29 clusters. This is quite reasonable, given that there are more than 96,000 observations. Moreover, we note that although there are 29 clusters, many are effectively the same cluster, thanks to the small Euclidean distances between them. This reduction of the effective number of clusters finds reasons more than statistical within the astro-physics paradigm. Indeed, astro-physicists (for example, [3]) have tried to split the data into 2 characteristics only, namely Radio loud and Radio quiet.

Driven by the above observations and discussions, we merge those clusters with Euclidean distances less than a prefixed limit. Table 1 shows how the number of clusters change if the prefixed limit is changed. The original central clustering con-

Table 1: Table showing the variation in the number clusters with change in the prefixed limit

Value of prefixed limit	Number of clusters after merger
0.05	23
0.1	21
0.3	10
0.5	9
0.65	5
0.7	4
0.9	2

sisting of 29 clusters and the merged central clustering consisting of two components only (which corresponds to the prefixed limit being 0.9) are shown in Figures 1b and 1c respectively. The latter is provided to make our analysis comparable to the clustering done by astro-physicists on the basis of Radio loud and Radio quiet. To make the figure easily visible we have sampled 2000 observations and plotted them. Since colour pictures are not allowed we are forced to do so.

We have repeated the same analysis taking the maximum number of components, $M = 50$. No notable difference between the results of the two analyses were observed. Indeed, even with $M = 30$, the posterior probability of 30 components turned out to be negligible (about 0.001).

5. Possible Implications in Cosmology

The above statistical analyses shed new light on the following issues in quasar redshift and cosmology.

- The redhsift-magnitude relation for quasars.
- Clustering of quasars at low and high redshifts.
- Change point analysis and physical conditions for quasar redshifts.

The observed non-linearity of the distance-magnitude relation at high redshift might be due to the following reasons :

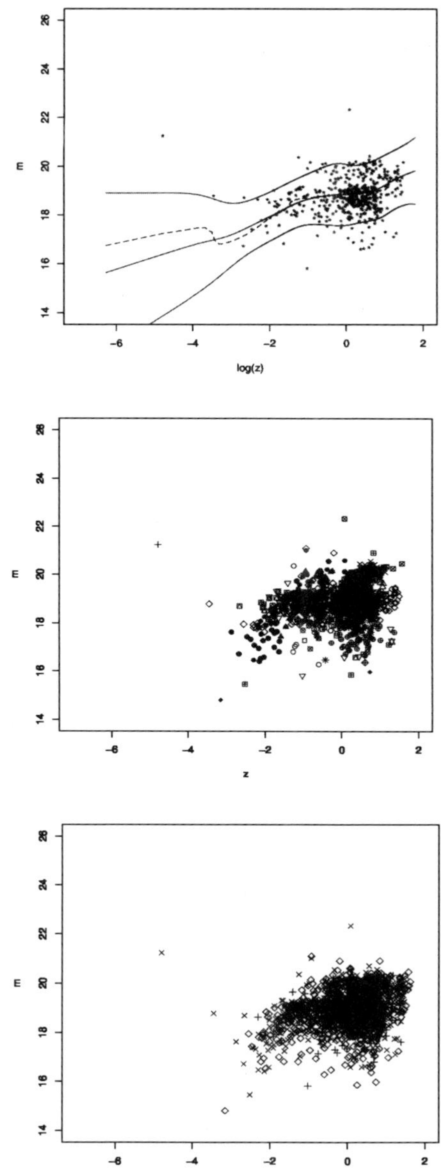

Fig. 1: In the figure of upper panel, the fitted curve is shown in solid line in middle, the change point curve in dotted, and the pointwise 95% credible intervals are shown in solid lines above and below; figure in middle panel shows central clustering with 29 clusters: different shapes indicate different clusters, figure in lower panel shows merged central clustering with 2 clusters.

- The presence of rapid luminosity evolution at high redshift may be one of the possible causes for the observed non-linearity.
- The effect of environment should be considered as a contributing factor to the Doppler effect. This would have important astronomical and, consequently, cosmological implications.

The fitte curve indicates that the non-linearity becomes prominent at intermediate and high redshifts. We have studied non-linearity of the curve using change-point analysis and found four most probably change points for various redshifts. These non-linearities can be approximated by linear curves and the details have already been described in the main text. The non-linearity may be due to evolution, acceleration or deceleration of the universe. The detailed cluster analysis shows the presence of 29 clusters and we can merge the very close components to form fewer number of clusters. The degree of reduction depends upon the prefixed threshold for merging the clusters. The environments around quasars might play a significant role in understanding the clusterings. For example, the environments around radio-loud and radio-quite quasars may be different. The detailed analysis of radio-loud and radio-quite quasars will shed new light in this regard and will be considered in subsequent publications.

References

[1] B. Efron and V. Petrosian, *The Astrophysical Journal* **399**, 345 (1992).

[2] B. Efron and V. Petrosian, *Journal of the American Statistical Association* **94**, 824 (1999).

[3] S. Roy, D. Dutta, J. Ghosh, M. Roy and M. Kafatos, Non-parametric tests for quasar data and hubble diagram, in *Modeling and Simulation in Science: 6th International Workshop on Data Analysis in Astronomy*, eds. V. Di Gesú, G. Lo Bosco and M. Maccarone (World Scientific, NJ, 2007).

[4] S. Mukhopadhyay, S. Roy and S. Bhattacharya, Fast and Efficient Bayesian Semi-Parametric Curve-Fitting and Clustering in Massive Data. (submitted)

[5] S. R. Dalal and W. J. Hall, *Journal of the Royal Statistical Society. Series B* **45**, 278 (1983).

[6] P. Diaconis and D. Ylvisaker, Quantifying prior opinion, in *Bayesian Statistics 2*, eds. J. M. Bernardo, J. O. Berger, A. P. Dawid and A. F. M. Smith (Oxford University Press, 1985).

[7] M. D. Escobar and M. West, *Journal of the American Statistical Association* **90**, 577 (1995).

[8] S. Richardson and P. J. Green, *Journal of the Royal Statistical Society. Series B* **59**, 731 (1997).

[9] P. J. Green, *Biometrika* **82**, 711 (1995).

AUTHOR INDEX

PARTICIPANTS

Sebastian E. AHNERT
Cavendish Laboratory
University of Cambridge
JJ Thomson Avenue
Cambridge, CB3 0HE, UK
sea31@cam.ac.uk

Gustav BERNROIDER
University of Salzburg,
Dept. Organismic Biology
A-5020 Salzburg, AUSTRIA
Gustav.Bernroider@sbg.ac.at

Luciano BOI
CAMS - UMR 8557
54, bd Raspail
F75270 Paris Cedex 06, FRANCE
Luciano.Boi@ehess.fr

Alfred M. BRUCKSTEIN
Department of Computer Science
Technion IIT
Technion City Haifa 32000, ISRAEL
freddy@cs.technion.ac.il

Stefano CAVUOTI
Dipartimento di Scienze Fisiche
Università di Napoli Federico II
via Cinthia, I, 80126, Napoli, ITALY
stefano.cavuoti@gmail.com

Dalia CHAKRABARTY
Department of Statistic, University of Warwick
Coventry CV4 7AL, UK
D.Chakrabarty@warwick.ac.uk

Oliver DORN
School of Mathematics
The University of Manchester
Oxford Rd
Manchester M13 9PL, UK
Oliver.Dorn@manchester.ac.uk

Jean-Claude HEUDIN
Institut International du Multimédia
Pôle Universitaire Léonard de Vinci
92916 PARIS LA DEFENSE, FRANCE
Jean-Claude.Heudin@devinci.fr

Maxim Pavlovich IGAEV
Institute of Applied Physics of Rusian
Accademy of Science
46 Uljanov Street
603950 Nizhny Novgorod, Russia
mooz@list.ru

Antonio INSOLIA
Dipartimento di Fisica Universita' di Catania
Istituto Nazionale di Fisica Nucleare
Via S. Sofia 64, 95123 Catania, ITALY
antonio.insolia@ct.infn.it

Robert M. KIEHN
University of Houston, TX USA
410 Summer Rd, Georgetown
TX, 78633, USA
RKiehn2352@aol.com

Giosuè LO BOSCO
Dipartimento di Matematica ed Informatica
Università degli Studi di Palermo
lobosco@unipa.it

Hans LILJENSTRÖM
Dept. of Energy and Technology, SLU
P.O. Box 7032
750 07 Uppsala, SWEDEN
Hans.Liljenstrom@et.slu.se

Maria Concetta MACCARONE
Istituto Nazionale di Astrofisica
Via Ugo La Malfa 153
90146 Palermo, ITALY
Cettina.Maccarone@ifc.inaf.it

Rafael MOLINA
Departamento de Ciencias de la Computación e I. A.
E.T.S. Ing. Informática
Universidad de Granada
18071 Granada, SPAIN
rms@decsai.ugr.es

Fionn MURTAGH
Dept. Computer Science
Royal Holloway, University of London
Egham, Surrey TW20 0EX, UK
fmurtagh@acm.org

Vladimir I. NEKORKIN
Institute of Applied Physics RAS
46 Ul'yanov Street
603950 Nizhny Novgorod, RUSSIA
vnekorkin@neuron.appl.sci-nnov.ru

Antonio PAGLIARO
Istituto Nazionale di Astrofisica
Via Ugo La Malfa 153
90146 Palermo, ITALY
pagliaro@ifc.inaf.it

Sankar K. PAL
Indian Statistical Institute
203, B.T.Road
Kolkata 700108, INDIA
sankar@isical.ac.in

Tim N. PALMER
European Centre for Medium-Range
Weather Forecasts
Shinfield Park, Reading RG2 9AX, UK
tim.palmer@ecmwf.int

Carlos PEREIRA
Instituto de Matemàtica e Estatistica
Universidade de SaoPaulo Brasilia
Rua do Matão, 1010
Cidade Universitária
São Paulo, SP, BRASIL, CEP 05508-090
cadebp@gmail.com

Sisir ROY
Physics and Applied Mathematics Unit
Indian Statistical Institute
203 B.T. Road, Kolkata 700108, INDIA
sisir.sisirroy@gmail.com

Antonio SAGGION
Dipartimento di Fisica G. Galilei
Università di Padova
Via Marzolo 8, 35131 Padova, ITALY
saggion@pd.infn.it

Emanuele SALERNO
ISTI-CNR
Via Moruzzi 1
I-56124 Pisa, Italy
emanuele.salerno@isti.cnr.it

Gaetano SCANDARIATO
Osservatorio Astrofisico di Catania
via S. Sofia 78, 95123 Catania
gas@oact.inaf.it

Mario SCUDERI
Università di Catania
Dipartimento di Fisica e Astronomia
Via S. Sofia 64, 95123 Catania
mario.scuderi@ct.infn.it

Rossana SPORTELLI
Dipartimento di Matematica ed Informatica
Università degli Studi di Palermo
sportell@math.unipa.it

Marco Elio TABACCHI
Università degli Studi di Palermo
Dipartimento di Matematica ed Informatica
Via Archirafi 34, 90123 Palermo
e Istituto Nazionale di Ricerche Demopolis
tabacchi@unipa.it—demopolis.it

Bertrand ZAVIDOVIQUE
Institute of Electronics Fundamentals
University Paris Sud, Orsay, France
bertrand.zavidovique@u-psud.fr